Graphene from Natural Sources

Graphene from Natural Sources

Synthesis, Characterization, and Applications

Edited by
Amir Al-Ahmed and Inamuddin

CRC Press
Taylor & Francis Group
Boca Raton London New York

CRC Press is an imprint of the
Taylor & Francis Group, an **informa** business

First edition published 2023
by CRC Press
6000 Broken Sound Parkway NW, Suite 300, Boca Raton, FL 33487-2742

and by CRC Press
4 Park Square, Milton Park, Abingdon, Oxon, OX14 4RN

CRC Press is an imprint of Taylor & Francis Group, LLC

ISBN: 978-0-367-77091-4 (hbk)
ISBN: 978-0-367-77093-8 (pbk)
ISBN: 978-1-003-16974-1 (ebk)

DOI: 10.1201/9781003169741

Typeset in Times
by KnowledgeWorks Global Ltd.

Contents

Preface

Graphene (graphene oxide and reduced graphene oxide) is a two-dimensional layered carbon material. It is finding uses in many applications due to its fascinating electrical, thermal, and mechanical properties. Bulk graphene is prepared by chemical vapor deposition (CVD), epitaxial growth, and an arc discharge method. In these approaches, graphene is obtained through a series of physical and chemical changes using high temperatures and carbon precursor gas (e.g., methane, ethane) and metal (e.g., copper, nickel) substrates; silicon carbide; or gas (e.g., hydrogen, helium) and graphite sources, respectively. Graphene can also be obtained by chemical treatment of pure graphite using the Hummers' method. All these methods have higher yields but expensive and multistaged methods and require strong conditions. However, graphene can also be obtained from natural and industrial carbonaceous wastes and even from food items and insects through mild conditions. There are many individual pieces of literature available on these topics; however, there is no comprehensively compiled book, so this book will be one of its kind in this field. This edition will provide a state-of-the-art overview of graphene synthesis using different natural raw materials including agricultural waste, vegetable waste, biowastes, and animal waste. Their methods of characterization and some important applications are also discussed.

Chapter 1 discusses the synthesis, characterization, and applications of graphene synthesized from sugar and sugarcane extracts. Its primary focus is on the quality/quantity of graphene obtained by various routes using sugarcane extracts as the precursor and reducing agent. Furthermore, it includes the present applications, prospects, and challenges.

Chapter 2 describes the honey-mediated green synthesis of graphene. The chemical composition and various properties of honey are discussed in detail. Various characterization techniques and applications of honey-mediated green synthesized graphene are discussed in brief. The chapter is mainly focused on the possibility of the replacement of inorganic chemicals from honey for the synthesis of graphene along with its advantages, drawbacks, and future perspective.

Chapter 3 details the various methods of extracting graphene from agro-waste including animal waste. Several potential applications of bio-graphene are also discussed in detail such as biomedical, electronics, batteries, conductive ink, etc.

Chapter 4 presents the different types of graphene, traditional methods of graphene synthesis, and some applications of carbon-based nanomaterial. In addition, this chapter describes the composition, properties, and application of essentials oils. The use of essential oils in the fabrication of graphene is also discussed.

Chapter 5 details the production of graphene from biowastes, which is a captivating idea particularly when the petroleum and allied resources are continuously depleting. It also demonstrates an overview of various implausible biowastes that have been utilized for the production of graphene. It suggests these miraculous biowastes as a hidden treasure of precursors for graphene production.

Chapter 6 discusses the chemical composition and properties of rice husk (RH), graphene, and RH nanocomposite manufacture. The main goal is to highlight the benefits and future viability of available green synthesis processes for creating graphene from biomass precursors as a sustainable energy source.

Chapter 7 discusses the facile synthesis of graphene from vegetable waste. The search for natural precursors for the cost-effective and environmentally friendly synthesis of graphene is increasing nowadays. Thus, several vegetable wastes that have the potential to be used as a precursor for graphene synthesis are discussed in brief. Also, several vegetable wastes that can be used as a reducing agent to obtain graphene are briefly discussed.

Chapter 8 details the history and green synthesis of graphene oxide. Applications of graphene oxide are discussed in chemistry, physics, medicine, and food packaging. Additionally, environmentally safe agriculture purposes are mentioned, such as growth promoters, slow-release fertilizers, pesticides, and heavy metal adsorbents that effectively stop environmental destruction and reduce toxic chemicals.

Chapter 9 begins with a general introduction to graphene followed by the composition of sugarcane bagasse and why the disposal of this waste material is a problem. Acidic method, synthesis from organic matter, Hummer's method, and using Urea are discussed. Various characterization techniques like FTIR, UV, XRD, and SEM are mentioned along with the application of sugarcane bagasse-derived graphene in the field of membranes, biomedical, sensors coatings.

Chapter 10 reviews recent developments, synthesis methods, characterization, and application of produced graphene from leaf wastes and their derivatives as eco-friendly and low-cost sources. The effects of temperature and synthesis conditions on the characteristics of produced graphene are discussed. The synthesis of graphene quantum dots from leaves is also discussed.

Chapter 11 encompasses the synthesis and characterization of graphene and its derivatives using leaf wastes. These bio-based and unused waste materials can provide significant results in producing graphene by avoiding the usage of toxic chemicals. Adapting inexpensive starting materials will pave the way for sustainable processing and recycling of materials.

Chapter 12 briefly focuses on the biosynthesis of reduced graphene oxide (rGO), functionalization, rGO nanocomposite fabrication, and characterization and its functionality as an antibacterial template. The antibacterial activity of the produced rGO nanocomposites on a vast number of bacteria and its mode of action are detailed.

Chapter 13 elucidates the advances of graphene for supercapacitor application. Initially, a detailed description of the supercapacitor principle, fabrication, and characteristics is given. An outline of modified graphene materials like fullerene, carbon onions is presented. The focus is mainly on various graphene composites and their efficiency as electrode material for supercapacitors.

Editors

Amir-Al-Ahmed, PhD, is working as a Research Scientist-II (Associate Professor) in the Interdisciplinary Research Center for Renewable Energy and Power Systems (IRC-REPS), at King Fahd University of Petroleum & Minerals (KFUPM), Saudi Arabia. He graduated in chemistry from the Department of Chemistry, Aligarh Muslim University (AMU), India. Then completed his MPhil (2001) and PhD (2004) in Applied Chemistry from the Department of Applied Chemistry, AMU, India, followed by three consecutive postdoctoral fellowships in South Africa and Saudi Arabia. During this period, he worked on various multidisciplinary projects, in particular, conducting polymers, electrochemical sensors, nano-materials, polymeric membranes, electro-catalysis and solar cells. At present, his research activity is fundamentally focused on the 3rd generation solar cell devices, such as, low band gap semiconductors, quantum dots, perovskites, and tandem cells. At the same time, he is also working on energy storage technologies, such as heat storage, evaluation of electricity storage devices and dust repellent coating for PVs. He has worked on different NSTIP, KACST and Saudi Aramco funded projects in the capacity of a principle and co-investigator. Dr. Amir has eight US patents, over 60 journal articles, invited book chapters and conferences publications. He has edited ten books with Trans Tech Publication, Springer, and Elsevier and several other books are in progress. He is also the Editor-in-Chief of an international journal *Nano Hybrids and Composites* along with Professor Y. H. Kim.

Inamuddin, PhD, is an Assistant Professor in the Department of Applied Chemistry, Aligarh Muslim University, Aligarh, India. He earned his MSc in organic chemistry at Chaudhary Charan Singh (CCS) University, Meerut, India, in 2002. He earned his MPhil and PhD in applied chemistry at Aligarh Muslim University (AMU), India, in 2004 and 2007, respectively. He has extensive research experience in the multidisciplinary fields of analytical chemistry, materials chemistry, electrochemistry, and, more specifically, renewable energy and environment. He has worked on different research projects as project fellow and senior research fellow funded by the University Grants Commission (UGC), Government of India, and the Council of Scientific and Industrial Research (CSIR), Government of India. He has received the Fast Track Young Scientist Award from the Department of Science and Technology, India, to work in the area of bending actuators and artificial muscles. He has completed four major research projects sanctioned by the University Grant Commission, Department of Science and Technology, the Council of Scientific and Industrial Research, and the Council of Science and Technology, India. He has published 196 research articles in international journals of repute and 19 book chapters in knowledge-based book editions published by renowned international publishers. He has published 145 edited books with Springer (UK), Elsevier, Nova Science Publishers, Inc. (USA), CRC Press – Taylor & Francis Asia Pacific, Trans Tech Publications Ltd. (Switzerland), IntechOpen Limited (UK), Wiley-Scrivener (USA), and Materials Research Forum LLC (USA). He is a member of various journals' editorial boards. He is an Associate Editor for several journals (*Environmental Chemistry Letter, Applied Water Science* and *Euro-Mediterranean Journal for Environmental Integration*, Springer-Nature), Frontiers Section Editor (*Current Analytical Chemistry*, Bentham Science Publishers), Editorial Board Member (*Scientific Reports*, Nature), Editor (*Eurasian Journal of Analytical Chemistry*), and Review Editor (*Frontiers in Chemistry*, Frontiers, UK). He has also guest-edited various thematic special issues to the journals of Elsevier, Bentham Science Publishers, and John Wiley & Sons, Inc. He has attended as well as chaired sessions at various international and national conferences. He has worked as a Postdoctoral Fellow, leading a research team at the Creative Research Initiative Center for Bio-Artificial Muscle, Hanyang University, South Korea, in the field of renewable energy, especially biofuel cells. He has also worked as a Postdoctoral Fellow at the Center of Research

Excellence in Renewable Energy, King Fahd University of Petroleum and Minerals, Saudi Arabia, in the field of polymer electrolyte membrane fuel cells and computational fluid dynamics of polymer electrolyte membrane fuel cells. He is a life member of the *Journal of the Indian Chemical Society*. His research interests include ion exchange materials, a sensor for heavy metal ions, biofuel cells, supercapacitors, and bending actuators.

Contributors

Nadia Akram
Department of Chemistry
Government College University Faisalabad
Faisalabad, Pakistan

Vivian C. Akubude
Department of Agricultural and Bioresource Engineering
Federal University of Technology
Owerri, Nigeria

Telli Alia
Laboratoire de protection des écosystèmes en zone aride and semi aride
Université de Kasdi Merbah
Ghardaia, Algérie

Fozia Anjum
Department of Chemistry
Government College University Faisalabad
Faisalabad, Pakistan

Hamidreza Bagheri
Department of Chemical Engineering
Faculty of Engineering
Shahid Bahonar University of Kerman
Kerman, Iran

Surender Duhan
Department of Physics
Deenbandhu Chhotu Ram University of Science and Technology
Murthal Sonipat, Haryana, India

Marzieh Fatehi
Department of Chemical Engineering
Faculty of Engineering
Shahid Bahonar University of Kerman
Kerman, Iran

Eksha Guliani
Department of Chemistry
Amity Institute of Applied Sciences
Amity University
Noida, India

Amal I. Hassan
Radioisotope Department
Nuclear Research Center
Atomic Energy Authority
Dokki, Giza, Egypt

R. Imran Jafri
Department of Physics and Electronics
Christ University (Deemed to be University)
Bengaluru, Karnataka, India

Christine Jeyaseelan
Department of chemistry
Amity Institute of Applied Sciences
Amity University
Noida, India

Rita Joshi
Centre of Excellence: Nanotechnology
Indian Institute of Technology Roorkee
Roorkee, India

Akbar Karami
Department of Horticultural Science, School of Agriculture
Shiraz University
Shiraz, Iran

Aruna Jyothi Kora
National Centre for Compositional Characterisation of Materials (NCCCM)
Bhabha Atomic Research Centre (BARC)
Hyderabad, India
and
Homi Bhabha National Institute (HBNI)
Anushakti Nagar, Mumbai, India

Atul Kumar
Department of Physics
Deenbandhu Chhotu Ram University of Science and Technology
Murthal Sonipat, Haryana, India

Spandana Bhat Kuruveri
Department of Mechanical Engineering
National Institute of Technology Karnataka
Srinivasnagar, Surathkal, India

Udaya Bhat Kuruveri
Department of Metallurgical and Materials Engineering
National Institute of Technology Karnataka
Srinivasnagar, Surathkal, India

Indranil Lahiri
Centre of Excellence: Nanotechnology
and
Department of Metallurgical and Materials Engineering
Indian Institute of Technology Roorkee
Roorkee, India

Priyadharshini Madheswaran
Smart Materials Interface Laboratory
Department of Physics
Periyar University
Salem, Tamil Nadu, India

Ali Mohebbi
Department of Chemical Engineering
Faculty of Engineering
Shahid Bahonar University of Kerman
Kerman, Iran

Seyyed Sasan Mousavi
Department of Horticultural Science
School of Agriculture
Shiraz University
Shiraz, Iran

Akshaya S. Nair
Department of Physics and Electronics
Christ University (Deemed to be University)
Bengaluru, Karnataka, India

Victor C. Okafor
Department of Agricultural and Bioresource Engineering
Federal University of Technology
Owerri, Nigeria

Jelili A. Oyedokun
Engineering and Scientific Services Department
National Centre for Agricultural Mechanization
Ilorin, Nigeria

Devadas Bhat Panemangalore
Department of Metallurgical and Materials Engineering
National Institute of Technology Karnataka
Srinivasnagar, Surathkal, India

Hosam M. Saleh
Radioisotope Department
Nuclear Research Center
Atomic Energy Authority
Dokki, Giza, Egypt

Athul Satya
Department of Physics and Electronics
Christ University (Deemed to be University)
Bengaluru, Karnataka, India

Muhammad Shahbaz
Department of Chemistry
Government College University Faisalabad
Faisalabad, Pakistan

Adona Vallattu Soman
Department of Physics and Electronics
Christ University (Deemed to be University)
Bengaluru, Karnataka, India

Sourav Melethethil Surendran
Department of Physics and Electronics
Christ University (Deemed to be University)
Bengaluru, Karnataka, India

K.S. Suresh
Department of Metallurgical and Materials Engineering
Indian Institute of Technology Roorkee
Roorkee, India

Pazhanivel Thangavelu
Smart Materials Interface Laboratory
Department of Physics
Periyar University
Salem, Tamilnadu, India

Akanksha R. Urade
Centre of Excellence: Nanotechnology
Indian Institute of Technology Roorkee
Roorkee, India

Gunjan Varshney
Department of Chemistry
Amity Institute of Applied Sciences
Amity University
Noida, India

Khalid Mahmood Zia
Department of Chemistry
Government College University Faisalabad
Faisalabad, Pakistan

1 Graphene from Sugar and Sugarcane Extract

Synthesis, Characterization, and Applications

Akanksha R. Urade, Rita Joshi, K.S. Suresh, and Indranil Lahiri

CONTENTS

1.1 INTRODUCTION

Graphene, a name given to a basic structure of graphite, carbon nanotubes, and fullerene, is arranged in honeycomb structure (Novoselov et al. 2004). Since the discovery of graphene in 2004, this two-dimensional single atomic layer of sp^2-bonded carbon atoms has shown a lofty promise for future electronic and optical devices owing to its distinctive properties such as high thermal conductivity (Lancellotti et al. 2020; Lin et al. 2020), high current carrying capability (Deb, Seriani, and Sarkar 2021), tunable electronic band gap (Mahdavifar, Shekarforoush, and Khoeini 2021), optical transmittance (Bae et al. 2010; Koo et al. 2020), and ultra-high mechanical strength (Geim and Novoselov 2007).

Novoselov et al. (2004) reported the preparation of the first graphene films by mechanical exfoliation of graphite. This approach was easy and promised high-quality graphene production, but the process was time consuming (Yi and Shen 2015). The high mobility of graphene makes it ideal for electronic applications requiring high mobility and fast response times (Giannazzo et al. 2020; Sood et al. 2015). The high transparency and low resistance of graphene make it promising for transparent conductive coating for photonic devices (Li et al. 2020). Graphene has a lot of capability for further

DOI: 10.1201/9781003169741-1

applications such as anticorrosion coatings (Punith et al. 2020) and paints (Fu et al. 2020), efficient and precise sensors (Fu et al. 2020), wearable (Zheng et al. 2020), flexible displays (Muralee et al. 2020), efficient solar panels (Taheri et al. 2018), faster DNA sequencing (Rani and Ray 2021), drug delivery (Ghamkhari et al. 2020), and more. It is eminent noting that graphene's unrealized significance can depend on morphology and structure, mostly determined by the synthesis techniques and subsequent processing methods.

To scale up the graphene-based devices from a laboratory to an industry, graphene's mass production is the foremost concern (Novoselov et al. 2012). Moreover, most of the graphene applications, as mentioned earlier, require a high-quality single to bilayer graphene (Lin and Gai 2016). However, currently available commercial graphene materials are expensive, resulting in significantly slowing down their potential usage in these applications (Avouris and Dimitrakopoulos 2012; Edwards and Coleman 2013). The growing demand for graphene mass production has triggered a worldwide focus on graphene synthesis from natural sources (Berktas et al. 2020; Ikram, Jan, and Ahmad 2020). The worldwide research interest is currently focused on green technology-based graphene synthesis to commercialize graphene applications (Akhavan, Bijanzad, and Mirsepah 2014; Ruan et al. 2011). This chapter initiates with a short epitome of the importance of graphene synthesis from natural resources with a detailed discussion of graphene synthesis from sugar and sugarcane extract and its potential applications in all possible emerging fields.

1.2 IMPORTANCE OF SYNTHESIZING GRAPHENE FROM NATURAL SOURCES

The use of biomass waste in synthesizing graphene has been considered green to pact with the muddles associated with worldwide pollutions (Ikram, Jan, and Ahmad 2020). Recently, lots of evaluations have reported on the use of natural sources for the synthesis of graphene (Akhavan, Bijanzad, and Mirsepah 2014; Kumar, Singh, and Singh 2016; Saikia et al. 2020). Carbon is the primary source of graphene and is present abundantly in nature. Recent studies have proposed biomass as a hugely relevant alternative build-up material for devising beneficial carbonaceous materials due to its climate-friendly nature, worldwide accessibility, and imperishable production in bulk quantities as well as low cost and temperature demand (Ansari et al. 2018; Goswami et al. 2017; Shi et al. 2016). For example, agricultural waste biomass gradually attracted worldwide attention as a low-cost resource for nanomaterials synthesis (Kumar, Singh, and Singh 2016; Mohan, Manoj, and Panicker 2019). Goswami et al. (2017) have reported the use of rice straw and agricultural waste for synthesizing graphene oxide nanoplatelets (GONPs). They noted that the GONPs were similar regarding structure and function to those prepared from graphite in the literature. In another approach, Tamilselvi et al. (2020) transmuted coconut shell and coir into reduced graphene oxide (rGO) via simple catalytic oxidation. Likewise, Hashmi et al. (2020) reported the use of orange peel constituents, sugarcane bagasse, and rice bran for synthesizing the GO. The disposal of polyethylene terephthalate (PET) bottle waste has become a significant challenge worldwide, especially in developing countries for waste and environmental management. Essawy et al. (2017) have reported the potential of utilizing PET bottle waste by the synthesis of graphene. Their results revealed that the prepared graphene has a relatively high surface area. Furthermore, Ding et al. (2020) reported a climate-friendly, scalable, and straightforward process to synthesize transparent graphene films from black liquor.

Sugar and sugarcane waste is hugely disposed of from juice restaurants and industries every day. Considering the large trash accessible and based on several works earlier regulated in turning waste to wealth by procuring waste biomass into usable materials, there is a probability of converting sugar and sugarcane into graphene (Singh et al. 2018; Wadhwa et al. 2020). Green nanotechnology has come out as a flexible platform capable of bestowing efficient, cost-effective, and environmental-friendly solutions to our society's worldwide sustainability debate. Recently, He et al. (2020) reported a simple glucose-blowing approach for a simple, systematic production of graphene-like materials with a large specific surface area and good conductivity. In another different approach,

Suvarna and Binitha (2020) converted graphite via jaggery into graphene using a cost-effective and simple ball milling technique. The material displayed excellent adsorbent behavior to Cr(VI), one of the most toxic and common pollutants from industrial effluent. In this chapter, sugar and sugarcane bagasse potential as an agro-industrial waste is discussed for application as a carbon source in graphene and its derivatives synthesis.

1.3 CHARACTERIZATION OF GRAPHENE-BASED MATERIALS

Sugar, sugarcane extract, and its derivatives are major agricultural wastes, especially in Brazil, India, and China. A novel strategy can be evolved through its proper utility that can be beneficial both economically and environmentally. As sugarcane is having an appreciable amount of cellulose and hemicellulose content, it is used to produce carbon-related materials. Due to the unique bulk and surface structure of carbon materials, the characterization techniques are different from traditional ones. These include X-ray diffraction (XRD), scanning electron microscopy (SEM), transmission electron microscopy (TEM), high-resolution TEM (HRTEM), Raman spectroscopy, Fourier transform infrared spectroscopy (FTIR), X-ray photoelectron spectroscopy (XPS), atomic force microscopy (AFM), ultraviolet-visible spectroscopy (UV-Vis), thermogravimetric analysis (TGA), inductively coupled plasma mass spectroscopy (ICP), X-ray fluorescence (XRF), and scanning tunneling microscopy (STM). Out of all the techniques mentioned here, some of the most widely used ones are discussed in detail that emphasizes graphene identification.

1.3.1 X-Ray Diffraction

XRD is a crucial tool to analyze the crystalline structure of any material. Additionally, the XRD technique is used to estimate the crystallite size and interplanar spacing. Usually, the high-intensity peak at $2\theta = 26.6°$ for a CuKα X-ray source corresponds to pristine graphitic structure, as shown in Figure 1.1(a) (Bo et al. 2014). This peak confirmed the well-arranged-layered structure of graphite, having an interlayer spacing of 0.34 nm along (002) orientations. Another most important function of XRD in characterization is to ensure the reduction of GO to graphene. As the graphite is oxidized to GO, the 2θ peak is shifted left to ~10.02°, further increasing the interplanar distance to ~0.9 nm. The interplanar distance is significantly larger than that of graphite due to intercalation of oxygen-containing functional groups, namely, hydroxyl, epoxy, carbonyl, and carboxyl groups at the carbon basal planes between the GO layers (Lin et al. 2010). After the successful reduction, oxygen-containing functional groups were eliminated, and the peak at ~10.02° completely disappears, whereas

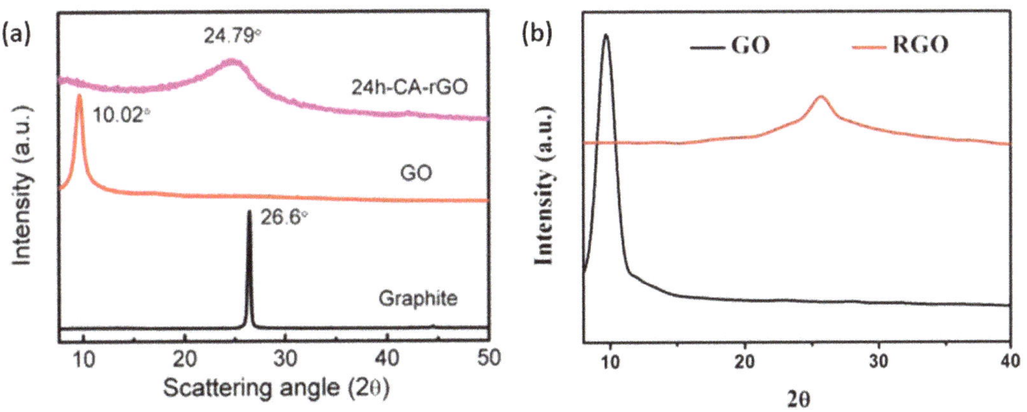

FIGURE 1.1 (a) XRD patterns of pristine graphite, GO, and caffeic acid reduced GO (Bo et al. 2014). (b) XRD patterns of GO and sugarcane bagasse reduced GO (Li et al. 2018b). (Reprinted with permission.)

a broader peak at ~24–26° corresponding to the (002) plane of reduced GO appears (Bo et al. 2014). This indicates the restoration of the π-conjugated structure of graphene due to reduction (Thakur and Karak 2012). Additionally, the interplanar spacing of rGO was reduced to 0.36 nm, which is an evidence for removing oxygen-containing functional groups. The broad peak for rGO implies that the crystal phase (002) is arranged randomly compared to highly crystallized graphite.

Li, Jin et al. (2018) reduced GO using sugarcane bagasse and used it for the removal of cadmium in an aqueous solution. They used XRD analysis to examine GO and rGO. For GO, a sharp intensity peak at $2\theta = 10.8°$ appears corresponding to an interplanar distance of 0.83 nm, whereas a broad peak at $2\theta = 26.5°$ with reduced interplanar distance is observed for rGO (Figure 1.1(b)).

1.3.2 Raman Spectroscopy

Raman spectroscopy is a fast and nondestructive versatile tool to study and characterize different graphitic structures, such as graphite, diamond, carbon nanotubes, graphene, and fullerenes, as shown in Figure 1.2(a). It is basically used to analyze the number of layers, edge chirality, and doping levels. The Raman spectra of all these graphitic structures show different prominent features at D, G, and 2D peaks around 1350, 1580, and 2700 cm^{-1}, respectively. The position, intensity, and line shapes of these peaks give much information regarding their structure, defects, strain, chirality, stacking order, number of layers, and quality.

The G band is common to all sp^2 systems that arise due to the stretching of C–C bond. While the position of G band is insensitive to the excitation frequency, it is highly sensitive to the doping and strain in the layers. Additionally, the D peak is induced by defects, like disordered graphene, nanographene, GO, and rGO. The number of layers in graphene, i.e., monolayer to few-layer graphene to bulk graphite, can be distinguished from the 2D peak. Both D and 2D bands are highly sensitive to the excitation frequency and their position and shape vary with the same. Therefore, all the measurements related to D and 2D bands should be done with the same excitation frequency. The intensity ratio of D and G peaks (I_D/I_G) is used to analyze the level of disorder or defect density and is inversely proportional to the average crystallite size of the sp^2 clusters (La), using Tuinstra and Koenig equation (as shown in Equation 1.1) (Cançado et al. 2006)

$$\mathrm{La(nm)} = (2.4 \times 10^{-10})\lambda^4 (I_D/I_G)^{-1} \tag{1.1}$$

where λ is the wavelength of the laser beam used in Raman experiment.

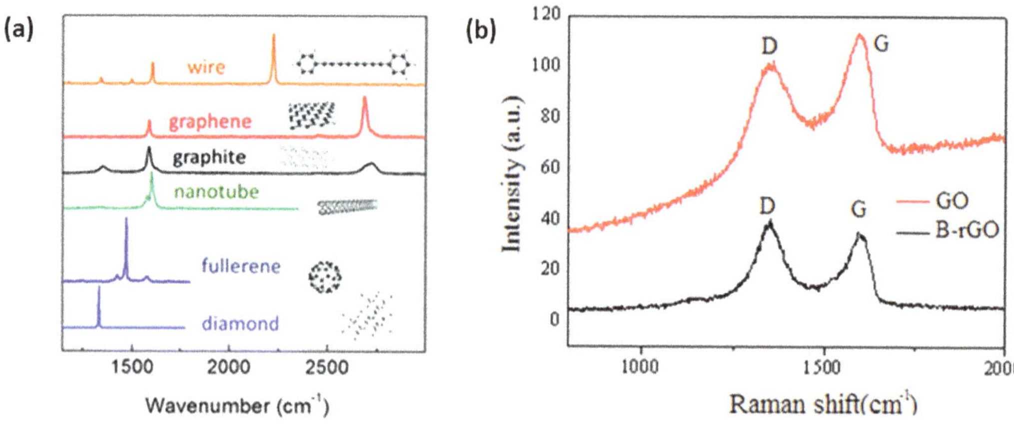

FIGURE 1.2 (a) Raman spectra of various carbonaceous materials. (b) Raman spectra of GO and sugarcane bagasse reduced GO. (Reprinted with permission from Gan et al. 2019.)

Gan et al. (2019) used bagasse rGO for the application in methyl blue dye removal. When the GO is analyzed by Raman spectroscopy, the G band is broadened and blue shifted to 1581 cm^{-1}. Furthermore, a prominent D band is visible at 1353 cm^{-1}. In contrast, the rGO shows the presence of D band at 1354 cm^{-1} and G band at 1584 cm^{-1}. It can be observed that the I_D/I_G ratio successively increases on reduction that indicates the decline in the sp^2 domain size (Figure 1.2(b)).

1.3.3 Electron Microscopy

SEM and TEM are used to visualize the structure and morphology of graphene-based materials. Usually, GO shows wrinkled and folded structures due to a significant amount of oxygen present in hydroxyl, epoxy, or carboxyl groups. These groups are covalently bonded to carbon atoms, destroying the sp^2 network (Lee and Kim 2014). When the GO is reduced to form rGO, some structure is recovered, leading to the removal of few wrinkles and folds. Li, Jin et al. (2018) showed the wrinkled and folded structure of GO (Figure 1.3i(a and b)), which was improved on the successful reduction by sugarcane bagasse. Apart from the morphology, field emission scanning electron microscopy (FESEM) is helpful in determining the elemental composition. The elemental content of carbon and oxygen is plotted by energy dispersive spectroscopy (EDS) analysis. The C/O ratio increased from 1.02 for GO to 4.27 for rGO. While the conventional SEM has a focus on the wrinkles and folded structures of graphene, recent advances in low-voltage SEM combined with an interplay with the working distance have helped to unravel the graphene domains having a different number of layers or thickness (Huang et al. 2018; Huang et al. 2019; Mikmeková et al. 2016). Huang et al. (2018) concomitantly tuned the accelerating voltage (to a lower value) and working distance (to a medium value) to enhance the collection efficiency of secondary electrons originating from the graphene layers. Furthermore, it was also proposed that the tilting of the sample could be effectively

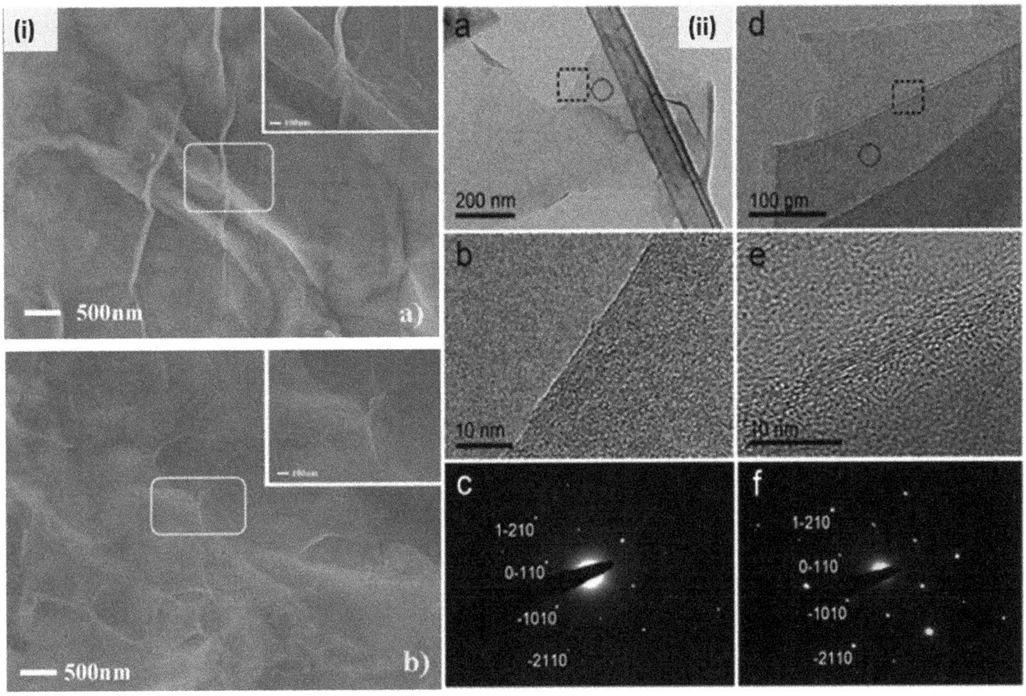

FIGURE 1.3 (i) FESEM images and EDS spectra of wrinkled GO and rGO (Li et al. 2018b), (ii) TEM characterization of graphene: (a) graphene sheet, (b) magnified image, (c, f) the electron diffraction patterns, (d) few-layer graphene sheet, (e) magnified image (Geng et al. 2010). (Reprinted with permission.)

employed to minimize the channeling contrast arising from different orientations of the substrate (Huang et al. 2019). Atomic resolution lattice images of graphene have also been observed within a SEM using a bright field detector of scanning TEM (Sunaoshi et al. 2016). With such advances, large area characterization of graphene at the atomic scale becomes much easier.

The TEM is high-potential microscopy that can observe the morphology and helps to determine the thickness and surface defects of graphene and related materials (Hernandez et al. 2008; Vij, Tiwari, and Kim 2016). By cross-sectional view of graphene, a number of layers present can be measured using TEM (Takamura et al. 2015). It is often associated with the electron diffraction pattern, which clearly shows the hexagonal pattern of graphene crystal structure (Geng et al. 2010) (Figure 1.3ii). Disordered or defective edges of graphene result in variation in the bandgap that could in turn degrade the carrier mobility, limiting their potential applications (Jia et al. 2011). In-situ heating inside an HRTEM makes it easier to study the defects along the edges of nanoribbons of graphene (Tan et al. 2009).

1.3.4 X-Ray Photoelectron Spectroscopy

XPS is a powerful surface-sensitive spectroscopic technique used to determine the elemental composition and chemical bonding, crucial for a better understanding of graphene derivatives. XPS was not a major technique for characterizing graphene until the recent preparation of graphene via GO and then rGO came to existence. The most important parameter to study the extent of oxidation/reduction is sp^2 carbon fraction, which is affected by functional groups' presence. XPS spectra are obtained by measuring the number of electrons that escape from the material's surface (<10 nm) and the kinetic energy when it is exposed to X-rays. The GO and rGO survey-scan spectra show two high-intensity peaks at binding energies of about 284 and 530 eV, popularly known as C1s and O1s peaks, respectively (Al-Gaashani et al. 2019). The C1s peak is further deconvoluted for oxygen-containing functional groups at 284.5 eV (C-C), 286.4 eV (C-O), 287.8 eV (C=O), and 289.0 eV (COOH) (Zhang et al. 2010) (Figure 1.4). The area covered by the peaks determines the percentage of respective functional groups present.

1.4 GRAPHENE FROM SUGAR

1.4.1 Sugar (Table Sugar) as a Precursor

Yearly a massive amount of wastes are generated from the agro-processing industries and becomes one of the severe environmental troubles. The carbon content of these wastes is more than 65%.

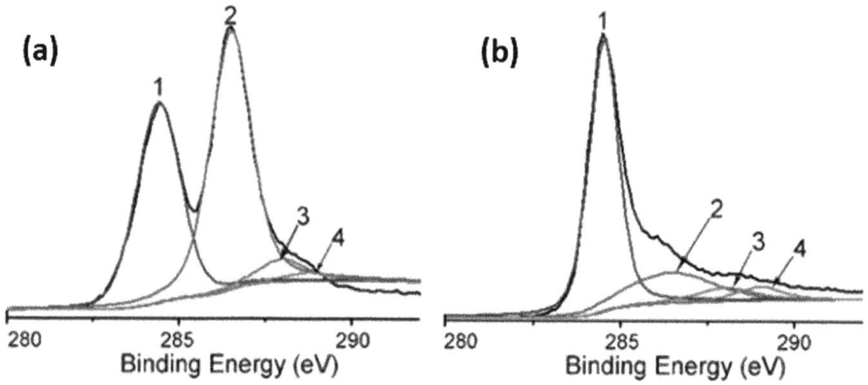

FIGURE 1.4 (a) XPS spectra of C 1s of GO and (b) GO reduced by L ascorbic acid. (Reprinted with permission from J. Zhang et al. 2010.)

These carbon sources are sustainable and affordable. In this aspect, sugar and sugarcane bagasse are significant agro-industries wastes studied recently as substitute prime precursors for the development of graphene and its derivatives (Wadhwa et al. 2020).

The synthesis of carbon-based materials from sugar was started as early as 1996 when Xing, Xue, and Dahn (1996) showed carbon materials formation by pyrolysis of table sugar. In this experiment, sugar precursors were put in a container and inserted into a quartz tube of the horizontal tube furnace. Sugar precursors were warmed up under an argon atmosphere at a heating rate of 25°C/minute from 100°C to pyrolyzing temperature, held for 1 hour, and further cooled down to room temperature. The pyrolyzing temperature was varied from 600 to 1600°C for different runs. The obtained final product exhibited a high capacity for Li intercalation, better than that of graphite. In 2004, Novoselov et al. (2004) reported the successful synthesis of monocrystalline graphitic films. This few atoms thick two-dimensional (2D) carbon material is referred to as few-layer graphene. After the discovery of graphene, its demand started increasing for large-scale production. Since there was already an attempt of pyrolyzing sugar for the synthesis of carbon materials, researchers began experimenting with sugar and its derivatives for the mass-scale production of graphene. Sugar being a less expensive and readily available precursor has gained lots of attention in recent time. The first report on graphene synthesis from sucrose (table sugar) was found in 2011 (Ruiz-Hitzky et al. 2011). In this work, a natural, eco-friendly, low-cost sucrose precursor was first transformed into caramel in the presence of sepiolite. This nanocomposite was further thermally treated at 800°C in an N_2 atmosphere. The resulting carbonaceous materials were found to exhibit typical characteristics of graphene layers. Their work opened the direction for the possibility of producing graphene-like materials from sucrose. In 2013, a similar kind of approach was used by Pan et al. (2013). They used edible sugar to synthesize nitrogen-doped graphene (NG) by pyrolysis of a homogenous mixture comprising urea and crystal sugar. This mixture was firstly dissolved in water to form a homogenous solution and warmed to remove excess water. This mixture was dried in the furnace at 800, 900, and 1000°C for 1 hour named NG800, NG900, and NG1000, respectively. The resultant nitrogen-doped graphene was of high surface area and pore volume. Figure 1.5(a and b) shows the nitrogen-doped graphene's TEM images at 1000°C. In a similar approach, Liu and Antonietti (2013) reported the synthesis of highly porous N-doped graphene nanoplatelets by exposing a mixture of sugar and nitrate. The methodology is extended to S-doped porous carbon sheets by replacing the nitrate with sulfate. With increasing attention in the synthesis of nitrogen-doped graphene, Shi et al. (2016) came up with a new way of utilizing N-doped graphene for catalytic application. They demonstrated controlled synthesis of Fe_3C-based CNT composites encapsulated with N-doped graphene layers via direct pyrolysis of glucose, melamine, and $FeCl_3$ at varied temperature. They claimed

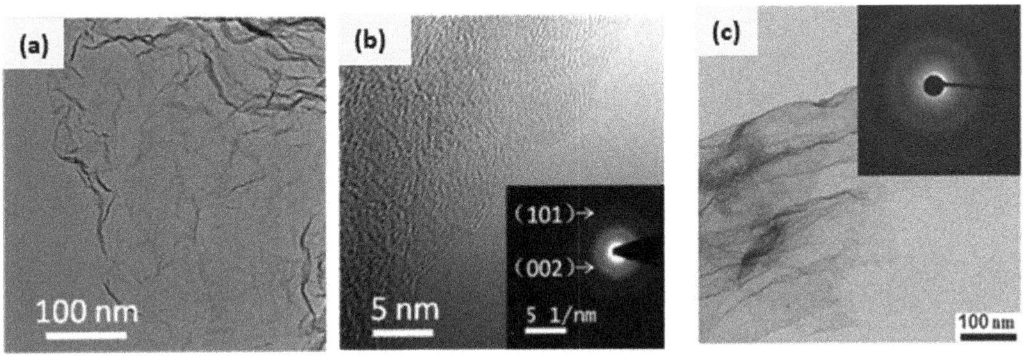

FIGURE 1.5 (a) TEM and (b) HRTEM images of N-doped graphene; the inset of (b) indicates SEAD pattern. (c) TEM images of the MS-derived graphene. (Reprinted with permission from Pan et al. 2013 and Liu, Giordano, and Antonietti 2014.)

that doping a graphitic structure into an N-doped graphene led to a non-uniform electron distribution leading to a significant enhancement of catalytic activity of carbon surface. In a little different carbonization approach, Liu, Giordano, and Antonietti (2014) proposed using an ionic molten-salt (MS) medium to convert glucose into nanoporous carbon. The MS-derived graphene scaffolds are hydrophobic and exhibit high selectivity and capacity for absorption of organics. Figure 1.5(c) shows the TEM image of the MS-derived graphene.

In an almost similar process, B. Zhang et al. (2014) have manufactured graphene on a large scale using glucose. It was claimed that the graphene produced in this method had electrical conductivity approximately equal that of graphene film produced by CVD. In this method, glucose and $FeCl_3$ were dissolved in water and vaporized at 80°C in the air to remove water. This prepared carbonized glucose was calcined at 700°C under the flow of argon. Graphene sheets were acquired after washing with solvents and drying. Raman and HRTEM confirmed the formation of graphene. Figure 1.6(a and b) presents TEM images of graphene produced by this method. In further development of the process, Seo et al. (2013) described a fresh approach for the growth of vertical graphene nanosheets (VGNSs) using five different kinds of palatable products, i.e., sugar, honey, butter, milk, and cheese. VGNS was found to be synthesized by a low-temperature (450°C) plasma technique. However, it was reported that the VGNSs grown by sugar showed more defects as compared to other products.

In a further progress in graphene growth, direct thermal decomposition of sugar has confirmed superior quality graphene and reproducibility at a large scale. Chithaiah et al. (2020) recently reported a rapid method for synthesizing rGO nanosheets by direct thermal decomposition of sugar at 475°C, without any hazardous chemicals. Figure 1.7(a) shows SEM images of the rGO nanosheets produced through this method. The rGO nanosheets produced in this method were composed of a large number of nanosheets. The TEM image (Figure 1.7(b and c)) shows that the rGO sample comprises nanosheets with a smooth surface. By a similar process called solvothermal method, Huang et al. (2012) demonstrated the synthesis of graphene with tunable edge defects. Oxygenated groups when attached to tunable mount defects could be useful as a humidity sensor by adsorbing humidity gas molecules. In a quite comparable process, Choi et al. (2019) synthesized sugar-based magnetic pseudo--GO (SGO). This was also prepared by pyrolyzing the sugar at 750°C for 1 hour and then oxidizing the obtained pyrolyzed sugar using ozone as an oxidant. Upcoming research works need to be pointed to acknowledge the mechanisms of both the growth processes of graphene from sugar and applying the knowledge in developing practical devices.

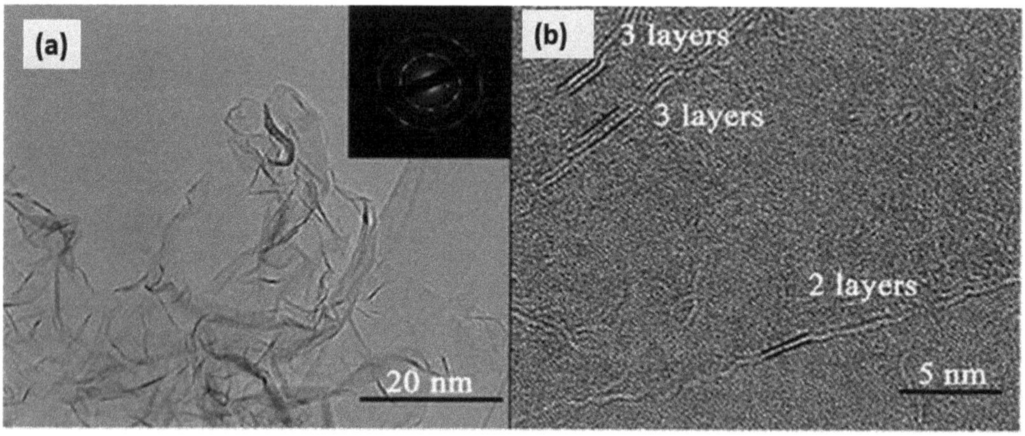

FIGURE 1.6 (a) TEM image of the graphene sheets prepared from glucose (inset SAED pattern). (b) The TEM image of the graphene sheets. (Reprinted with permission from Li et al. 2014.)

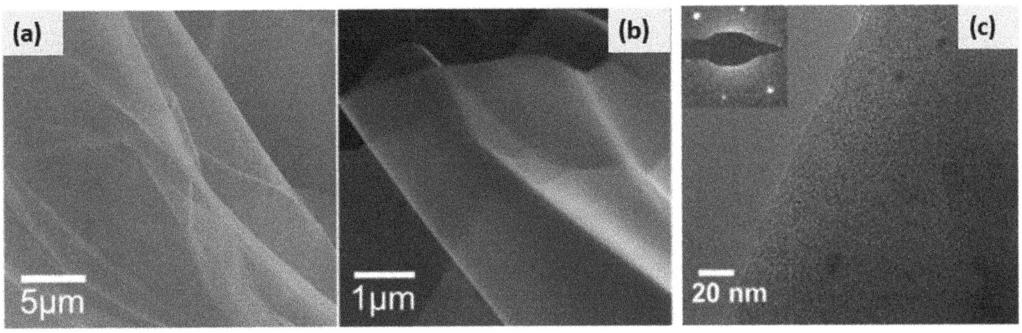

FIGURE 1.7 (a) SEM Image. (b and c) TEM and SAED patterns (inset) of the rGO. (Reprinted with permission from Chithaiah et al. 2020.)

1.4.2 Sugar as a Reducing Agent

One of the most efficient and widely used methods for the synthesis of graphene is the Hummers method. But the chemical reduction of GO is generally carried off using hydrazine/hydrazine derivatives as the reducing agent. Woefully, the use of highly pernicious hydrazine to reduce GO is hazardous. However, other potential reducing agents such as hydroquinone and $NaBH_4$ may produce environmental unacceptable products that can hamper its larger-scale production potential. Sugar, being non-toxic and environmentally friendly, can be used as a reducing agent due to its lenient reductive ability (Li, Xing, and Wang 2012; Qian, Lu, and Gao 2011; Shen et al. 2011a; Shen et al. 2011b; Wang et al. 2011; Wojtoniszak and Mijowska 2012). In 2010, Zhu et al. (2010) first reported the use of glucose as a reducing agent for an ecofreak and leisurely technique toward the synthesis of water-soluble graphene nanosheets (Figure 1.8(a)). The process used the modified Hummers method to synthesize GO. Then, glucose was added into the aqueous dispersion of GO with ammonia solution and stirred for 60 min at 95°C. Ultimately, the final stable black dispersion was centrifuged and washed with water. They observed that GO reduced rapidly in the presence of glucose. Graphene nanosheets (GNSs), produced in this method, was confirmed by the TEM, as shown in Figure 1.8(b), and Raman spectroscopy results (see Figure 1.8(c)). In a similar process, SuryaBhaskaram, Cheruku, and Govindaraj (2014) have synthesized rGO using glucose as a reducing agent. In recent years, the simplicity of this process has attracted much attention for using sugar as a reducing agent, as this process may be a feasible technique in mass-scale production of rGO (Bhargava et al. 2019; Lau et al. 2019; Salazar et al. 2019; Shang et al. 2019).

An individualistic study by Kamisan et al. (2016) has also reported related properties by reducing GO with glucose as a reducing agent with ultrasonication. They synthesized GO by the modified Hummers method, and glucose was added to GO following ultrasonic irradiation. The procured samples were centrifuged and washed repeatedly with distilled and vacuum dried at 50°C for further use. In a slight modification to this technique, Narayanan, Park, and Han (2020) have proclaimed the synthesis of electrospun poly(vinyl alcohol) (PVA)/rGO nanofibrous scaffold. Here the GO was produced by the modified Hummers method. Glucose was added to the homogeneous solution of GO along with NH_4OH for reduction. The mixture was then stirred powerfully for a few minutes and heated to 90°C for 3 hours. Ultimately, the mixture was centrifuged, washed, and dried to accumulate the glucose-reduced GO (GrGO). It was claimed that the presence of GrGOin PVA nanofibers enhanced the thermal properties of scaffolds. The glucose-based reduction is one of the most suitable one-pot synthesis processes for rGO production. The simplicity of the procedure captivated the heed of the scientific society, and the same kind of process to produce rGO by using glucose was followed by many research groups globally (Bhargava and Khan 2018; Casa et al. 2018; Çıplak 2018; Hossain, Das, and Park 2017; Shuvra, Das, and Barui 2020).

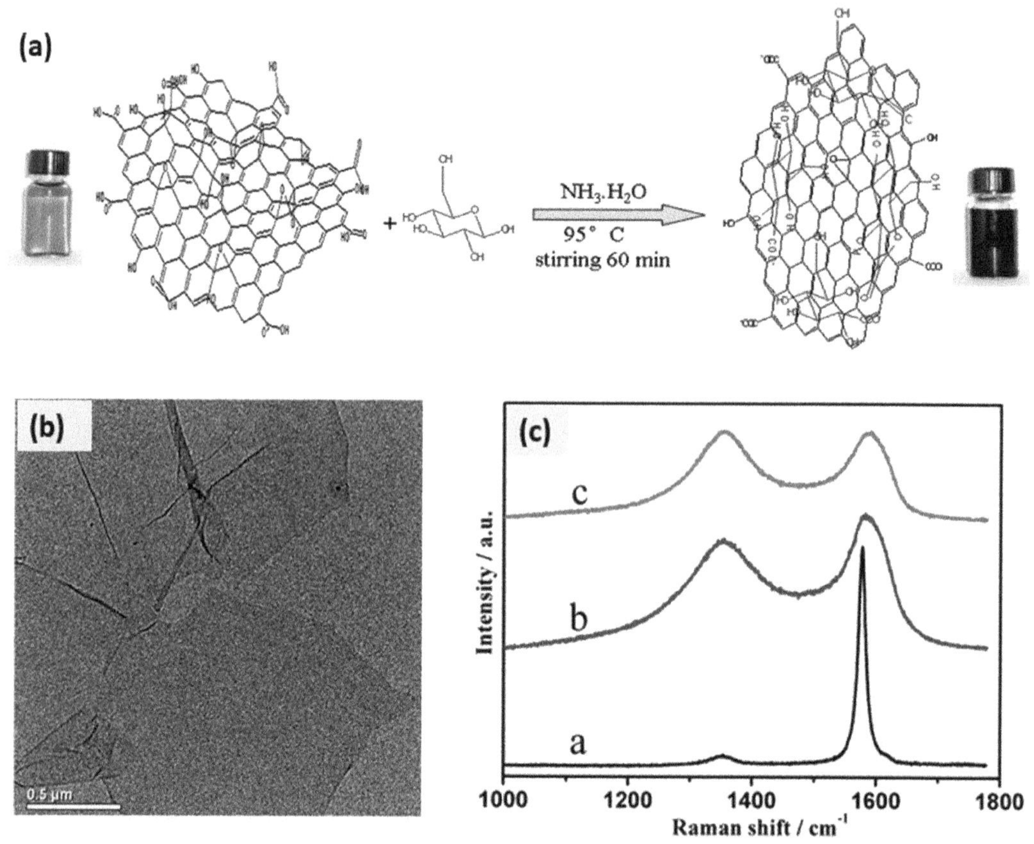

FIGURE 1.8 (a) TEM of the GNS (c) Raman spectra of pristine graphite (a), GO (b), and GNS (c) after its reduction with glucose. (Reprinted with permission from Zhu et al. 2010.)

1.4.3 Three-Dimensional Graphene from Sugar

Three-dimensional graphene is a propitious material because of its potential applications in environmental protection, energy storage, and conversion. Recently, significant attempts have been dedicated to designing and manufacturing 3D graphene from sugar as a precursor or a reducing agent (Cheng, Du, and Zhu 2012; He et al. 2020; Zhang et al. 2017). A recent finding by Wang et al. (2013) showed that the sugar-blowing method could be used to synthesize 3D strutted graphene (SG) (Figure 1.9(a)). In this approach, sugar was mixed with ammonium salts (NH_4Cl) and treated at 1350°C for 3 hours under an Ar atmosphere in a tube furnace. A black foam-like product, i.e., SG, was collected. The SG has hierarchical porous structures at different scales: bubbles, ripples, and concaves (Figure 1.9(b)). Graphitic membranes circumscribe the bubbles joined onto graphitic struts (Figure 1.9(c and d)). The yield of final product was 16 ± 5 wt% related to raw glucose. Also, the estimated cost found to be only \$0.5 per gram (laboratory). A similar approach was reported by Jiang et al. (2015) for the fabrication of SG by ammonium-assisted chemical blowing. In this approach, a household sugar was directly mixed with NH_4Cl and $(NH_4)_2CO_3$ and heated in a horizontal furnace at 1400°C and annealed for 3 hours under an Ar atmosphere (Figure 1.9(e–g)).

Eventually, several research groups ameliorate the sugar-blowing-assisted method. For a moment, a gas-foaming technique was flourished to formulate 3D graphene layers to reduce GO. The advantage of this technique lies in averting the restacking of graphene sheets. Nevertheless, 3D graphene still has frail connections that are inseparable (Hao et al. 2015). Then, an advanced practice was used to assemble graphene-like carbon nanosheets with glucose as a carbon precursor, Fe species as a graphitization catalyst, and NH_4Cl as a blowing agent. But the catalyst contaminations

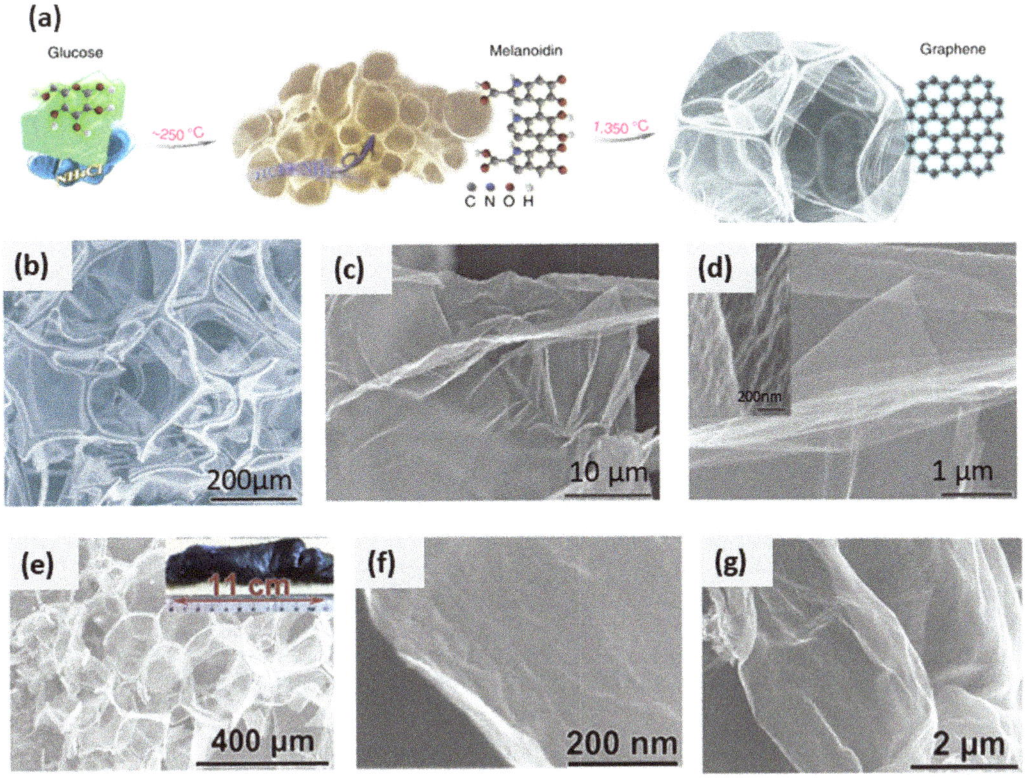

FIGURE 1.9 (a) Growth process of sugar-blowing production. (b–d) SEM images of the SG. (e–g) Photograph and SEM image of the SG produced by ammonium-assisted chemical blowing at different scale. (Reprinted with permission from Wang et al. 2013 and Jiang et al. 2015.)

could hamper the purity of 3D graphene (Lei et al. 2015). All these problems were sorted by Han et al. (2018). They reported a catalyst-free simple approach for producing high-quality 3D porous graphene (3D-PG) by the sugar-blowing method (Han et al. 2018). Olszowska et al. (2017) have propounded a comprehensive review on the synthesis and application of three-dimensional nanostructured graphene.

Aside from these main advancing routes, there have been efforts to produce free-standing 3D graphene foams (GFs) through the (3D-printing method (Sha et al. 2017). In this approach, a mixture of Ni and sucrose was manually placed onto a platform, and a commercial CO_2 laser was used to convert the Ni/sucrose mixture into 3D GFs. It was claimed that the 3D-printed GFs showed multilayered graphene with low density (~0.015 g/cm^3) and high porosity (~99.3%). Figure 1.10 shows the SEM images of the 3D-printed GFs after the three-step removal of the Ni scaffolds.

1.4.4 Graphene Composite from Sugar

Current research has observed expeditious advances in the synthesis of graphene-based composites due to their diversity of applications in numerous technologies (Hoa et al. 2017; Hossain, Das, and Park 2017; Shang et al. 2019). The graphene-based composites were generally developed using GO as the starting precursor and glucose as a reducing agent to convert GO to rGO. The metal oxide blossom on graphene performs a vital role in modifying the numerous properties of the composite (Çıplak 2018; Wadhwa et al. 2020). Particularly, these nanocomposites exhibit superior electrochemical performance due to a synergistic effect arising from the high surface area of graphene and redox reactions from the transition metal oxides. Recently, Bhargava and Khan (2018) have produced

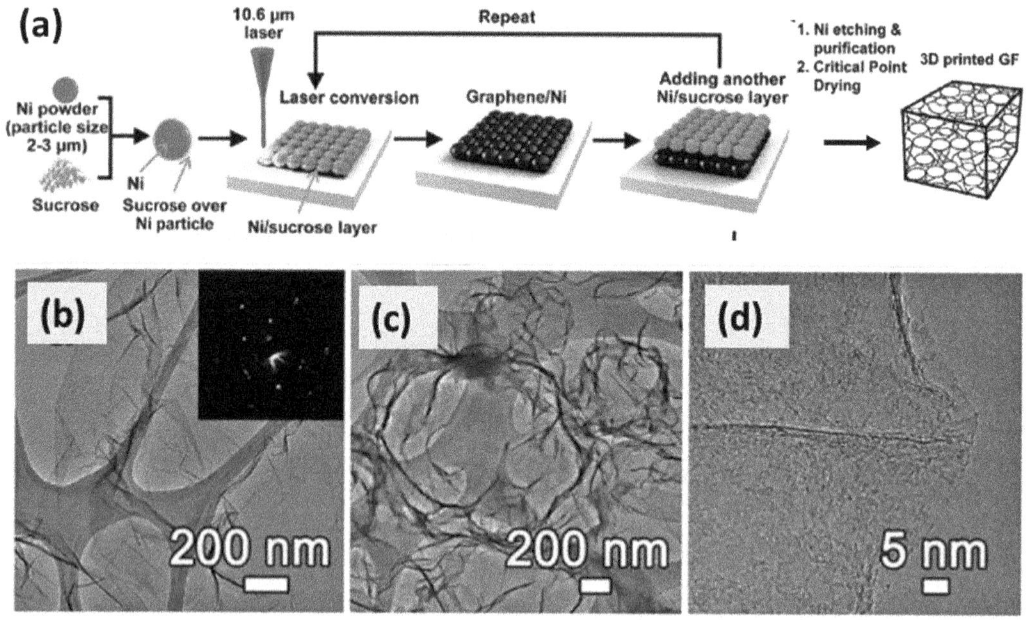

FIGURE 1.10 (a) Schematic of synthesis of 3D GF using 3D printing. (b–d) TEM images of GFs after removing Ni scaffolds. The inset in (b) is the SAED pattern. (Reprinted with Permission from Sha et al. 2017.)

graphene-metal composites by chemical reduction. Initially, the solution of GO and $CuSO_4$ was mixed and stirred, followed by dropwise addition of NaOH into the solution. Once the homogeneous solution formed, glucose powder was added and heated at 70°C. Finally, the resultant precipitates were washed and dried. The obtained nanocomposites are called Cu_2O/rGO. Figure 1.11(a) shows spherical Cu_2O nanoparticles grown on the rGO sheets. In a similar approach, Zhigang (2014)

FIGURE 1.11 (a) FE-SEM images Cu_2O/rGO nanocomposites (Bhargava and Khan 2018). (b, c) TEM image for rGO/Au. (d, e) TEM image for rGO/Ag nanocomposites (Wu et al. 2017). (f) and (g) TEM images of G-rGO-PB and corresponding SAED pattern of G-rGO-PB (Zhang et al. 2017). (Reprinted with permission.)

also reported an efficient way to synthesize a cuprous oxide-graphene composite using glucose. In a similar process, Wu et al. (2017) reported the synthesis of an rGO/metal (oxide) (e.g., rGO/Au and rGO/Ag) composite using glucose as the reducing agent and the stabilizer (Figure 1.11(b–e)). In a further development of the process, graphene-Prussian Blue hybrid composites were developed by using glucose as a reducing agent (G-rGO-PB). This hybrid composite was reported to exhibit as a high-performance material for biosensors and supercapacitor electrodes (Figure 1.11(f and g)) (Zhang et al. 2017). In recent years, lots of research efforts have been devoted to glucose-mediated metal-graphene composite synthesis for a large-scale growth of the graphene composite for enhanced application (Ge et al. 2016; Li et al. 2016; Ma et al. 2016).

Predominantly, graphene synthesis techniques from sugar accessible to date have been reviewed in this section. The methodologies range widely from the pyrolysis of sugar to the sugar-blowing method, the direct thermal decomposition of sugar, and the plasma technique. After the discovery of graphene in 2004, recent years have seen massive growth in mass-scale graphene synthesis from sugar. Upcoming research attempts on graphene synthesis should govern large-scale graphene production to escort a new uprising in the electronics and semiconductor industries.

1.5 GRAPHENE FROM SUGARCANE EXTRACT

All biomasses (including sugarcane) consist of three main components: cellulose, hemicellulose, and lignin (Kim and Day 2011). Among all the features, cellulose shares the most considerable portion and attracts significant research interest in the field of carbon-based nanomaterials. After extracting out the valuable sugar from the plant, the leftover part is known as sugarcane bagasse. It serves as a material of choice to synthesize graphene-based materials due to its easy availability, low cost, and biodegradable nature. There are various routes for synthesizing graphene from sugarcane extracts, which are discussed in the next subsections.

1.5.1 Sugarcane Extract as a Precursor

In recent years, various graphene derivatives were produced using graphite as a precursor material. But using agro-wastes as the precursor could help produce graphene and a pollution-free environment. Usually, sugarcane bagasse is the most suitable agro-waste widely explored due to its high economic value and the best environmental and ecological benefits. Long et al. (2019) converted cellulose bundle crystal extracted from sugarcane leftover to few-layer cellulose (two-dimensional) by a specific technique, as shown in Figure 1.12. The two-dimensional cellulose was converted into few-layer graphene by pyrolysis, whereas the cellulose bundle can produce graphite.

Xiao et al. (2017) synthesized a graphene-like material (FZS900) by sugarcane carbonization and activation. Initially, $ZnCl_2$ and $FeCl_3$ were added to bulk sugarcane biomass and kept for stirring at 80°C for 2 days. Then the mixture was dried at 80°C for 5 days. Afterward, it was heated in the furnace at 900°C for 1 hour and washed in HCl and Milli-Q water several times. The obtained graphene material exhibited abundant micropores and high surface area (2280 m^2/g), supporting its high adsorption capacities for naphthalene, 1-naphthol, and phenanthrene. Suvarna and Binitha (2020) prepared holey graphene by jaggery using the ball milling technique. Holey graphene is a particular category of functionalized graphene where functional groups are mostly attached at the edges and holes, providing it high surface area, making it suitable for various applications (Lin et al. 2017).

Graphitic carbon is also the most suitable candidate as an anode material in rechargeable lithium ion batteries (RLiBs), and it is claimed that RLiBs are the best choices as environment-friendly energy effective devices. Alvarez et al. (2017) prepared carbonaceous materials using palygorskite and sugarcane molasses and examined its electrical, thermal, and electrochemical properties

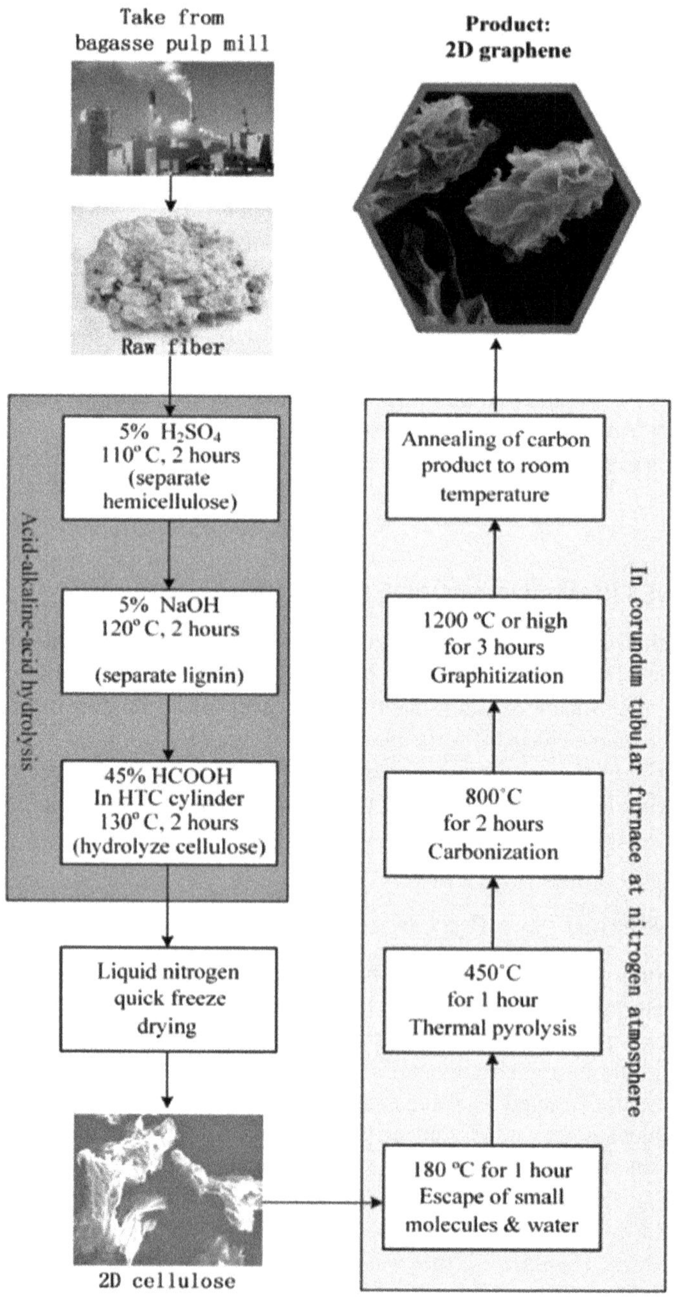

FIGURE 1.12 Synthesis technique of two-dimensional graphene from sugarcane bagasse. (Reprinted with permission from Long et al. 2019.)

suitable as an anode material in RLiBs. The palygorskite-graphene-like carbon nanocomposite was further oxidized to GO like carbon (GOLC) by a two-step treatment with acids. GOLC had a good combination of a large specific surface area (467 m^2/g), high electrical conductivity (1 S/cm in the range of 173–293 K), and good thermal conductivity (1.1 W/m/K) that makes it an excellent candidate for both as an anode and a cathode in RLiBs. The morphology of both the graphene-like materials was studied by SEM and TEM images (Figure 1.13).

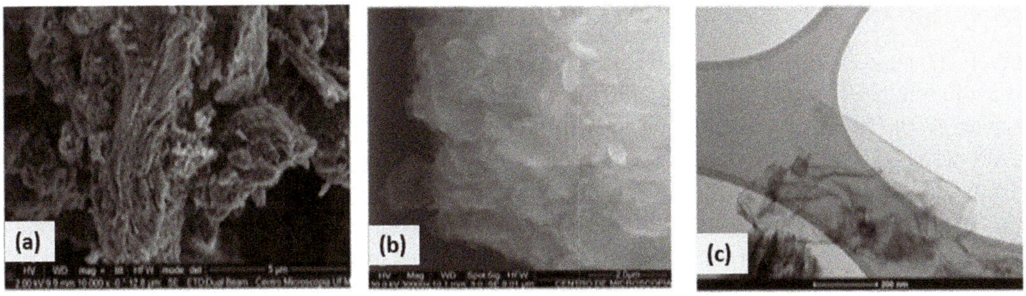

FIGURE 1.13 (a) SEM images of the palygorskite-graphene-like carbon and (b) GOLC. (c) TEM image of GOLC. (Reprinted with permission from Alvarez et al. 2017.)

In the present era of carbon and related materials, the research community is working to explore different carbonaceous materials. Tang et al. (2018) fabricated a different type of glassy carbon known as graphene microcrystal (GMC) from lignin. Usually, petroleum, coal, or natural gases are the traditional precursors of glassy carbon. But the overconsumption of these fossil fuels is an important issue. To deal with the growing environmental and ecological problems, biomass is the perfect replacement. The lignin is the renewable biomass that can be extracted from sugarcane bagasse. Tang et al. (2018) refined lignin from sugarcane bagasse and further fabricated GMC by two techniques: the pyrolysis of lignin in a tubular furnace and hydrothermal carbonization of lignin, followed by pyrolysis. The TEM, XRD, and Raman spectra of GMC are very close to previously reported rGO images and are shown in Figure 1.14.

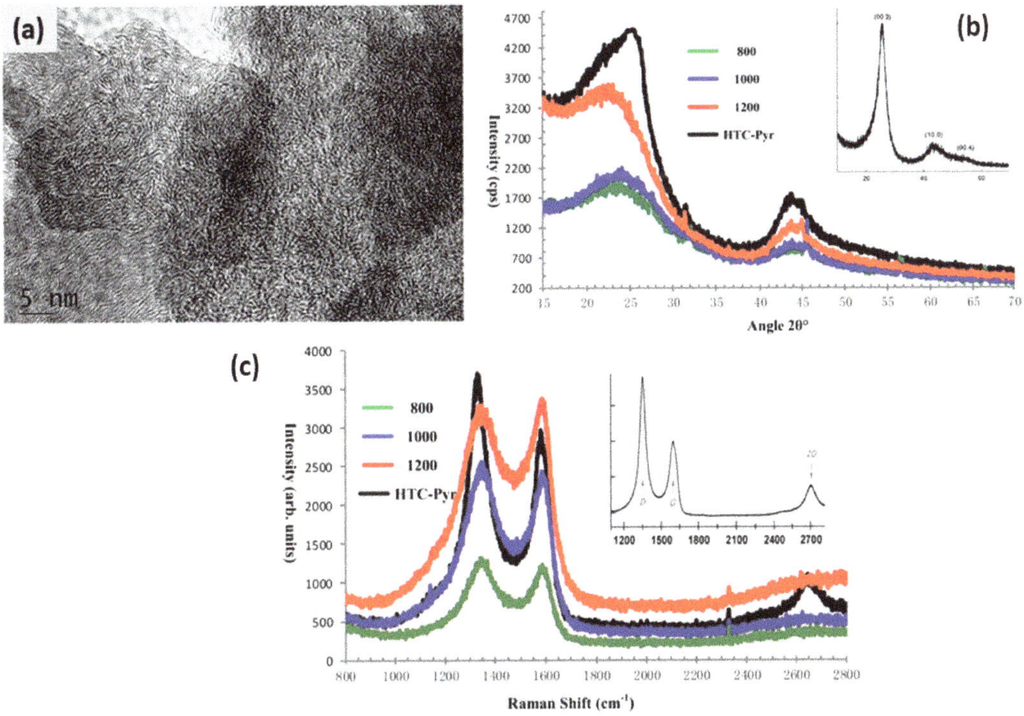

FIGURE 1.14 (a) TEM, (b) XRD, and (c) Raman of GMC, prepared at different temperatures. (Reprinted with permission from Tang et al. 2018.)

1.5.2 Sugarcane Extract as a Reducing Agent

Graphene in large quantities is usually synthesized by reducing GO by various reducing agents such as hydrazine. Still, the high cost and the hazards caused to the environment restrict the approach. Replacing the harmful reducing agents with green reductants is an environmental-friendly way to produce graphene in high quantity. In recent years, many green reductants such as vitamin C (Fernández-Merino et al. 2010), green tea (Weng et al. 2019), aloe vera (Ramanathan et al. 2017), etc. have been employed for this purpose. Other than these widely used green reducing agents, sugarcane bagasse is an appropriate reductant to reduce graphene oxide. Graphene oxide is synthesized by the Hummers method and further reduced by grinded sugarcane bagasse powder. Other than reduction, sugar extracts play an important role as a stabilizer (Li, Jin et al. 2018). To investigate the extent of reduction, Li, Jin et al. (2018) characterized the obtained rGO by XRD, SEM, Raman spectroscopy, and XPS. All the observations were in the agreement of previously reported rGO. The XPS analysis of GO and rGO gives the evidence for better reduction as the C/O ratio increased from 2.03 to 3.37 for GO to rGO, respectively. Additionally, rGO synthesized using sugarcane bagasse was considered suitable for removing cadmium from aqueous solution.

Excessive use of dyes in industries like paper, textile, food, and plastics is a serious threat to present ecosystem. Various techniques like reverse osmosis, ozone treatment, photochemical oxidation, and activated carbon adsorption have been adopted for the removal of harmful dyes from water. Recently, Naveen et al. (2015) used the graphene silica composite (GSC) synthesized by sugarcane juice for the adsorption of harmful industrial dyes such as Eriochrome black T, Methyl red and Congo red. The adsorption capacity of different dyes was investigated by UV-Vis spectra by varying the dye concentration. Gan et al. (2019) also used GO reduced by bagasse for the removal of methyl blue dye.

Other than applications in supercapacitors, water purification, and dye removal, the chemically reduced graphene has an important role to reduce environmental pollution. Many toxic and hazardous polycyclic aromatic hydrocarbons (PAHs) are being released in the environment by incomplete burning of fossil fuels. Singh et al. (2018) synthesized a sheet and disc-shaped graphene by two approaches (rG1 and rG2) using sugarcane extract and used them for the photodegradation of PAHs under the UV irradiation. The schematic of synthesis steps of rG1 and rG2 are shown in Figure 1.15(a). The large surface area and π-π conjugation of rGO make it a better candidate as a photocatalyst material. The XRD peaks of rG1 and rG2 appeared at $2\theta = 24.75°$ and $2\theta = 21.52°$, respectively, as shown in Figure 1.15(b). This change in interplanar distance can be related to change of their structures due to different approaches. Figure 1.15(c and d) shows the TEM images of rG1 and rG2.

Li, Gan et al. (2018) developed a new biomaterial from sugarcane bagasse that contained rGO and Burkholderia cepacia. This new biomaterial was able to adsorb and biodegrade malachite green, a harmful triphenylmethane dye extensively used in the textile industries. An aqueous solution of sugarcane bagasse was added to the GO suspension and ammonia was added to the mixture. Sugarcane bagasse consists of fructose that is transformed into reducing sugars due to ketonal tautomerism in the presence of ammonia base (Zhu et al. 2010). It acts as a better reducing agent for the transformation of GO to B-rGO. The Raman spectra of both GO and B-rGO show two main characteristic peaks (G peak at 1612 cm^{-1} and D peak at 1356 cm^{-1}), as shown in Figure 1.16. The I_D/I_G ratio of GO (0.86) significantly increases for B-rGO (1.32), which indicates that the graphene sheets in B-rGO are disordered, and the reduction causes a decrease in the size of the in-plane sp^2 domain.

1.6 APPLICATIONS

One of the potential applications of graphene is in energy conversion devices (Çıplak 2018). To prosper a futuristic energy device, an active electrode material with high capacity is indispensable. Graphene is transpiring as a distinctive morphology of carbon materials with capability for

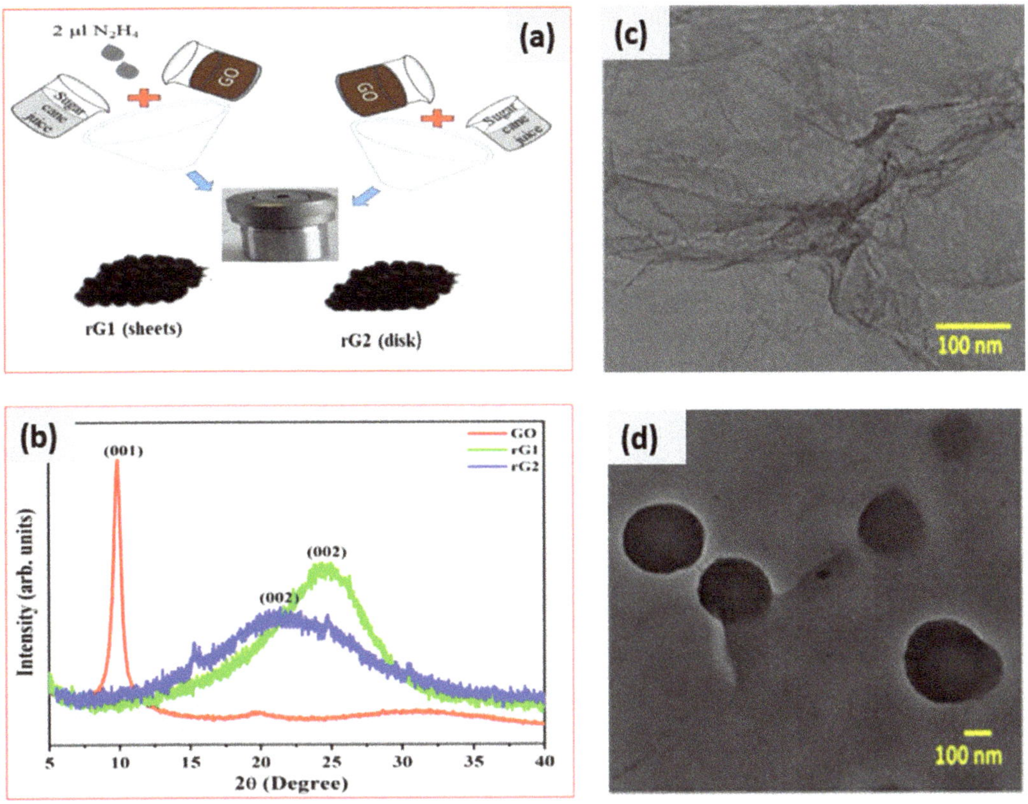

FIGURE 1.15 (a) The schematic diagram of synthesis steps for rGO sheets and disk-type structures (rG1 and rG2) using sugarcane juice. (b) XRD pattern of GO, rG1, and rG2. (c) TEM of rG1 and (d) TEM of rG2. (Reprinted with permission from Singh et al. 2018.)

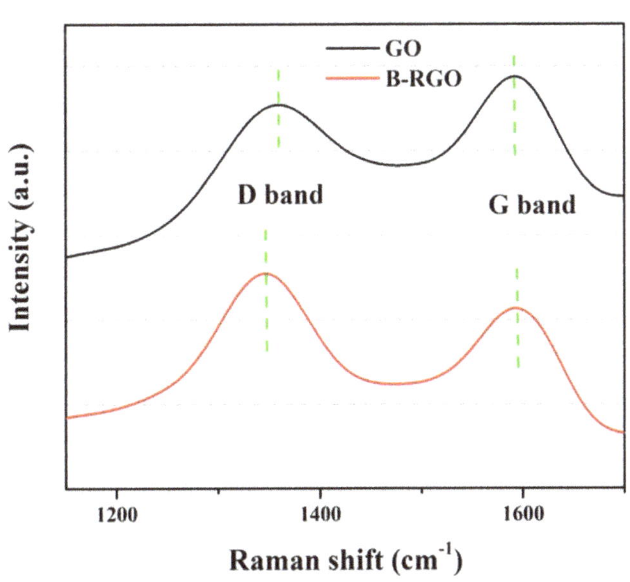

FIGURE 1.16 Raman spectra of GO and sugarcane bagasse reduced GO. (Reprinted with permission from Li et al. 2018a.)

electrochemical energy storage device applications to sensors, drug delivery, biomedical applications, batteries, etc.

He et al. (2020) recently fabricated a graphene-like foam/NiO (GLF/NiO) composite for high-performance electrochemical energy storage. The asymmetric supercapacitor by He et al. (2020) was prepared from the glucose-blowing approach. The detailed process of graphene/NiO composite blend could be perceived in the study of He et al. (2020). The asymmetric supercapacitor constructed using GLF/NiO showed a high specific capacitance up to 152 F/g and an energy density of 47.6 Wh/kg at the power density of 750 W/kg with 83.6% capacitance retention after 6000 cycles (Figure 1.17(a–c)). In another work by Chithaiah et al. (2020), hydrogen-treated rGO (H-rGO) was prepared by thermal decomposition of glycine and sucrose. The graphene supercapacitor prepared by this method demonstrated the specific capacitance of 203 F/g at 0.5 A/g. Although these research works have shown techniques to prepare graphene, these methods may not be acceptable to attain high specific surface area since van der Waals interactions easily restack graphene sheets during the preparation processes that can affect the loss of ultrahigh specific surface area.

This problem of restacking may probably be addressed by constructing graphene hydrogel demonstrated by Shang et al. (2019). The restacking of graphene sheets could be prevented by creating graphene hydrogel (GH) (Shang et al. 2019). The as-prepared asymmetric supercapacitor showed a 5.6-Wh/kg energy density at a power density of 226.5 W/kg. Woefully, to avert graphene sheets from restacking, graphene hydrogels had to encounter an unfavorable and stagnant freeze-drying process. The reduction was brought out in liquid surroundings and the ultimate outcome suffered from low electrical conductivity due to the high intersheet resistance. To address this issue, Han et al. (2018) demonstrated the synthesis of three-dimensional porous graphene film (3D-PG) by the sugar-blowing method. Their supercapacitor device displayed a high capacitance of about 115 F/g at a scan rate of 10 mV/s with a 90% excellent cycling stability after 10,000 cycles with 12-ohm contact interface resistance (Figure 1.17(d–f)).

In another approach by Ma, Xue, and Qin (2014), sugar-derived carbon/graphene composite materials were used as electrodes for supercapacitor application. Their electrodes for supercapacitors

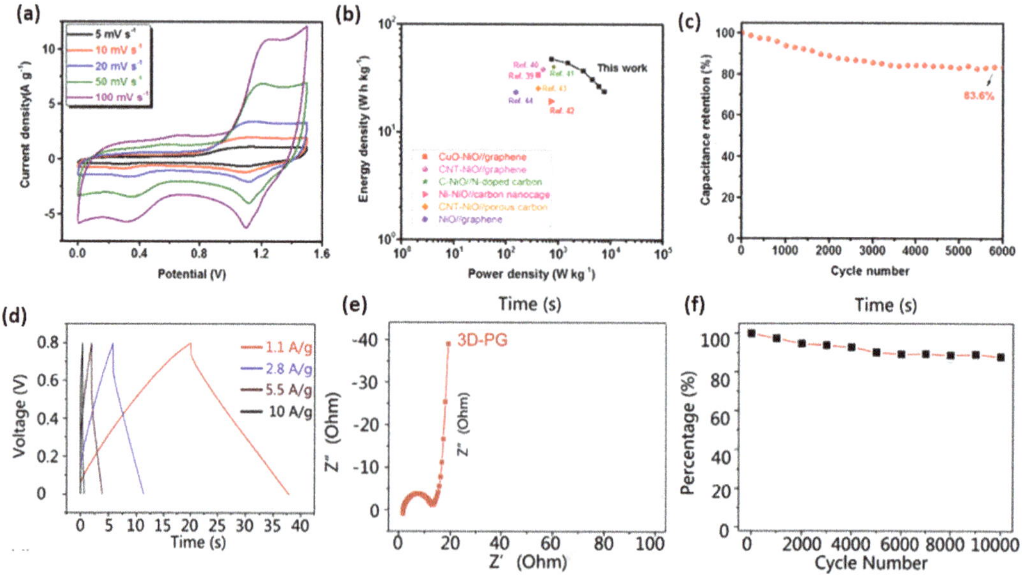

FIGURE 1.17 (a) CV curves of the asymmetric supercapacitor at various scan rates, (b) Ragone plots, (c) cycling stability (He et al. 2020), (d) galvanostatic charge-discharge (GCD) curves at different current densities of a 3D-PG electrode densities, (e) Nyquist plots, (d) cycling stability (Han et al. 2018). (Reprinted with permission.)

exhibit a high specific capacitance of 273 F/g as well as excellent electrochemical stability. They claimed that inserting sugar-derived carbon into graphene nanosheets (GNSs) can stabilize the electrode structure and prevents the agglomeration due to the van der Waals interactions, which enhanced the electrochemical activity of GNSs during the charge-discharge process. Works by Qian Lu, and Gao (2011), Wang et al. (2011), Zhang et al. (2017), and Li et al. (2016) have also demonstrated the use of glucose-reduced graphene composite for high-performance energy device applications.

Another most optimistic application of graphene is in sensors, covering bio and electrochemical sensors (Salazar et al. 2019; Zhang and Li 2017). In work by Hoa et al. (2017), highly porous 3D rGO showed a good sensing property toward glucose sensors. Furthermore, it was demonstrated that anchoring Ag NPs to the 3D rGO networks could enhance the antibacterial ability of the fabricated glucose sensor. The elevated sensor revealed an excellent sensitivity of 725.0 μA/cm^2/mM with a rapid response time of 11 s. In the same year, Hossain, Das, and Park (2017) also reported the graphene electrochemical sensor for detecting glucose. This sensor exhibited a short response time (4 s), high sensitivity (69.44 μA/mM/cm^2), and a wide detection range (0.002–12 mM) for glucose sensing (Figure 1.18(a)).

In a similar approach, Li et al. (2013) also demonstrated the use of silver nanoparticles-graphene oxide nanocomposite to detect tryptophan. This electrochemical sensor demonstrated the detection limit up to 2.0 nM. In another work by Zhu et al. (2020), hydrogen peroxide (H_2O_2) found in cells was also detected. This H_2O_2 biosensor prepared from a sugar-derived 3D graphene framework (GFs) electrode exhibits an ultrasensitive detection limit of 0.032 ± 0.005 μM at a working potential of −0.55 V in 0.01 M N_2-saturated phosphate-buffered saline (PBS, pH = 7.4) (Figure 1.18(b)). Furthermore, it was proposed that this biosensor can also detect H_2O_2 released from living tumorigenic cells in real time.

Recently, graphene has been used extensively in water purification. Gupta et al. (2012) demonstrated a green method for synthesizing a graphene material from cane sugar. It was claimed that their composite could effectively remove contaminants from water. Li, Jin, et al. (2018) reported the use of rGO extracted from sugarcane as an adsorbent for the eviction of cadmium in an aqueous solution. Table 1.1 shows the synthesis, properties, and applications of graphene prepared from sugar and sugarcane extract.

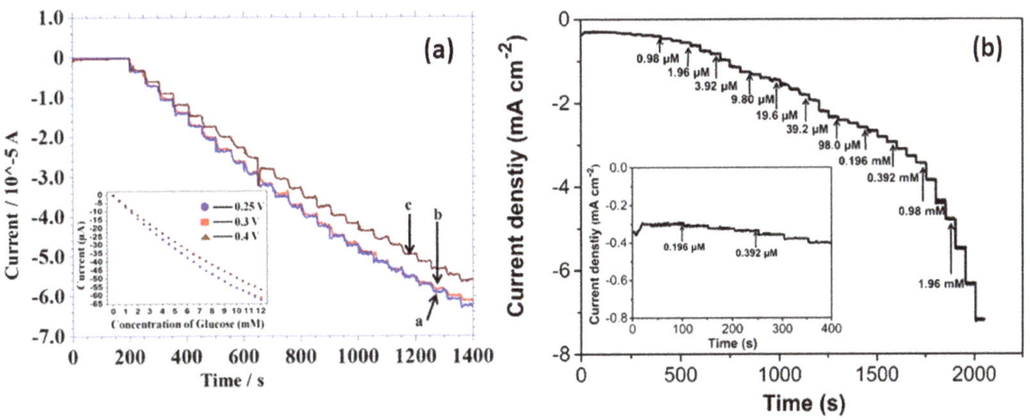

FIGURE 1.18 (a) Amperometric response of biosensor with a successive injection of the concentration of glucose in 0.25 V, 0.3 V, and 0.4 V, respectively, with inset showing calibration curve (Hossain, Das, and Park 2017). (b) Amperometric response of biosensors with the successive addition of different concentrations of H_2O_2 at the potential of −0.55 V and inset is the amperometric response of GFs/GCE toward low concentrations of H_2O_2 (Zhu et al. 2020). (Reprinted with permission.)

TABLE 1.1
Graphene Properties Prepared from Sugar and Sugarcane Extract

Method	Precursor/ Reducing Agent	Synthesis Temp.	Final Product	Properties	Raman	Graphene Layer	Application	Refs.
3D printing	Sucrose	80°C	Graphene foam	Electrical conductivity ~8.7 S/m Porosity ~99.3%	I_D/I_G 0.44 I_G/I_{2D} 2.28	Multilayer	–	Sha et al. (2017)
Sugar blowing	Glucose	1350°C	3D strutted graphene	Electrical conductivity 20,000 S/m Surface area 1005 m^2/g	$2D_{FWHM}$ 48 cm^{-1}	Mono-/few-layer	Supercapacitor	Wang et al. (2013)
Chemical blowing	Sucrose	1400°C	Graphene foam	Porosity 99.5–99.8%	$2D_{FWHM}$ 55 cm^{-1}	Mono-/few-layer	Supercapacitors	Jiang et al. (2015)
Thermolysis	Glucose	750°C	Patched graphene	Density 0.0077 g/cm^3 specific surface area of 820 m^2/g	I_D 1360 cm^{-1} I_G 1578 cm^{-1}	Nanocrystalline graphene flakes	Oxygen reduction reaction	Li et al. (2012)
Thermal decomposition	Sucrose	475 °C	rGO and N-rGO	–	–	Nanosheets	Supercapacitor	Chithaiah et al. (2020)
Glucose-blowing	Glucose	650 °C	Graphene-like foam	Surface area 323 m^2/g	I_D/I_G 0.80	–	Supercapacitor	He et al. (2020)
Modified Hummers method	Glucose	90°C	PVA/rGO	Tensile strengths 5.96 ± 0.966 MPa	I_D 1350 cm^{-1} I_G 1590 cm^{-1}	–	Skin tissue engineering	Narayanan, Park, and Han (2020)
Hydrothermal method	Glucose	RT	GQD-Au hybrid	Particle size~5–100 nm	–	–	Luminescent coating	Wadhwa et al. (2020)
Pyrolysis	Glucose	–	3D graphene frameworks	Surface area 853.19 m^2/g	–	Few layers of graphene	sensor	Zhu et al. (2020)
Chemical exfoliation	Sugarcane bagasse	60°C	rGO	–	I_D 1354 cm^{-1} I_G 1584 cm^{-1}	Wrinkles	Dye removal	Gan et al. (2019)
Chemical reduction	Glucose	70°C	Cu_2O/rGO	Optical band gap 2.01 eV	I_D/I_G 1.03	Crumbled graphene	–	Bhargava and Khan (2018)

(Continued)

TABLE 1.1 (*Continued*)
Graphene Properties Prepared from Sugar and Sugarcane Extract

Method	Precursor/ Reducing Agent	Synthesis Temp.	Final Product	Properties	Raman	Graphene Layer	Application	Refs.
Chemical and heat treatment	Sugarcane bagasse	300°C	Graphene quantum dots	Highly fluorescent		–	–	Baweja and Jeet (2019)
Thermal and chemical	Sugar molasses	1073K	Palygorskite-graphene like carbon nanocomposite	Electrical conductivity 1 S/m Thermal conductivity 1.1 W/mK	–	–	Batteries	Alvarez et al. (2017)
Pyrolysis	Sugarcane bagasse	1200°C	Graphene micro-crystal	–	–	–	–	Tang et al. (2018)
Carbonization	Sugarcane biomass	900°C	Graphene-like material	–	–	–	Adsorption of organic pollutants from water	Xiao et al. (2017)
Thermal and chemical	Sugarcane	1200°C	Graphene two dimensional crystal	–	–	Multilayer	–	Long et al. (2019)
Chemical route	Sugarcane bagasse	60°C	Reduced graphene oxide	–	–	Multilayer	Dye removal	Gan et al. (2019)
Chemical route	Sugarcane bagasse	90°C	Nanobiomaterial		I_D/I_G 1.32	Multilayer	Water purification	Li et al. (2018a)
Chemical route	Sugarcane juice	350°C	Graphene silica composite	–	–	Multilayer	Dye adsorption	Naveen et al. (2015)
Chemical route	Sugarcane bagasse	95°C	Reduced graphene oxide	–	I_D/I_G 1.16	Multilayer	Removal of cadmium	Li et al. (2018b)

1.7 FUTURE PROSPECTIVE AND CHALLENGE

This chapter gives detailed reports on the quest to find scalable and cost-efficient pathways to produce graphene-based materials using sugar and sugarcane waste. High-carbon content in sugar waste makes it promising for graphitization to be applied in the production of high-quality graphene sheets. All the above-discussed graphene preparation routes are rapid, reliable, efficient, scalable, and cost-effective. Yet, most of the studies have not dealt with systematic characterization, graphene purity, and application. For instance, the effect of sugar as a reducing agent on the edge and defect distribution of graphene has not been explained. The impact of sugar purity and concentration on the growth mechanism of graphene has not been addressed plainly. Furthermore, the quality level of the resulting graphene is not discussed for any intended application. Additionally, more light needs to be thrown on the electrical conductivity of graphene obtained from sugar. Hence, more investigation is needed to optimize the preparation methods. Apart from all these challenges, we believe this chapter will open a window of opportunity for a breakthrough in graphene development not only from sugar waste, but also all the natural wastes for several useful applications such as supercapacitor, next-generation water filter, fuel cell, H_2 generation, etc.

REFERENCES

Akhavan, O., K. Bijanzad, and A. Mirsepah. 2014. "Synthesis of Graphene from Natural and Industrial Carbonaceous Wastes." *RSC Advances* 4: 20441–8.

Al-Gaashani, R., A. Najjar, Y. Zakaria, S. Mansour, and M. A. Atieh. 2019. "XPS and Structural Studies of High Quality Graphene Oxide and Reduced Graphene Oxide Prepared by Different Chemical Oxidation Methods." *Ceramics International* 45: 14439–48.

Alvarez, E., Y. Laffita, L. Montoro, et al. 2017. "Electrical, Thermal and Electrochemical Properties of Disordered Carbon Prepared from Palygorskite and Cane Molasses." *Journal of Solid State Chemistry* 246: 404–11.

Ansari, M., M. Shoeb, P. Nayab, M. Mobin, Rahisuddin, I. Khan, and W. Siddiqi. 2018. "Honey Mediated Green Synthesis of Graphene Based NiO2/Cu2O Nanocomposite (Gr@NiO2/Cu2O NCs): Catalyst for the Synthesis of Functionalized Schiff-Base Derivatives." *Journal of Alloys and Compounds* 738: 56–71.

Avouris, P., and C. Dimitrakopoulos. 2012. "Graphene: Synthesis and Applications." *Materials Today* 15: 86–97.

Bae, S., H. Kim, Y. Lee, X. Xu, J. Park, Y. Zheng, J. Balakrishnan, et al. 2010. "Roll-to-Roll Production of 30-Inch Graphene Films for Transparent Electrodes." *Nature Nanotechnology* 5: 574–8.

Baweja, H., and K. Jeet. 2019. "Economical and Green Synthesis of Graphene and Carbon Quantum Dots from Agricultural Waste." *Materials Research Express* 6: 0850g8.

Berktas, I., M. Hezarkhani, L. Poudeh, and B. Okan. 2020. "Recent Developments in the Synthesis of Graphene and Graphene-like Structures from Waste Sources by Recycling and Upcycling Technologies: A Review." *Graphene Technology* 5: 59–73.

Bhargava, R., and S. Khan. 2018. "Enhanced Optical Properties of Cu_2O Anchored on Reduced Graphene Oxide (RGO) Sheets." *Journal of Physics Condensed Matter* 30: 35703.

Bhargava, R., S. Khan, M. Ansari, and N. Ahmad. 2019. "Green Synthesis Approach for the Reduction of Graphene Oxide by Using Glucose." *AIP Conference Proceedings* 2115: 030075.

Bo, Z., X. Shuai, S. Mao, H. Yang, J. Qian, J. Chen, J. Yan, and K. Cen. 2014. "Green Preparation of Reduced Graphene Oxide for Sensing and Energy Storage Applications." *Scientific Reports* 4: 4684.

Caņcado, G., K. Takai, T. Enoki, M. Endo, Y. A. Kim, H. Mizusaki, et al. 2006. "General Equation for the Determination of the Crystallite Size La of Nanographite by Raman Spectroscopy." *Applied Physics Letters* 88: 163106.

Casa, M., M. Sarno, R. Liguori, et al. 2018. "Conductive Adhesive Based on Mussel-Inspired Graphene Decoration with Silver Nanoparticles." *Journal of Nanoscience and Nanotechnology* 18: 1176–85.

Cheng, J., J. Du, and W. Zhu. 2012. "Facile Synthesis of Three-Dimensional Chitosan-Graphene Mesostructures for Reactive Black 5 Removal." *Carbohydrate Polymers* 88: 61–7.

Chithaiah, P., M. Raju, G. Kulkarni, and C. N. R. Rao. 2020. "Simple Synthesis of Nanosheets of RGO and Nitrogenated RGO." *Beilstein Journal of Nanotechnology* 11: 68–75.

Choi, Y., J. Choi, L. Lingamdinne, et al. 2019. "Removal of U(VI) by Sugar-Based Magnetic Pseudo–Graphene Oxide and Its Application to Authentic Groundwater Using Electromagnetic System." *Environmental Science and Pollution Research* 26: 22323–37.
Çıplak, Zafer, and Nuray Yıldız. 2018. "A Parametric Study for the Synthesis of Graphene–AgAu Nanocomposites: Performances as Electrode Material." *Journal of Materials Science: Materials in Electronics* 29 (12): 10411–26.
Deb, J., N. Seriani, and U. Sarkar. 2021. "Ultrahigh Carrier Mobility of Penta-Graphene: A First-Principle Study." *Physica E: Low-Dimensional Systems and Nanostructures* 127: 114507.
Ding, Z., T. Yuan, J. Wen, et al. 2020. "Green Synthesis of Chemical Converted Graphene Sheets Derived from Pulping Black Liquor." *Carbon* 158: 690–7.
Edwards, R., and K. Coleman. 2013. "Graphene Synthesis: Relationship to Applications." *Nanoscale* 5: 38–51.
Essawy, N., S. Ali, H. Farag, A. Konsowa, M. Elnouby, and H. Hamad. 2017. "Green Synthesis of Graphene from Recycled PET Bottle Wastes for Use in the Adsorption of Dyes in Aqueous Solution." *Ecotoxicology and Environmental Safety* 145: 57–68.
Fernández-Merino, M., L. Guardia, J. I. Paredes, et al. 2010. "Vitamin C Is an Ideal Substitute for Hydrazine in the Reduction of Graphene Oxide Suspensions." *Journal of Physical Chemistry C* 114 (14): 6426–32.
Fu, P., G. Teri, X. Chao, J. Li, et al. 2020. "Modified Graphene-Feve Composite Coatings: Application in the Repair of Ancient Architectural Color Paintings." *Coatings* 10: 1162.
Gan, L., B. Li, Y. Chen, B. Yu, and Z. Chen. 2019. "Green Synthesis of Reduced Graphene Oxide Using Bagasse and Its Application in Dye Removal: A Waste-to-Resource Supply Chain." *Chemosphere* 219: 148–54.
Ge, H., T. Hao, H. Osgood, et al. 2016. "Advanced Mesoporous Spinel Li4Ti5O12/RGO Composites with Increased Surface Lithium Storage Capability for High-Power Lithium-Ion Batteries." *ACS Applied Materials and Interfaces* 8: 9162–9.
Geim, A., and K. Novoselov. 2007. "The Rise of Graphene PROGRESS." *Nature Materials* 6: 183–91.
Geng, J., B. Kong, S. Yang, and H. Jung. 2010. "Preparation of Graphene Relying on Porphyrin Exfoliation of Graphite." *Chemical Communications* 46: 5091–3.
Ghamkhari, A., S. Ravasjani, M. Alebi, H. Hamishehkar, and M. Hamblin. 2020. "Development of a Graphene Oxide-Poly Lactide Nanocomposite as a Smart Drug Delivery System." *International Journal of Biological Macromolecules* 169: 521–31.
Giannazzo, F., R. Dagher, E. Schilirò, S. E. Panasci, G. Greco, G. Nicotra, F. Roccaforte, et al. 2020. "Nanoscale Structural and Electrical Properties of Graphene Grown on AlGaN by Catalyst-Free Chemical Vapor Deposition." *Nanotechnology* 32: 015705.
Goswami, S., P. Banerjee, S. Datta, A. Mukhopadhayay, and P. Das. 2017. "Graphene Oxide Nanoplatelets Synthesized with Carbonized Agro-Waste Biomass as Green Precursor and Its Application for the Treatment of Dye Rich Wastewater." *Process Safety and Environmental Protection* 106: 163–72.
Gupta, S., T. Sreeprasad, S, Maliyekkal, S. Das, and T. Pradeep. 2012. "Graphene from Sugar and Its Application in Water Purification." *ACS Applied Materials and Interfaces* 4: 4156–63.
Han, F., O. Qian, B. Chen, H. Tang, and M. Wang. 2018. "Sugar Blowing-Assisted Reduction and Interconnection of Graphene Oxide into Three-Dimensional Porous Graphene." *Journal of Alloys and Compounds* 730: 386–91.
Hao, J., Y. Liao, Y. Zhong, D. Shu, et al. 2015. "Three-Dimensional Graphene Layers Prepared by a Gas-Foaming Method for Supercapacitor Applications." *Carbon* 94: 879–87.
Hashmi, A., A. Singh, B. Jain, and A. Singh. 2020. "Muffle Atmosphere Promoted Fabrication of Graphene Oxide Nanoparticle by Agricultural Waste." *Fullerenes Nanotubes and Carbon Nanostructures* 28: 627–36.
He, C., Y. Jiang, X. Zhang, X. Cui, and Y. Yang. 2020. "A Simple Glucose-Blowing Approach to Graphene-Like Foam/NiO Composites for Asymmetric Supercapacitors." *Energy Technology* 8: 1900923.
Hernandez, Y., V. Nicolosi, M. Lotya, et al. 2008. "High-Yield Production of Graphene by Liquid-Phase Exfoliation of Graphite." *Nature Nanotechnology* 3: 563–8.
Hoa, L., N. Linh, J. Chung, and S. Hur. 2017. "Green Synthesis of Silver Nanoparticle-Decorated Porous Reduced Graphene Oxide for Antibacterial Non-Enzymatic Glucose Sensors." *Ionics* 23: 1525–32.
Hossain, F., P. Das, and J. Park. 2017. "Development of High Performance Electrochemical and Physical Biosensors Based on Chemically Modified Graphene Nanostructured Electrodes." *Journal of the Electrochemical Society* 164: B391–6.
Huang, L., D. Zhang, F. Zhang, Z. Feng, Y. Huang, and Y. Gan. 2018. "High-Contrast SEM Imaging of Supported Few-Layer Graphene for Differentiating Distinct Layers and Resolving Fine Features: There Is Plenty of Room at the Bottom." *Small* 14: 1704190.

Huang, L., D. Zhang, F. Zhang, Y. Huang, and Y. Gan. 2019. "Graphene to Graphene, and Substrate to Substrate: How to Reliably Differentiate Supported Graphene from Polycrystalline Substrates Using SEM?" *Materials Research Express* 6: 085604.

Huang, Q., D. Zeng, S. Tian, and C. Xie. 2012. "Synthesis of Defect Graphene and Its Application for Room Temperature Humidity Sensing." *Materials Letters* 83: 76–9.

Ikram, R., B. Jan, and W. Ahmad. 2020. "Advances in Synthesis of Graphene Derivatives Using Industrial Wastes Precursors; Prospects and Challenges." *Journal of Materials Research and Technology* 9: 15924–51.

Jia, X., J. Delgado, M. Terrones, V. Meunier, and M. Dresselhaus. 2011. "Graphene Edges: A Review of Their Fabrication and Characterization." *Nanoscale* 3: 86–95.

Jiang, X., X. Wang, P. Dai, et al. 2015. "High-Throughput Fabrication of Strutted Graphene by Ammonium-Assisted Chemical Blowing for High-Performance Supercapacitors." *Nano Energy* 16: 81–90.

Kamisan, A., L. Zainuddin, A. Kamisan, et al. 2016. "Ultrasonic Assisted Synthesis of Reduced Graphene Oxide in Glucose Solution." *Key Engineering Materials* 708: 25–9.

Kim, M., and D. Day. 2011. "Composition of Sugar Cane, Energy Cane, and Sweet Sorghum Suitable for Ethanol Production at Louisiana Sugar Mills." *Journal of Industrial Microbiology and Biotechnology* 38: 803–7.

Koo, D., S. Jung, J. Seo, et al. 2020. "Flexible Organic Solar Cells Over 15% Efficiency with Polyimide-Integrated Graphene Electrodes." *Joule* 4: 1021–34.

Kumar, R., R. Singh, and D. Singh. 2016. "Natural and Waste Hydrocarbon Precursors for the Synthesis of Carbon Based Nanomaterials: Graphene and CNTs." *Renewable and Sustainable Energy Reviews* 58: 976–1006.

Lancellotti, L., E. Bobeico, M. Noce, et al. 2020. "Graphene as Non Conventional Transparent Conductive Electrode in Silicon Heterojunction Solar Cells." *Applied Surface Science* 525: 146443.

Lau, K., R. Ginting, S. Tan, et al. 2019. "Sodium Cholate as Efficient Green Reducing Agent for Graphene Oxide via Flow Reaction for Flexible Supercapacitor Electrodes." *Journal of Materials Science: Materials in Electronics* 30: 19182–8.

Lee, G., and B. Kim. 2014. "Biological Reduction of Graphene Oxide Using Plant Leaf Extracts." *Biotechnology Progress* 30: 463–9.

Lei, H., T. Yan, H. Wang, L. Shi, J. Zhang, and D. Zhang. 2015. "Graphene-like Carbon Nanosheets Prepared by a Fe-Catalyzed Glucose-Blowing Method for Capacitive Deionization." *Journal of Materials Chemistry A* 3: 5934–41.

Li, B., L. Gan, G. Owens, and Z. Chen. 2018. "New Nano-Biomaterials for the Removal of Malachite Green from Aqueous Solution via a Response Surface Methodology." *Water Research* 146: 55–66.

Li, B., X. Jin, J. Lin, and Z. Chen. 2018. "Green Reduction of Graphene Oxide by Sugarcane Bagasse Extract and Its Application for the Removal of Cadmium in Aqueous Solution." *Journal of Cleaner Production* 189: 128–34.

Li, J., D. Kuang, Y. Feng, F. Zhang, Z. Xu, M. Liu, and D. Wang. 2013. "Green Synthesis of Silver Nanoparticles-Graphene Oxide Nanocomposite and Its Application in Electrochemical Sensing of Tryptophan." *Biosensors and Bioelectronics* 42: 198–206.

Li, L., P. Gao, S. Gai, F. He, Y. Chen, M. Zhang, and P. Yang. 2016. "Ultra Small and Highly Dispersed Fe3O4 Nanoparticles Anchored on Reduced Graphene for Supercapacitor Application." *Electrochimica Acta* 190: 566–73.

Li, S., Y. Xing, and G. Wang. 2012. "A Graphene-Based Electrochemical Sensor for Sensitive and Selective Determination of Hydroquinone." *Microchimica Acta* 176: 163–8.

Li, X., S. Kurasch, U. Kaiser, and M. Antonietti. 2012. "Synthesis of Monolayer-Patched Graphene from Glucose." *Angewandte Chemie International Edition* 51: 9689–92.

Li, Y., K. Tantiwanichapan, A. Swan, and R. Paiella. 2020. "Graphene Plasmonic Devices for Terahertz Optoelectronics." *Nanophotonics* 9: 1901–20.

Lin, K., H. Lin, T. Yang, and B. Jia. 2020. "Structured Graphene Metamaterial Selective Absorbers for High Efficiency and Omnidirectional Solar Thermal Energy Conversion." *Nature Communications* 11: 1–10.

Lin, X., and J. Gai. 2016. "Synthesis and Applications of Large-Area Single-Layer Graphene." *RSC Advances* 6: 17818–44.

Lin, Y., Y. Liao, Z. Chen, and J. Connell. 2017. "Holey Graphene: A Unique Structural Derivative of Graphene." *Materials Research Letters* 5: 209–34.

Lin, Z., Y. Yao, Z. Li, Y. Liu, Z. Li, and C. Wong. 2010. "Solvent-Assisted Thermal Reduction of Graphite Oxide." *Journal of Physical Chemistry C* 114: 14819–25.

Liu, X., and M. Antonietti. 2013. "Moderating Black Powder Chemistry for the Synthesis of Doped and Highly Porous Graphene Nanoplatelets and Their Use in Electrocatalysis." *Advanced Materials* 25: 6284–90.
Liu, X., C. Giordano, and M. Antonietti. 2014. "A Facile Molten-Salt Route to Graphene Synthesis." *Small* 10: 193–200.
Long, S., Q. Du, S. Wang, P. Tang, D. Li, and R. Huang. 2019. "Graphene Two-Dimensional Crystal Prepared from Cellulose Two-Dimensional Crystal Hydrolysed from Sustainable Biomass Sugarcane Bagasse." *Journal of Cleaner Production* 241: 118209.
Ma, H., J. Zeng, S. Harrington, L. Ma, M. Ma, X. Guo, and Y. Ma. 2016. "Hydrothermal Fabrication of Silver Nanowires-Silver Nanoparticles-Graphene Nanosheets Composites in Enhancing Electrical Conductive Performance of Electrically Conductive Adhesives." *Nanomaterials* 6: 119.
Ma, J., T. Xue, and X. Qin. 2014. "Sugar-Derived Carbon/Graphene Composite Materials as Electrodes for Supercapacitors." *Electrochimica Acta* 115: 566–72.
Mahdavifar, M., S. Shekarforoush, and F. Khoeini. 2021. "Tunable Electronic Properties and Electric-Field-Induced Phase Transition in Phosphorene/Graphene Heterostructures." *Journal of Physics D* 54: 095108.
Mikmeková, E., L. Frank, I. Müllerová, B. W. Li, R. S. Ruoff, and M. Lejeune. 2016. "Study of Multi-Layered Graphene by Ultra-Low Energy SEM/STEM." *Diamond and Related Materials* 63: 136–42.
Mohan, A., Manoj B, and S. Panicker. 2019. "Facile Synthesis of Graphene-Tin Oxide Nanocomposite Derived from Agricultural Waste for Enhanced Antibacterial Activity against Pseudomonas Aeruginosa." *Scientific Reports* 9: 1–12.
Muralee, C., R. Vinodh, S. Sambasivam, I. Obaidat, and H. Kim. 2020. "Recent Progress of Advanced Energy Storage Materials for Flexible and Wearable Supercapacitor: From Design and Development to Applications." *Journal of Energy Storage* 27: 101035.
Narayanan, K., G. Park, and S. Han. 2020. "Electrospun Poly(Vinyl Alcohol)/Reduced Graphene Oxide Nanofibrous Scaffolds for Skin Tissue Engineering." *Colloids and Surfaces B* 191: 110994.
Naveen, P., K. Govindaraju, G. Kumar, U. Suganya, and K. Vijai Anand. 2015. "Graphene from Sugarcane Extract and Its Application on Organic Dyes Adsorption." *Pollution Research* 34: 803–8.
Novoselov, K., A. Geim, S. Morozov et al. 2004. "Electric Field Effect in Atomically Thin Carbon Films." *Science* 306: 666–9.
Novoselov, K. S., V. I. Fal'Ko, L. Colombo, P. R. Gellert, M. G. Schwab, and K. Kim. 2012. "A Roadmap for Graphene." *Nature* 490: 192–200.
Olszowska, K., J. Pang, P. Wrobel, et al. 2017. "Three-Dimensional Nanostructured Graphene: Synthesis and Energy, Environmental and Biomedical Applications." *Synthetic Metals* 234: 53–85.
Pan, F., J. Jin, X. Fu, Q. Liu, and J. Zhang. 2013. "Advanced Oxygen Reduction Electrocatalyst Based on Nitrogen-Doped Graphene Derived from Edible Sugar and Urea." *ACS Applied Materials and Interfaces* 5: 11108–14.
Punith M., P. Laxmeesha, S. Ray, and C. Srivastava. 2020. "Enhancement in the Corrosion Resistance of Nanocrystalline Aluminium Coatings by Incorporation of Graphene Oxide." *Applied Surface Science* 533: 147512.
Qian, Y., S. Lu, and F. Gao. 2011. "Preparation of MnO_2/Graphene Composite as Electrode Material for Supercapacitors." *Journal of Materials Science* 46: 3517–22.
Ramanathan, S., E. Elanthamilan, A. Obadiah, et al. 2017. "Aloe Vera (L.) Burm.f. Extract Reduced Graphene Oxide for Supercapacitor Application." *Journal of Materials Science* 28: 16648–57.
Rani, S., and S. J. Ray. 2021. "DNA and RNA Detection Using Graphene and Hexagonal Boron Nitride Based Nanosensor." *Carbon* 173: 493–500.
Ruan, G., Z. Sun, Z. Peng, and J. Tour. 2011. "Growth of Graphene from Food, Insects, and Waste." *ACS Nano* 5: 7601–7.
Ruiz-Hitzky, E., M. Darder, F. Fernandes, E. Zatile, F. Palomares, and P. Aranda. 2011. "Supported Graphene from Natural Resources: Easy Preparation and Applications." *Advanced Materials* 23: 5250–5.
Saikia, B., S. Benoy, M. Bora, J. Tamuly, M. Pandey, and D. Bhattacharya. 2020. "A Brief Review on Supercapacitor Energy Storage Devices and Utilization of Natural Carbon Resources as Their Electrode Materials." *Fuel* 282: 118796.
Salazar, P., I. Fernández, M. Rodríguez, et al. 2019. "One-Step Green Synthesis of Silver Nanoparticle-Modified Reduced Graphene Oxide Nanocomposite for H_2O_2 Sensing Applications." *Journal of Electroanalytical Chemistry* 855:113638.
Seo, D., A. Rider, Z. Han, S. Kumar, and K. Ostrikov. 2013. "Plasma Break-down and Re-Build: Same Functional Vertical Graphenes from Diverse Natural Precursors." *Advanced Materials* 25: 5638–42.
Sha, J., Y. Li, R. Salvatierra, T. Wang, P. Dong, Y. Ji, S. Lee, et al. 2017. "Three-Dimensional Printed Graphene Foams." *ACS Nano* 11: 6860–7.

Shang, Y., L. Xu, L. Cai, and X. Jiang. 2019. "Synthesis of Graphene Hydrogel and Graphene Oxide/Polyaniline Composites for Asymmetric Supercapacitor." *IOP Conference Series* 562: 012105.

Shen, J., M. Shi, H. Ma, B. Yan, N. Li, and M. Ye. 2011a. "Hydrothermal Synthesis of Magnetic Reduced Graphene Oxide Sheets." *Materials Research Bulletin* 46: 2077–83.

Shen, J., B. Yan, M. Shi, H. Ma, N. Li, and M. Ye. 2011b. "One Step Hydrothermal Synthesis of TiO2-Reduced Graphene Oxide Sheets." *Journal of Materials Chemistry* 21: 3415–21.

Shi, J., Y. Wang, W. Du, and Z. Hou. 2016. "Synthesis of Graphene Encapsulated Fe_3C in Carbon Nanotubes from Biomass and Its Catalysis Application." *Carbon* 99: 330–37.

Shuvra, S., A. Das, and A. Barui. 2020. "Surface Functionalization of Green-Synthesized Reduced Graphene Oxide with PPIX Enhances Photosensitization of Cancer Cells." *Photochemistry and Photobiology* 96: 1283–93.

Singh, A., B. Ahmed, A. Singh, and A. Ojha. 2018. "Photodegradation of Phenanthrene Catalyzed by RGO Sheets and Disk like Structures Synthesized Using Sugar Cane Juice as a Reducing Agent." *Spectrochimica Acta – Part A* 204: 603–10.

Sood, A., I. Lund, Y. Puri, et al. 2015. "Review of Graphene Technology and Its Applications for Electronic Devices." *Graphene - New Trends and Developments*. London, Intech Publication.

Sunaoshi, T., K. Kaji, Y. Orai, C. Schamp, and E. Voelkl. 2016. "STEM/SEM, Chemical Analysis, Atomic Resolution and Surface Imaging at ≤ 30 KV with No Aberration Correction for Nanomaterials on Graphene Support." *Microscopy and Microanalysis* 22: 604–5.

SuryaBhaskaram, D., R. Cheruku, and G. Govindaraj. 2014. "Sugar Assisted Graphene: A Green Synthesis Approach." *International Journal of ChemTech Research* 6: 3291–3.

Suvarna, K.S., and N. Binitha. 2020. "Graphene Preparation by Jaggery Assisted Ball-Milling of Graphite for the Adsorption of Cr(VI)." *Materials Today* 25: 236–40.

Taheri, B., N. Nia, A. Agresti, et al. 2018. "Graphene-Engineered Automated Sprayed Mesoscopic Structure for Perovskite Device Scaling-Up." *2D Materials* 5: 045034.

Takamura, M., K. Furukawa, H. Okamoto, S. Tanabe, H. Yamaguchi, and H. Hibino. 2015. "Epitaxial Graphene Resonators Obtained by Electrochemical Etching," *International Conference on Solid State Devices and Materials*.

Tamilselvi, R., M. Ramesh, G. Lekshmi, et al. 2020. "Graphene Oxide – Based Supercapacitors from Agricultural Wastes: A Step to Mass Production of Highly Efficient Electrodes for Electrical Transportation Systems." *Renewable Energy* 151: 731–9.

Tan, Y., H. Stormer, P. Kim, K. Novoselov, M. Cohen, S. Louie, X. Wang, et al. 2009. "Graphene at the Edge: Stability and Dynamics." *Science* 323: 1705–8.

Tang, P., Q. Du, D. Li, et al. 2018. "Fabrication and Characterization of Graphene Microcrystal Prepared from Lignin Refined from Sugarcane Bagasse." *Nanomaterials* 8: 565.

Thakur, S., and N. Karak. 2012. "Green Reduction of Graphene Oxide by Aqueous Phytoextracts." *Carbon* 50: 5331–9.

Vij, V., J. Tiwari, and K. Kim. 2016. "Covalent versus Charge Transfer Modification of Graphene/Carbon-Nanotubes with Vitamin B1: Co/N/S-C Catalyst toward Excellent Oxygen Reduction." *ACS Applied Materials and Interfaces* 8: 16045–52.

Wadhwa, S., A. John, A. Mathur, M. Khanuja, G. Bhattacharya, S. Roy, and S. Ray. 2020. "Engineering of Luminescent Graphene Quantum Dot-Gold (GQD-Au) Hybrid Nanoparticles for Functional Applications." *MethodsX* 7: 100963.

Wang, J., Z. Gao, Z. Li, B. Wang, Y. Yan, Q. Liu, T. Mann, M. Zhang, and Z. Jiang. 2011. "Green Synthesis of Graphene Nanosheets/ZnO Composites and Electrochemical Properties." *Journal of Solid State Chemistry* 184: 1421–7.

Wang, X., Y. Zhang, C. Zhi, X. Wang, D. Tang, Y. Xu, Q. Weng, et al. 2013. "Three-Dimensional Strutted Graphene Grown by Substrate-Free Sugar Blowing for High-Power-Density Supercapacitors." *Nature Communications* 4: 2905.

Weng, X., J. Wu, L. Ma, G. Owens, and Z. Chen. 2019. "Impact of Synthesis Conditions on Pb(II) Removal Efficiency from Aqueous Solution by Green Tea Extract Reduced Graphene Oxide." *Chemical Engineering Journal* 359: 976–81.

Wojtoniszak, M., and E. Mijowska. 2012. "Controlled Oxidation of Graphite to Graphene Oxide with Novel Oxidants in a Bulk Scale." *Journal of Nanoparticle Research* 14: 1248.

Wu, X., Y. Xing, D. Pierce, and J. Zhao. 2017. "One-Pot Synthesis of Reduced Graphene Oxide/Metal (Oxide) Composites." *ACS Applied Materials and Interfaces* 9: 37962–71.

Xiao, X., B. Chen, L. Zhu, and J. Schnoor. 2017. "Sugar Cane-Converted Graphene-like Material for the Superhigh Adsorption of Organic Pollutants from Water via Coassembly Mechanisms." *Environmental Science and Technology* 51: 12644–52.

Xing, W., J. Xue, and J. Dahn. 1996. "Optimizing Pyrolysis of Sugar Carbons for Use as Anode Materials in Lithium-Ion Batteries." *Journal of the Electrochemical Society* 143: 3046–52.

Yi, M., and Z. Shen. 2015. "A Review on Mechanical Exfoliation for the Scalable Production of Graphene." *Journal of Materials Chemistry A* 3: 11700–715.

Zhang, J., H. Yang, G. Shen, P. Cheng, J. Zhang, and S. Guo. 2010. "Reduction of Graphene Oxide Vial-Ascorbic Acid." *Chemical Communications* 46: 1112–4.

Zhang, B., J. Song, G. Yang, and B. Han. 2014. "Large-Scale Production of High-Quality Graphene Using Glucose and Ferric Chloride." *Chemical Science* 5: 4656–60.

Zhang, J., L. Xing, T. Liu, K. Qin, J. Zhou, H. Cui, S. Zhuo, and W. Si. 2017. "Three-Dimensional Reduced Graphene Hydrogels Using Various Carbohydrates for High Performance Supercapacitors." *Journal of Nanoscience and Nanotechnology* 17: 1099–107.

Zhang, M., C. Hou, A. Halder, J. Ulstrup, and Q. Chi. 2017. "Interlocked Graphene–Prussian Blue Hybrid Composites Enable Multifunctional Electrochemical Applications." *Biosensors and Bioelectronics* 89: 570–7.

Zhang, Y., and G. Li. 2017. "A Novel Electrochemical Sensor Based on a Graphene-Silver Platform for the Sensitive Determination of a Tumor-Supplied Group of Factors." *International Journal of Electrochemical Science* 12: 10095–106.

Zheng, C., K. Lu, Y. Lu, S. Zhu, Y. Yue, X. Xu, C. Mei, H. Xiao, Q. Wu, and J. Han. 2020. "A Stretchable, Self-Healing Conductive Hydrogels Based on Nanocellulose Supported Graphene towards Wearable Monitoring of Human Motion." *Carbohydrate Polymers* 250: 116905.

Zhigang, N. 2014. "Reduced Graphene Oxide-Cuprous Oxide Hybrid Nanopowders: Hydrothermal Synthesis and Enhanced Photocatalytic Performance under Visible Light Irradiation." *Materials Science in Semiconductor Processing* 23: 78–84.

Zhu, C., S. Guo, Y. Fang, and S. Dong. 2010. "Reducing Sugar: New Functional Molecules for the Green Synthesis of Graphene Nanosheets." *ACS Nano* 4: 2429–37.

Zhu, Y., K. Kang, Y. Jia, W. Guo, and J. Wang. 2020. "General and Fast Synthesis of Graphene Frameworks Using Sugars for High-Performance Hydrogen Peroxide Nonenzymatic Electrochemical Sensor." *Microchimica Acta* 187: 669.

2 Graphene from Honey

Atul Kumar and Surender Duhan

CONTENTS

2.1 INTRODUCTION

One of the most appealing nanomaterials is graphene, a two-dimensional (2D) carbon atoms layer stuck through sp^2 bonding [1]. Various methods for processing graphene have been developed over the years, including ultrasonic exfoliation [2, 3], chemical vapor deposition (CVD) [4], epitaxial growth [5], and chemical reduction of graphene [6]. Scientists used a variety of aqueous phytoextracts to reduce graphene oxide as reducing agents. Different naturally derived procedures are widely available to reduce graphene from graphite because of its less toxic [7], eco-friendly [8], and nonhazardous nature [9]. The plant extracts contain a high concentration of phenolic compounds such as vanillic acid, caffeic acid, gallic acid, protocatechuic acid, salicylic acid, chlorogenic acid, and others, as well as a high concentration of OH group, which provide them with slight reducing activities [10, 11]. Graphene is commercially produced from graphite in the form of multilayer nanoplates. The best quality graphene is monolayer, but it is not easy to produce.

Natural Honey, the earth's ancient foodstuff cradle, is an outstanding source of calories and nutrients [12]. *Apis mellifera* is an important honey bee species that produce honey by taking plant nectar [13, 14]. Since ancient times, honey has been used in beauty products and to form remedial to cure several diseases [15–18]. The special chemical characteristics of honey make it possible to be used in green nanoparticle synthesis. Consequently, honey-mediated synthesis has many advantages

DOI: 10.1201/9781003169741-2

over methods that mediate natural products, including a quicker mechanism than the other naturally synthesize techniques [19].

2.1.1 Properties of Honey

Natural Honey, based on water quality, is a sultry and glutinous liquid. It can also captivate and retain moistness in the natural atmosphere. Normally, honey possesses 18.8 percent water content, more than 60 percent relative humidity, and absorbs less moisture from the air [17].

Liquid honey ranges in color from pure and colorless to dark amber or black. Honey's shade ranges from yellow to yellowish-brown, which differs with plant source, period, and packing environment. However, purity or consistency is determined by the aggregation and composition of particles contained, such as pollen grains [20].

2.1.2 Chemical Composition of Honey

Since time immemorial, it has been regarded as one of the vigorous nutrition sources. Shugaba et al. analyzed honey and gave its "chemical composition as 80–85% saccharides, 15–17% water, 0.1–17% protein, 0.2% ash and amino acids, enzymes, vitamins and others, including phenolic antioxidants. Therefore, the same chemical composition and physical characteristics of natural honey differ according to the plant species that bees dry up and the climate and vegetation changes" [21].

Fructose falls from 32.56 and 38.2 percent and glucose between 28.54 and 31.3 percent makes up 85–95 percent of total Honey sugars and is readily absorbed into the gastrointestinal tract. There are some amounts of different types of disaccharides, and oligosaccharides are also available. Honey has fructooligosaccharides between 4 and 5 percent that act as probiotics [22].

Honey comprises proteins in amounts ranging from 0.1 to 0.5 percent, which vary considerably depending on the form and its source [17]. Vitamins C and B1 (thiamine) and B2 complex vitamins such as riboflavin, nicotine acid, B6, and pantothenic acid are also used in Natural Honey [23].

Mineral compounds are present in concentrations ranging from 0.1 to 1.0 percent. The most abundant metal is potassium, followed by calcium, magnesium, sodium, sulfur, and phosphorus. Gold, copper, zinc, and manganese are examples of trace elements [24]. Honey contains several enzymes, including oxidase, Invertase, amylase, and catalase; the Invertase, diastase, and glucose oxidase are the main enzymes. The action of the amylase enzyme produces long starch chains of dextrin and maltose [17]. These minor components are seen as having distinct medicinal or dietary properties, and Natural Honey has diverse and varied uses in the unique mixture. Table 2.1 represent the chemical composition of honey and Table 2.2. represent the different types of carbohydrates present in honey.

TABLE 2.1
Chemical Composition of Honey [22]

Composition	Percentage
Carbohydrates	80–85
Water	15–17
Protein	0.1–0.4
Ash	0.2
Minerals	0.1–1
Amino acid, enzymes, vitamins, and phenolic antioxidants	Traces

TABLE 2.2
Different Types of Carbohydrates Present in Honey [17]

Different Types of Carbohydrates	Percentage
Fructose	32.56–38.2
Glucose	28.54–31.3
Fructooligosaccharide	4–5

2.2 HONEY-MEDIATED NANOPARTICLES SYNTHESIS

The culture of nanoparticles by natural products must be very careful because it takes some time for nanoparticles to be converted. In addition, it can be difficult to separate nanoparticles from other natural resources.

Balasooriya et al. suggested the honey-mediated nanoparticle synthesis depends on several factors such as the reducing and stabilizing chemicals and the temperatures. Although gold, silver, and palladium nanoparticles were created using a generalized technique, carbon and platinum nanoparticles were created using unique conditions [25].

2.2.1 Graphene Synthesis

Each method for making or removing graphene, based on the preferred scale, clarity, and efflorescence of the particular component, is referred to as graphene synthesis. Graphene was first collected as small flakes of many microns in size by mechanical exfoliation of graphite with scotch tape [26, 27]. While this procedure of synthesizing graphene produces the best quality of graphene, a fabrication method capable of synthesizing wafer-scale graphene is needed for bulk synthesis. Several techniques for graphene synthesis have been established in recent years. Mechanical exfoliation [28], chemical exfoliation [29], chemical synthesis [30], and thermal CVD synthesis [31], on the other hand, are the most widely employed processes today. Other techniques, such as unwrapping of nanotubes and microwave synthesis, have also been published. When a mechanical exfoliation using an atomic force microscope (AFM) cantilever could produce graphs of just a few layers, the process problem was that graphs vary in thickness to 10 nm, equivalent to graphenes of 30 layers [3]. Graphene synthesis can be narrowly separated into two approaches: top-down and bottom-up. The Figure 2.1 and 2.2 illustrate schematic diagram of to-down and bottom-up approach of graphene synthesis respectively.

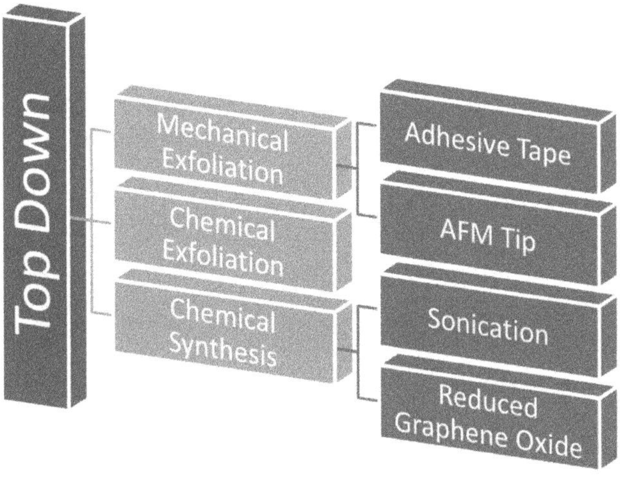

FIGURE 2.1 Schematic diagram of top-down approach of graphene synthesis [26–29].

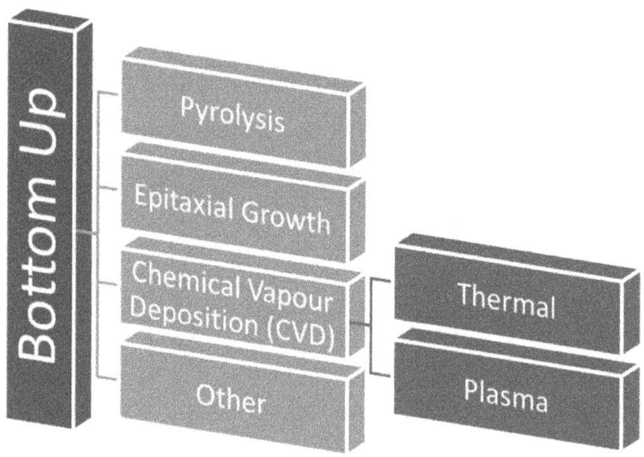

FIGURE 2.2 Schematic diagram of bottom-up approach of graphene synthesis [30, 31].

2.2.2 Honey-Mediated Graphene Synthesis

Ma et al. synthesized graphene from graphite powder by applying honey in 1:8 ratios. They then stirred it for 10 minutes at 2:00 pm. The mixture was pumped into the center roll of a three-roll mill. They kept the temperature and humidity steady. After 5 hours of testing, they washed the mixture with purified water and centrifuged the suspension to absorb the monolayer graphene. The team discovered that, in addition to producing high-quality graphene, the process provided a very high yield, with over 90 percent of the graphene being monolayer graphene [32]. The diagrammatic representation of graphene synthesis shown in the figure 2.3.

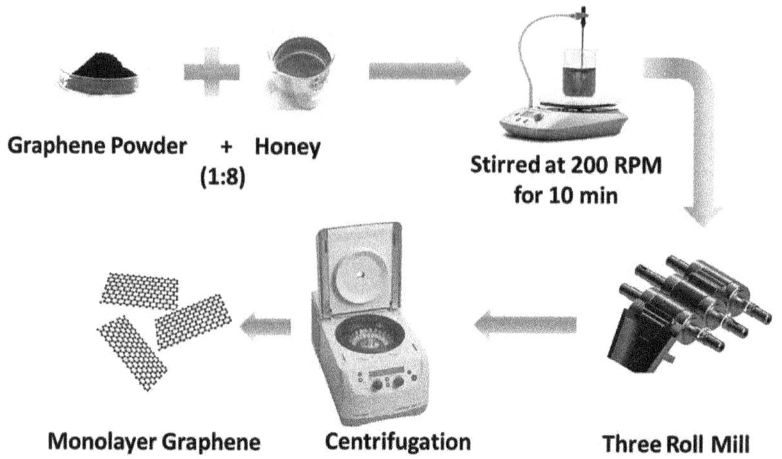

FIGURE 2.3 Schematic diagrammatic representation of graphene synthesis [32].

2.3 CHARACTERIZATION OF GRAPHENE

2.3.1 X-Ray Diffraction (XRD)

The detection of materials based on their division patterns is a key application of XRD research. XRD is a method for calculating the crystallographic structure of a substance used in the field of material science. XRD operates by radiation of incident X-rays and measurement of the intensity and dispersal angles of the X-rays leaving the material. The detection of materials based on their division patterns is a key application of XRD research. Figure 2.4 represents XRD diffraction patterns of pristine graphite, graphite oxide, and graphene. It can be easily detected at 25–26.6 degrees, a long and confined diffraction peak of pristine graphite. The peak moved to 13.9 degrees while the pristine graphite gets oxidized, because of oxygen molecules within graphite sheets [33].

No diffraction peak appeared whenever the graphite oxide becomes completely exfoliate, indicating the formation of the graphene layer.

2.3.2 Raman Spectroscopy

Raman microspectroscopy is an appropriate and accurate technique for determining any of these properties. Raman spectroscopy's high spectral and spatial resolutions, along with its high structural selectivity and nondestructive nature, allow it a standard characterization tool in the rapidly expanding field of graphene.

The major spectral characteristic of graphene is the G band, which occurs around 1580 cm^{-1}. It detects the number of graphene layers and is highly susceptible to strain effects. The shifting of the G band location toward lower frequencies in the diagram indicates more graphene layers. However, no significant difference in spectral form is detected. Furthermore, the G band is susceptible to doping, and the peak size and frequency of this pattern can be used to determine the degree of defects or doping concentration.

The D unit, also considered the disorder band, is affected by lattice movement away from the Brillouin core region, which can be found between 1270 and 1450 cm^{-1}.

The second-order two phonic phases are the 2D band, which is also called the G unit, which shows high-frequency laser excitement dependence because of a double resonance process that links an electronic band structure to a phonon wave vector. For a 514-nm laser excitation, this peak

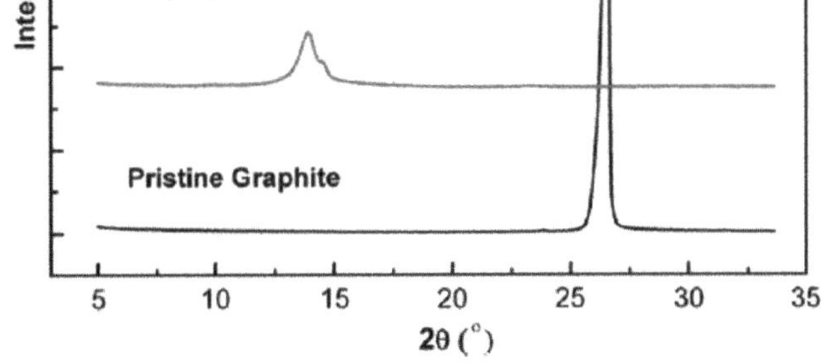

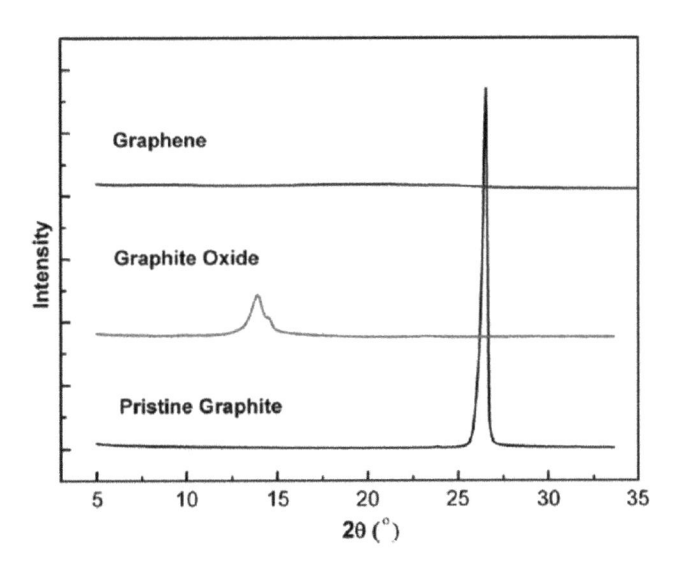

FIGURE 2.4 X-ray diffraction patterns of graphene, graphite oxide, and pristine graphite [34].

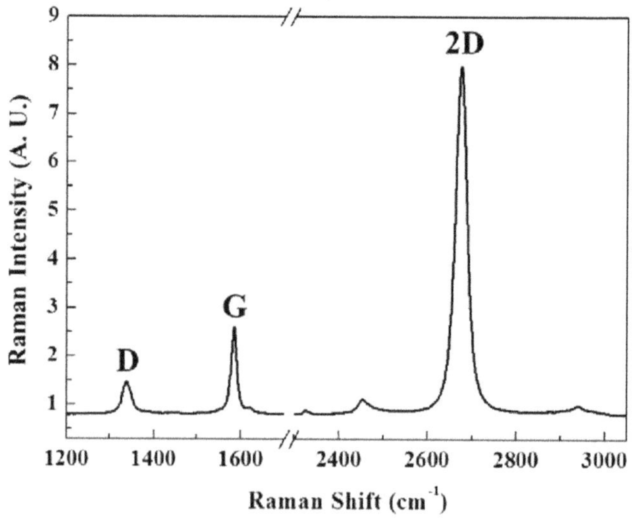

FIGURE 2.5 Raman spectrum of graphene [35].

occurs at about 2700 cm^{-1}. However, its behavior is more complex than the frequency shift seen in the G band. The Raman Spectrum of graphene is shown in the figure 2.5.

Recent research has shown that graphene can be used as a Raman substrate for testing complex samples. The substance will aid in the suppression of intense intrinsic fluorescence. Furthermore, when a sample is deposited onto a graphene layer, a Raman enhancement effect can be observed.

2.3.3 Scanning Electron Microscopy (SEM)

Nanomaterials are routinely imaged using electron microscopy, especially SEM. It has evolved into a fairly common, simple-to-use instrument capable of imaging even a single graphene layer.

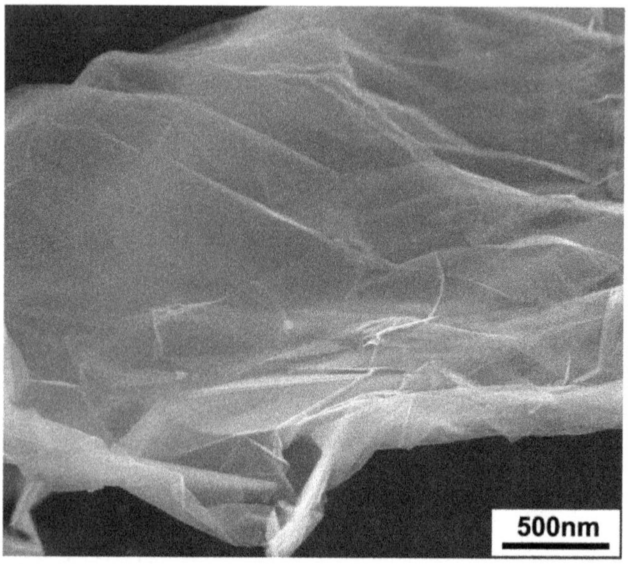

FIGURE 2.6 SEM image of graphene [37].

As an electron ray strikes a sample, it can create a range of scatterings that can be captured and collectively use to produce nanoscale images. Elastically dispersed electrons that produce backscattered electron signals with a high-energy incident sensitive to the atomic structure of the specimen are more effective in their dispersal of heavier elements [36]. Figure 2.6 represent SEM image of graphene.

Some incident electrons are often subjected to inelastic scattering to pass energy into the specimen atoms, resulting in lower energy secondary electrons from the energized atoms. High-resolution images can be obtained from secondary electron signals, but they are susceptible to the charged surface. Consequently, secondary electrons are harder to image isolation samples, such as graphenes.

2.3.4 Transmission Electron Microscope (TEM)

It is relatively easy to obtain TEM images of graphene, carbon nanotube (CNT), and other similar kinds of materials. In the electron wave projection, the interaction between the beam and the sample can be approached with a weak disturbance by the sample's electrostatic potential with only one or a few carbon atoms in the thickness. In addition, TEM allows the specimen to be shot in a series of fast consecutive images. In imaging, particularly in samples made up of light elements such as carbon, the effect of electron irradiation on TEM specimens should be considered [38]. TEM image of graphene can be seen in figure 2.7.

Although the radiation properties of electrons are usually an interference to attaining high-quality photographs, they also provide a possible chance for manipulating material characteristics on the nanoscale. It should be possible to customize various properties by controlled sputtered atomic availabilities and reconfiguring the framework. TEM is commonly used to investigate the structural consistency and layer number of graphene [40].

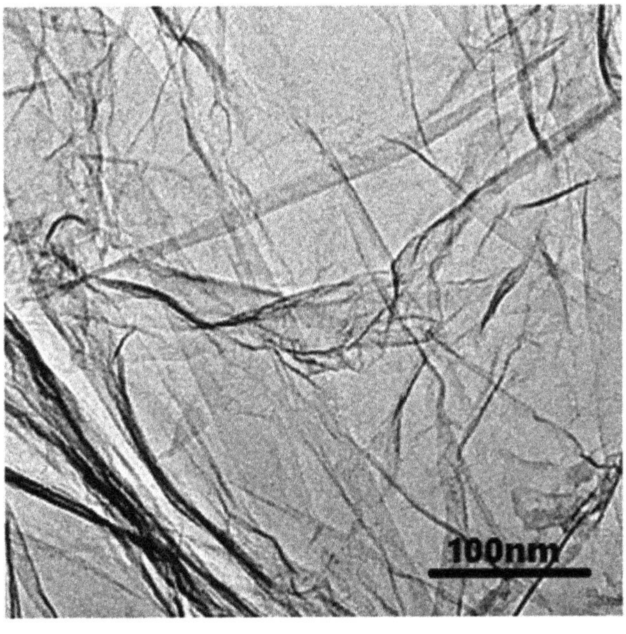

FIGURE 2.7 TEM image of graphene [39].

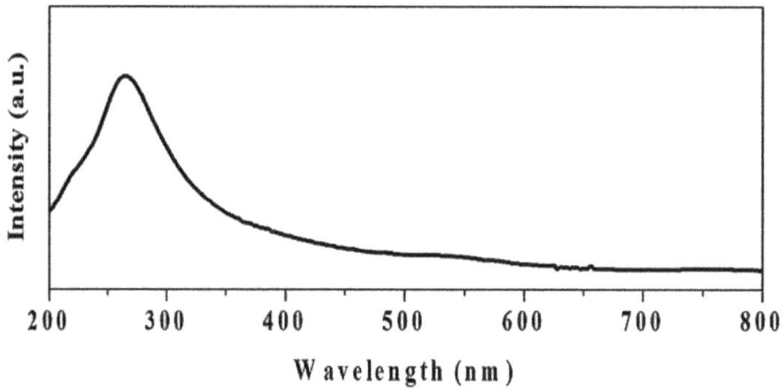

FIGURE 2.8 UV-VIS spectra of graphene [42].

2.3.5 UV-VIS Spectroscopy

Absorption spectroscopy in the UV-VIS range is convenient for identifying and characterizing graphene and other carbon allotropes. The absorption peak of graphene represents at 230 nm, which is correlated with the aromatic C-C bond's p-p* transition [41]. UV-VIS spectra of graphene ilustrated in figure 2.8.

A 300-nm shoulder due to n-p* transitions in the C=O bond is extremely critical in detecting oxidization with deep UV spectra. Another factor is the ability to calculate the number of layers in the sample quickly and non-destructively.

2.4 APPLICATIONS OF GRAPHENE

2.4.1 Medical

Graphene-based products, such as pristine-graphene tubes, graphene pellets, and graphene-oxide, provide a special, scalable, and tunable set of properties be creatively applied to biomedical applications [43]. There are various graphene technologies in biomedicine, divided into many categories: transport devices, sensors, tissue engineering, and biological agents [44–46].

The side dimensions of these 2D materials are adjustable between nanometers and millimeters. Their thickness can be varied from one to a hundred monolayers, and their bending rigidity can also be modulated. The flat surface is easy and operates to change the surface properties. It pioneered nanomaterial history, opening up huge design options for drug supply and ultrasensitive biosensors [45, 47], where it opens up huge possibilities.

2.4.2 Composites and Coatings

Graphene is one of the simplest and most effective ways to achieve its promise, combining it with existing components, called composite materials. Graphene-based composites would affect various fields, enhancing performance and extending their applications [48–51].

Researchers at The University of Manchester have also shown the opportunity for a rust-free future. A unique graphene coating is produced by combining graphs and paints that may signal the end of rust-related oxidation in ships and cars [52].

The same way can be used to avoid the movement of water and oxygen molecules that cause the spoilage of food through waterproof bricks and mortar [53], weatherproof buildings [54], and food packaging [55]. Additional benefits are derived from incorporating graphene-based composites into key industries, including production, transport, and aerospace. The schematic digram of different application areas of graphene depicted in Figure 2.9.

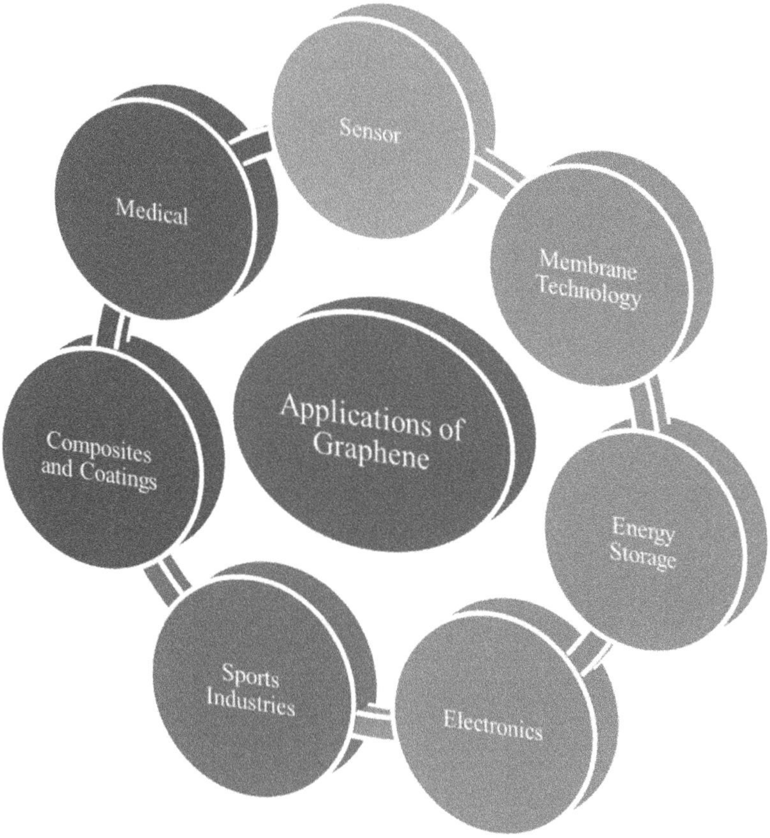

FIGURE 2.9 Schematic diagram of different application areas of graphene [43].

2.4.3 Sport Industries

Sporting products are also the first to adopt new material developments, as shown by Head's successful graphene-enhanced tennis racket. Graphene-based composites and laminates will be used to boost ski, cycling, and even Formula One sports equipment shortly [56].

2.4.4 Electronics

Graphene can be used as a coating to improve the efficiency of today's handset and tablet touch screens, etc. It can also be used to construct our computers' circuits, allowing them to run at breakneck speeds. These are only two of the ways graphene can improve today's electronics. Graphene can also be used to power the next wave of electronics [57].

The smaller the transistor, the better it performs inside circuits. The underlying threat that the electronics industry will face in the next 20 years is the more miniaturization of technology.

Graphene's peculiar properties of thinness and conductivity have sparked worldwide interest in its potential applications as a semiconductor. Graphene semiconductors, which are just one-atom thick and can process electricity at room temperature, have the potential to replace conventional computer chip technologies. According to research, graphene chips are now significantly quicker than silicon chips [58, 59].

2.4.5 Energy Storage

Graphene can greatly increase the longevity of a traditional lithium-ion battery so that batteries can be loaded quicker and power stored longer. The batteries can be so small and light that they can be sewn into clothing or body. It may have a significant effect on soldiers who bring up to 16 lb of battery at a time. They carry less battery weight and can recharge it through body heat, or the sun helps them spend more time on the field [60–62].

Graphene-based supercapacitors have the potential to deliver vast quantities of power while consuming much less energy than traditional systems. Since they are small, they can minimize the weight of automobiles or aircraft [63, 64].

2.4.6 Membrane Technology

When interacting with liquids and gases, graphene oxide membranes can form a perfect shield. They have an extraordinary ability to isolate organic solvent from water and extract water from a gas mixture. They have also been seen to avoid hydrogen, the most difficult gas to stop. The opportunity to open vast new markets and revolutionize many industrial processes means that a single plane can serve as an impenetrable shield [65].

The use of graphene coverings to avoid water movement and oxygen in the food and pharmaceutical packaging would enable food to remain fresher for a longer time. The absorption into the environment by power plants of toxic carbon dioxide, which graphene membranes could improve, is not currently being accomplished on a large scale [66].

2.4.7 Sensor

Graphene is a great material for the sensor. A graphene atom in its environment is revealed so that transition can be detected. The objective of chemical sensors is the identification of only one molecule of a potentially dangerous substance. Graphene also allows the creation of sensors in micrometer dimensions to track molecular events [67, 68].

Graphene oxide can be used for the production of intelligent food packaging. It reduces unnecessary food waste dramatically while also helping disease control. Graphene-coated packaging is capable of following changes in the atmosphere due to rotting food [69].

Graphene-based sensors could enhance the productivity of important crops tracked by the farming industry [70]. Farmers should monitor and take corrective action on any poisonous gasses that can affect crop areas. Given the sensitivity of the graphene sensors, the best areas for cultivation can be decided to depend upon atmospheric conditions [71, 72].

2.5 CONCLUSION

This chapter discusses the honey-meditated graphene synthesis, the characterization techniques, and the use of graphene. Graphene is the most used carbon-based material that has outstanding unique properties such as high conductivity, high tensile strength, high thermal conductivity, high carrier mobility, and transparency, which makes it a powerful candidate for a wide range of applications, such as sensors, electronics, sports equipment, storing of energy, membranes, and coatings. While it is very exciting to make advances in graph-based nanomaterials for applications, some of the key issues such as long-term stability, toxicity, and environmental implications must also be solved before commercialization. Continuing to improve graphic science and high-biological processing would allow for new nanomaterials based on graphene.

REFERENCES

[1] U. K. Sur, "Graphene: a rising star on the horizon of materials science," *Int J Electrochem*, vol. 2012, p. e237689, Sep. 2012, doi: 10.1155/2012/237689
[2] A. V. Tyurnina *et al.*, "Ultrasonic exfoliation of graphene in water: a key parameter study," *Carbon*, vol. 168, pp. 737–747, Oct. 2020, doi: 10.1016/j.carbon.2020.06.029
[3] Md. S. A. Bhuyan, Md. N. Uddin, Md. M. Islam, F. A. Bipasha, and S. S. Hossain, "Synthesis of graphene," *Int Nano Lett*, vol. 6, no. 2, pp. 65–83, Jun. 2016, doi: 10.1007/s40089-015-0176-1
[4] G. Kalita and M. Tanemura, *Fundamentals of Chemical Vapor Deposited Graphene and Emerging Applications*. IntechOpen, 2017. doi: 10.5772/67514
[5] F. Bonaccorso, A. Lombardo, T. Hasan, Z. Sun, L. Colombo, and A. C. Ferrari, "Production and processing of graphene and 2D crystals," *Mater Today*, vol. 15, no. 12, pp. 564–589, Dec. 2012, doi: 10.1016/S1369-7021(13)70014-2
[6] Y. Zhu, H. Ji, H.-M. Cheng, and R. S. Ruoff, "Mass production and industrial applications of graphene materials," *Natl Sci Rev*, vol. 5, no. 1, pp. 90–101, Jan. 2018, doi: 10.1093/nsr/nwx055
[7] A. M. Jastrzębska, P. Kurtycz, and A. R. Olszyna, "Recent advances in graphene family materials toxicity investigations," *J Nanopart Res*, vol. 14, no. 12, p. 1320, Nov. 2012, doi: 10.1007/s11051-012-1320-8
[8] L. Lu *et al.*, "Graphene oxide and H2 production from bioelectrochemical graphite oxidation," *Sci Rep*, vol. 5, no. 1, Art. no. 1, Nov. 2015, doi: 10.1038/srep16242
[9] S. Gambhir, R. Jalili, D. L. Officer, and G. G. Wallace, "Chemically converted graphene: scalable chemistries to enable processing and fabrication," *NPG Asia Mater*, vol. 7, no. 6, Art. no. 6, Jun. 2015, doi: 10.1038/am.2015.47
[10] B. K. Salunke and B. S. Kim, "Facile synthesis of graphene using a biological method," *RSC Adv*, vol. 6, no. 21, pp. 17158–17162, Feb. 2016, doi: 10.1039/C5RA25977K
[11] M. Y. Kim *et al.*, "Phenolic compounds and antioxidant activity in sweet potato after heat treatment," *J Sci Food Agric*, vol. 99, no. 15, pp. 6833–6840, Dec. 2019, doi: 10.1002/jsfa.9968
[12] M. Bhattacharya, "A history of evolution of the terms of carbohydrates coining the term 'glucogenic carbohydrates' and prescribing in grams per day for better nutrition communication," *J Public Health Nutr*, vol. 1, no. 4, 2018, Accessed: May 11, 2021. [Online]. Available: https://www.alliedacademies.org/abstract/a-history-of-evolution-of-the-terms-of-carbohydrates-coining-the-term-lsquoglucogenic-carbohydratesrsquo-and-prescribing-in-grams–10975.html
[13] E.-K. A. Taha, R. A. Taha, and S. N. AL-Kahtani, "Nectar and pollen sources for honeybees in Kafrelsheikh province of northern Egypt," *Saudi J Biol Sci*, vol. 26, no. 5, pp. 890–896, Jul. 2019, doi: 10.1016/j.sjbs.2017.12.010
[14] A. Al-Ghamdi, N. Adgaba, A. Getachew, and Y. Tadesse, "New approach for determination of an optimum honeybee colony's carrying capacity based on productivity and nectar secretion potential of bee forage species," *Saudi J Biol Sci*, vol. 23, no. 1, pp. 92–100, Jan. 2016, doi: 10.1016/j.sjbs.2014.09.020
[15] E. R. H. S. S. Ediriweera and N. Y. S. Premarathna, "Medicinal and cosmetic uses of Bee's Honey – a review," *Ayu*, vol. 33, no. 2, pp. 178–182, 2012, doi: 10.4103/0974-8520.105233
[16] F. Fratini, G. Cilia, S. Mancini, and A. Felicioli, "Royal Jelly: an ancient remedy with remarkable antibacterial properties," *Microbiol Res*, vol. 192, pp. 130–141, Nov. 2016, doi: 10.1016/j.micres.2016.06.007
[17] T. Eteraf-Oskouei and M. Najafi, "Traditional and modern uses of natural honey in human diseases: a review," *Iran J Basic Med Sci*, vol. 16, no. 6, pp. 731–742, Jun. 2013, Accessed: May 11, 2021. [Online]. Available: https://www.ncbi.nlm.nih.gov/pmc/articles/PMC3758027/
[18] S. Samarghandian, T. Farkhondeh, and F. Samini, "Honey and health: a review of recent clinical research," *Pharmacogn Res*, vol. 9, no. 2, pp. 121–127, 2017, doi: 10.4103/0974-8490.204647
[19] D. Philip, "Honey mediated green synthesis of silver nanoparticles," *Spectrochim Acta A Mol Biomol Spectrosc*, vol. 75, no. 3, pp. 1078–1081, Mar. 2010, doi: 10.1016/j.saa.2009.12.058
[20] P. B. Olaitan, O. E. Adeleke, and I. O. Ola, "Honey: a reservoir for microorganisms and an inhibitory agent for microbes," *Afr Health Sci*, vol. 7, no. 3, pp. 159–165, Sep. 2007, Accessed: May 12, 2021. [Online]. Available: https://www.ncbi.nlm.nih.gov/pmc/articles/PMC2269714/
[21] A. Shugaba, "Analysis of biochemical composition of honey samples from north-east Nigeria," *Biochem Anal Biochem*, vol. 2, no. 3, 2012, doi: 10.4172/2161-1009.1000139
[22] J. Tewari and J. Irudayaraj, "Quantification of saccharides in multiple floral honeys using Fourier transform infrared microattenuated total reflectance spectroscopy," *J Agric Food Chem*, vol. 52, no. 11, pp. 3237–3243, Jun. 2004, doi: 10.1021/jf035176+

[23] V. León-Ruiz, S. Vera, A. V. González-Porto, and M. P. San Andrés, "Analysis of water-soluble vitamins in honey by isocratic RP-HPLC," *Food Anal Methods*, vol. 6, no. 2, pp. 488–496, Apr. 2013, doi: 10.1007/s12161-012-9477-4
[24] V. Nanda, B. C. Sarkar, H. K. Sharma, and A. S. Bawa, "Physico-chemical properties and estimation of mineral content in honey produced from different plants in Northern India," *J Food Compos Anal*, vol. 16, no. 5, pp. 613–619, Oct. 2003, doi: 10.1016/S0889-1575(03)00062-0
[25] E. R. Balasooriya, C. D. Jayasinghe, U. A. Jayawardena, R. W. D. Ruwanthika, R. Mendis de Silva, and P. V. Udagama, "Honey mediated green synthesis of nanoparticles: new era of safe nanotechnology," *J Nanomater*, vol. 2017, p. e5919836, Mar. 2017, doi: 10.1155/2017/5919836
[26] A. A. Balandin *et al.*, "Superior thermal conductivity of single-layer graphene," *Nano Lett*, vol. 8, no. 3, pp. 902–907, Mar. 2008, doi: 10.1021/nl0731872
[27] H. P. Boehm, R. Setton, and E. Stumpp, "Nomenclature and terminology of graphite intercalation compounds (IUPAC Recommendations 1994)," *Pure Appl Chem*, vol. 66, no. 9, pp. 1893–1901, Jan. 1994, doi: 10.1351/pac199466091893
[28] M. Yi and Z. Shen, "A review on mechanical exfoliation for the scalable production of graphene," *J Mater Chem A*, vol. 3, no. 22, pp. 11700–11715, May 2015, doi: 10.1039/C5TA00252D
[29] M. Liu *et al.*, "One-step chemical exfoliation of graphite to ~100% few-layer graphene with high quality and large size at ambient temperature," *Chem Eng J*, vol. 355, pp. 181–185, Jan. 2019, doi: 10.1016/j.cej.2018.08.146
[30] S. Stankovich *et al.*, "Synthesis of graphene-based nanosheets via chemical reduction of exfoliated graphite oxide," *Carbon*, vol. 45, no. 7, pp. 1558–1565, Jun. 2007, doi: 10.1016/j.carbon.2007.02.034
[31] Z. Gao, Y. Zhang, Y. Fu, M. M. F. Yuen, and J. Liu, "Thermal chemical vapor deposition grown graphene heat spreader for thermal management of hot spots," *Carbon*, vol. 61, pp. 342–348, Sep. 2013, doi: 10.1016/j.carbon.2013.05.014
[32] Y. Ma *et al.*, "Robust and antibacterial polymer/mechanically exfoliated graphene nanocomposite fibers for biomedical applications," *ACS Appl Mater Interfaces*, vol. 10, no. 3, pp. 3002–3010, Jan. 2018, doi: 10.1021/acsami.7b17835
[33] Y. Seekaew, O. Arayawut, K. Timsorn, and C. Wongchoosuk, "Chapter Nine – Synthesis, Characterization, and Applications of Graphene and Derivatives," in *Carbon-Based Nanofillers and Their Rubber Nanocomposites*, S. Yaragalla, R. Mishra, S. Thomas, N. Kalarikkal, and H. J. Maria, Eds. Elsevier, 2019, pp. 259–283. doi: 10.1016/B978-0-12-813248-7.00009-2
[34] H.-B. Zhang *et al.*, "Electrically conductive polyethylene terephthalate/graphene nanocomposites prepared by melt compounding," *Polymer*, vol. 51, no. 5, pp. 1191–1196, Mar. 2010, doi: 10.1016/j.polymer.2010.01.027
[35] G. Heo, Y. Kim, S. Chun, and M. Seong, "Polarized Raman spectroscopy with differing angles of laser incidence on single-layer graphene," *Nanoscale Res Lett*, 2015, doi: 10.1186/s11671-015-0743-4
[36] J. Kim, F. Kim, and J. Huang, "Seeing graphene-based sheets," *Mater Today*, vol. 13, no. 3, pp. 28–38, Mar. 2010, doi: 10.1016/S1369-7021(10)70031-6
[37] D. Zhao, X. Wang, and D. Wu, "Enhanced thermoelectric properties of graphene/Cu_2SnSe_3 composites," *Crystals*, vol. 7, no. 3, Art. no. 3, Mar. 2017, doi: 10.3390/cryst7030071
[38] C. Mangler and J. C. Meyer, "Using electron beams to investigate carbonaceous materials," *C R Phys*, vol. 15, no. 2, pp. 241–257, Feb. 2014, doi: 10.1016/j.crhy.2013.10.011
[39] J. F. Dai, T. Xian, L. J. Di, and H. Yang, "Preparation of graphene nanocomposites and their enhanced photocatalytic activities," *J Nanomater*, vol. 2013, p. e642897, Nov. 2013, doi: 10.1155/2013/642897
[40] S. Mourdikoudis, R. M. Pallares, and N. T. K. Thanh, "Characterization techniques for nanoparticles: comparison and complementarity upon studying nanoparticle properties," *Nano-scale*, vol. 10, no. 27, pp. 12871–12934, Jul. 2018, doi: 10.1039/C8NR02278J
[41] M. Hu, Z. Yao, X. Wang, "Characterization techniques for graphene-based materials in catalysis," *AIMS Mater Sci*, vol. 4, no. 3, pp. 755–788, 2017, doi: 10.3934/matersci.2017.3.755
[42] K. Karthikeyan, V. Murugan, K. Gui-Shik, and J. K. Sang, "A one step hydrothermal approach for the improved synthesis of graphene nanosheets," *Curr Nanosci*, vol. 8, no. 6, pp. 934–938, Nov. 2012, Accessed: May 12, 2021. [Online]. Available: https://www.eurekaselect.com/104755/article
[43] K. Tadyszak, J. K. Wychowaniec, and J. Litowczenko, "Biomedical applications of graphene-based structures," *Nanomaterials (Basel)*, vol. 8, no. 11, Nov. 2018, doi: 10.3390/nano8110944
[44] X. Luo, C. L. Weaver, S. Tan, and X. T. Cui, "Pure graphene oxide doped conducting polymer nanocomposite for bio-interfacing," *J Mater Chem B*, vol. 1, no. 9, pp. 1340–1348, Mar. 2013, doi: 10.1039/C3TB00006K

[45] S. Goenka, V. Sant, and S. Sant, "Graphene-based nanomaterials for drug delivery and tissue engineering," *J Control Release*, vol. 173, pp. 75–88, Jan. 2014, doi: 10.1016/j.jconrel.2013.10.017
[46] J. K. Wychowaniec *et al.*, "Designing peptide/graphene hybrid hydrogels through fine-tuning of molecular interactions," *Biomacromolecules*, vol. 19, no. 7, pp. 2731–2741, Jul. 2018, doi: 10.1021/acs.biomac.8b00333
[47] J. Wu *et al.*, "Hierarchical construction of a mechanically stable peptide-graphene oxide hybrid hydrogel for drug delivery and pulsatile triggered release in vivo," *Nano-scale*, vol. 7, no. 5, pp. 1655–1660, Feb. 2015, doi: 10.1039/c4nr05798h
[48] M. Silva, I. S. Pinho, J. A. Covas, N. M. Alves, and M. C. Paiva, "3D printing of graphene-based polymeric nanocomposites for biomedical applications," *Funct Compos Mater*, vol. 2, no. 1, p. 8, Apr. 2021, doi: 10.1186/s42252-021-00020-6
[49] D. Verma, P. C. Gope, A. Shandilya, and A. Gupta, "Mechanical-thermal-electrical and morphological properties of graphene reinforced polymer composites: a review," *Trans Indian Inst Met*, vol. 67, no. 6, pp. 803–816, Dec. 2014, doi: 10.1007/s12666-014-0408-5
[50] A. Mirabedini, A. Ang, M. Nikzad, B. Fox, K.-T. Lau, and N. Hameed, "Evolving strategies for producing multiscale graphene-enhanced fiber-reinforced polymer composites for smart structural applications," *Adv Sci*, vol. 7, no. 11, p. 1903501, 2020, doi: https://doi.org/10.1002/advs.201903501
[51] Y. Li *et al.*, "Additive manufacturing high performance graphene-based composites: a review," *Composites, A: Appl Sci Manuf*, vol. 124, p. 105483, Sep. 2019, doi: 10.1016/j.compositesa.2019.105483
[52] Y. Su, V. G. Kravets, S. L. Wong, J. Waters, A. K. Geim, and R. R. Nair, "Impermeable barrier films and protective coatings based on reduced graphene oxide," *Nat Commun*, vol. 5, no. 1, Art. no. 1, Sep. 2014, doi: 10.1038/ncomms5843
[53] X. Chen, Y. Zhang, S. Li, Y. Geng, and D. Hou, "Influence of a new type of graphene oxide/silane composite emulsion on the permeability resistance of damaged concrete," *Coatings*, vol. 11, no. 2, Art. no. 2, Feb. 2021, doi: 10.3390/coatings11020208
[54] S. K. Tiwari, S. Sahoo, N. Wang, and A. Huczko, "Graphene research and their outputs: status and prospect," *J Sci: Adv Mater Devices*, vol. 5, no. 1, pp. 10–29, Mar. 2020, doi: 10.1016/j.jsamd.2020.01.006
[55] A. K. Sundramoorthy, T. H. Vignesh Kumar, and S. Gunasekaran, "Chapter 12 – Graphene-Based Nanosensors and Smart Food Packaging Systems for Food Safety and Quality Monitoring," in *Graphene Bioelectronics*, A. Tiwari, Ed. Elsevier, 2018, pp. 267–306. doi: 10.1016/B978-0-12-813349-1.00012-3
[56] W. Kong *et al.*, "Path towards graphene commercialization from lab to market," *Nat Nanotechnol*, vol. 14, no. 10, Art. no. 10, Oct. 2019, doi: 10.1038/s41565-019-0555-2
[57] Y. Sun, M. Sun, and D. Xie, "5 – Graphene Electronic Devices," in *Graphene*, H. Zhu, Z. Xu, D. Xie, and Y. Fang, Eds. Academic Press, 2018, pp. 103–155. doi: 10.1016/B978-0-12-812651-6.00005-7
[58] Y. Obeng and P. Srinivasan, "Graphene: is it the future for semiconductors? An overview of the material, devices, and applications," *Interface Mag*, vol. 20, no. 1, pp. 47–52, Jan. 2011, doi: 10.1149/2.F05111if
[59] S. Maharubin, X. Zhang, F. Zhu, H.-C. Zhang, G. Zhang, and Y. Zhang, "Synthesis and applications of semiconducting graphene," *J Nanomater*, vol. 2016, p. e6375962, Nov. 2016, doi: 10.1155/2016/6375962
[60] R. Raccichini, A. Varzi, S. Passerini, and B. Scrosati, "The role of graphene for electrochemical energy storage," *Nat Mater*, vol. 14, no. 3, Art. no. 3, Mar. 2015, doi: 10.1038/nmat4170
[61] L. Bai *et al.*, "Graphene for energy storage and conversion: synthesis and interdisciplinary applications," *Electrochem Energy Rev*, vol. 3, no. 2, pp. 395–430, Jun. 2020, doi: 10.1007/s41918-019-00042-6
[62] A. G. Olabi, M. A. Abdelkareem, T. Wilberforce, and E. T. Sayed, "Application of graphene in energy storage device – a review," *Renew Sustain Energy Rev*, vol. 135, p. 110026, Jan. 2021, doi: 10.1016/j.rser.2020.110026
[63] D. Majumdar, M. Mandal, and S. K. Bhattacharya, "Journey from supercapacitors to supercapatteries: recent advancements in electrochemical energy storage systems," *Emergent Mater*, vol. 3, no. 3, pp. 347–367, Jun. 2020, doi: 10.1007/s42247-020-00090-5
[64] Q. Ke and J. Wang, "Graphene-based materials for supercapacitor electrodes – a review," *J Mater*, vol. 2, no. 1, pp. 37–54, Mar. 2016, doi: 10.1016/j.jmat.2016.01.001
[65] Q. Xu and W. Zhang, *Next-Generation Graphene-Based Membranes for Gas Separation and Water Purifications*. IntechOpen, 2016. doi: 10.5772/64396
[66] P. Chaudhary, F. Fatima, and A. Kumar, "Relevance of nanomaterials in food packaging and its advanced future prospects," *J Inorg Organomet Polym*, vol. 30, no. 12, pp. 5180–5192, Dec. 2020, doi: 10.1007/s10904-020-01674-8
[67] X. Jin *et al.*, "Review on exploration of graphene in the design and engineering of smart sensors, actuators and soft robotics," *Chem Eng J Adv*, vol. 4, p. 100034, Dec. 2020, doi: 10.1016/j.ceja.2020.100034

[68] T. Wang *et al.*, "A review on graphene-based gas/vapor sensors with unique properties and potential applications," *Nano-Micro Lett*, vol. 8, no. 2, pp. 95–119, Apr. 2016, doi: 10.1007/s40820-015-0073-1
[69] J. K. Heising, G. D. H. Claassen, and M. Dekker, "Options for reducing food waste by quality-controlled logistics using intelligent packaging along the supply chain," *Food Addit Contam, A Chem Anal Control Expo Risk Assess*, vol. 34, no. 10, pp. 1672–1680, Oct. 2017, doi: 10.1080/19440049.2017.1315776
[70] A. K. Srivastava, A. Dev, and S. Karmakar, "Nanosensors and nanobiosensors in food and agriculture," *Environ Chem Lett*, vol. 16, no. 1, pp. 161–182, Mar. 2018, doi: 10.1007/s10311-017-0674-7
[71] M. Kodu *et al.*, "Graphene-based ammonia sensors functionalised with sub-monolayer V_2O_5: a comparative study of chemical vapour deposited and epitaxial graphene," *Sensors*, vol. 19, no. 4, Art. no. 4, Jan. 2019, doi: 10.3390/s19040951
[72] S. Bandi and A. K. Srivastav, "Chapter Seven – Graphene-Based Chemiresistive Gas Sensors," in *Comprehensive Analytical Chemistry*, vol. 91, C. M. Hussain, Ed. Elsevier, 2020, pp. 149–173. doi: 10.1016/bs.coac.2020.08.006

3 Graphene from Animal Waste

Vivian C. Akubude, Victor C. Okafor, and Jelili A. Oyedokun

CONTENTS

3.1 INTRODUCTION

Global population is on the increase on daily basis with rising energy demand. The use of fossil fuel is also on the increase with technological advancement. The activities of human beings at domestic, agricultural and industrial levels release pollutants into the environment. These pollutants are in solid, liquid or gaseous form. Conversion of waste to bio-based products of economic value will serve as control measure to environmental pollution and, in return, wealth creation strategy. Agro-waste has been utilized for the production of several bioproducts such as biogas, biofertilizer, biopolymers, biocomposites, and biochar, and these are all environment-friendly products.

Graphene is carbon allotrope with a single layer of two-dimensional, hexagonal lattice held together by σ bonds, comprising a network of sp^2-bonded carbon atoms as shown in Figure 3.1. It is the thinnest material that is stable in its free state and the strongest known material. It is characterized by notable electronic properties, thermal conductivity and large surface area, excellent mechanical, physical and chemical properties. The electronic properties of graphene are determined by the number of graphene layers and their arrangement. It is characterized by high carrier mobility, large electrical conductivity and superior transmittance (Biswas and Lee 2011; Li et al 2015). Derivates of graphene are graphene oxide (GO) and reduced graphene oxide (rGO). GO comprises carbon-based hexagonal rings with the hybridization of sp^2 and sp^3 carbon atoms, and it has a high surface area. It can be employed in electrode materials for batteries, sensors, transistors, solar cells, conducting polymer composites, photo degradation, biodegradation, clean energy devices, water treatment, etc.

Diverse techniques have been developed by researchers for the production of graphene and graphene-based products from renewable and eco-friendly carbon sources such as animal waste. In times past, graphene derivates were produced from graphite powders using a two-step process. Currently, research has shown that these derivates can be prepared using non-graphitic biomass and agro-waste materials, avoiding the graphite precursor. Also, a report has shown the possibilities of

DOI: 10.1201/9781003169741-3

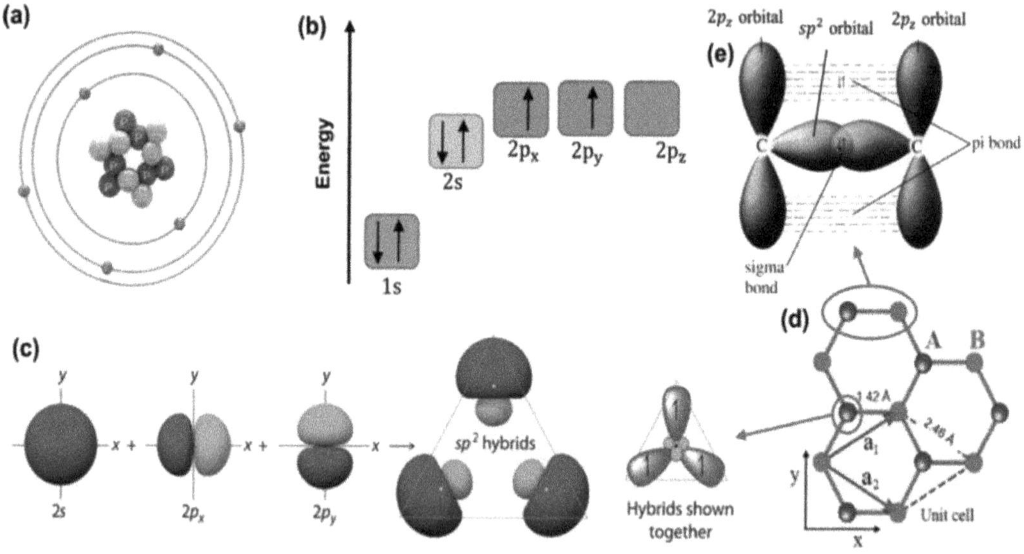

FIGURE 3.1 Graphene structure and its hybridization forms (Yang et al 2018).

preparing graphene directly without producing GO. Among the diverse methods used for graphene extraction, carbon vapour deposition (CVD) is the most researched one despite its limitations. The separation challenge associated with graphene dissociation from its substrate in this method may hinder its applications for commercial purposes (William et al 1958; Marcano et al 2010; Moo et al 2014; Shams et al 2015; Somanathan et al 2015; Chen et al 2016; Shi et al 2016; Goswami et al 2017; Purkait et al 2017; Singh et al 2017; Silpa and Sivmangai 2019).

3.2 METHODS OF EXTRACTING GRAPHENE FROM AGRO-WASTE, INCLUDING ANIMAL WASTE

Extraction methods of graphene are very vital as they determine their structures and properties. There are various forms of graphene from different extraction methods like single layer, double layer or multiple layers. And each of these forms of graphene serves different purposes. The method to be used for the extraction process is determined by the desired size, quantity and purity.

3.2.1 Carbonization

Carbonization is a thermal treatment process that involves the conversion of agro-waste into carbon materials. When this occurs in the absence of oxygen, it is described as pyrolysis which is a very common method of processing biomass into bio-based products like biochar, biogas and bio-oil. The end product is dependent on the pyrolysis process conditions such as temperature, heating rate and holding rate. For instance, high temperature favours biochar production, and fast heating rate favours bio-oil production. The possibility of deciding the end product via the manipulation of the process parameter makes it an interesting technique for generating diverse bioproducts. Also, in the presence of oxygen, oxidized carbon can be produced and, when exfoliated, gives GO. There are several research outputs in this area where agro-waste is converted into graphene. Usually, the reduction process starts at a temperature of 300°C. There are accounts of conversion of sugarcane bagasse into graphene derivatives, oil palm empty bunch into graphene (Widiatmoko et al 2019). Also, the issue of moisture in the biomass raw material can be tackled using hydrothermal carbonization. This has been used together with pyrolysis in producing graphene from wheat straw (Chen et al 2016).

3.2.2 Exfoliation Method

Exfoliation method applies mechanical and chemical mechanisms in breaking down the weak bonds and separates them into distinctive grapheme sheets held together by van der Waals forces. Research results point out that this process in graphene production is applied by employing potassium metal for the separation of pure graphene sheets, where ethanol was applied to cast them off in different layers. Several techniques have been applied to this method for generating graphene and its derivates from biomass, like sonication, Hummer's principle and Marcano's oxidation techniques. Some of the agro-wastes that have been utilized as carbon sources in this method for graphene extraction include oil palm leaves palm kernel shells (Nasir et al 2017), empty fruit bunches (Nasir et al 2017), sugarcane bagasse, rice husks, wheat straws (Chen et al 2016) and camphor (Shams et al 2015).

3.2.3 Carbon Growth

Carbon growth method uses the principle of chemical vapour deposition in growing carbon-like materials unto a metal surface. This method tends to produce high-quality graphene or graphene/metal composites. Application of pyrolysis to agro-waste tends to break down the long chains of carbon, hydrogen and oxygen compounds within them into smaller molecules. Carbon growth shares lots of similarities with CVD where carbon molecules released from pyrolysis process is left on a metal surface. This method has been employed by several researchers for graphene extraction from biomass (Bhuyan et al 2016; Suk et al 2011; Yang et al 2013; Lavin-Lopez et al 2014; Leng et al 2016; Mamat et al 2018; Yan et al 2018; He et al 2019; Liu et al 2019). Usually, it involves a mixing of the agro-waste with a metal precursor that serves as a catalyst. Iron powder was combined with Kraft lignin biomass to produce graphene pyrolysis which allows amorphous carbon to wrap up on the metal. Also, when the mixture is further heated to a very high temperature, graphene growth is observed on the metal surface (Liu et al 2019). In another study, the impact of varying atmospheric conditions on graphene extraction was investigated. Result shows that under welding gas atmospheric condition, more graphene was produced using lignocellulosic biomass. This method has been employed in converting animal waste such as cockroach leg, dog faeces, cow dung chicken bone and chitosan to graphene and its derivatives (Ruan et al 2011; Akhavan et al 2014; Hao et al 2015), as shown in Table 3.1. Results from this research show graphene/graphene derivates of different morphologies, structures and sizes.

3.3 POTENTIAL APPLICATIONS OF BIO-GRAPHENE AND ITS DERIVATIVES

Graphene and its derivatives can be used in technological fields due to their unique attributes. The discovery of graphene is the beginning of major breakthroughs in several areas such as medical field, electronics, battery system, solar cells, water treatment process and sensors.

3.3.1 Biomedical Applications

In medical field, graphene has been used in biosensors, bioimaging, drug delivery, tissue engineering as shown in Figure 3.2. It serves as a vehicle for drug delivery because of its hydrophobic properties. For instance, it has been used for drug delivery process for antibiotics, peptides, antibodies, genes, poorly soluble drugs and drugs for cancer treatment. The treatment of cancer using graphene and graphene-based substances has been extensively researched (Patel et al 2016).

TABLE 3.1
Utilization of Animal Waste in Production of Graphene and Its Derivatives

			Graphene Attributes		Graphene Form	
S/n	Animal Waste Type	Thermal Treatment	No. of layers	Thickness (nm)	Graphene	References
1	Chitosan	Chemical activation with KOH	4	1.5		Hao et al (2015), Safian et al (2020)
2	Chicken bone	Mixing with $FeCl_3$ and $ZnCl_2$	3	3	Graphene oxide	Akhavan et al (2014), Safian et al (2020)
3	Cow dung	Mixing with $FeCl_3$ and $ZnCl_2$	3	3	Graphene oxide	Akhavan et al (2014), Safian et al (2020)
4	Dog faeces	CVD with a Cu s ubstrate	1	–	Graphene	Ruan et al (2011), Safian et al (2020)
5	Cockroach legs	CVD with a Cu substrate	1	–	Graphene	Ruan et al (2011), Safian et al (2020)

Also, their intrinsic mechanical properties make graphene a good option in tissue-engineering scaffolds. Presently, they are employed in stem cell engineering and musculskeletal tissue engineering as tissue damage is a key issue that can result in death of human lives (Zhu et al 2010; Ziyad 2010; Li et al 2013; Guo et al 2017). The use grapheme in bio-imaging has also received reasonable research efforts (Ziyad 2010). Bio-imaging is part of regenerative medicine that deals with capturing selected biological entities. This is quite useful in monitoring and tracking issues within the entire body system. Abnormal processes such as cancerous cells and necrosis can be detected using graphene-based imaging systems like fluorescence/confocal imaging, surface-enhanced Raman scattering, coherent anti-Stokes Raman scattering imaging, magnetic resonance imaging, positron-emission tomography, ultrasound imaging, photoacoustic imaging and electron paramagnetic resonance imaging (Krishna et al 2001; Huff and Cheng 2007; Matsumoto et al 2007; Evans and Xie 2008; Feleppa et al 2011; Huang et al 2012a, b; Shi et al 2013; Chen et al 2014; Wang et al 2014; Jakubovic et al 2018; Jang et al 2018).

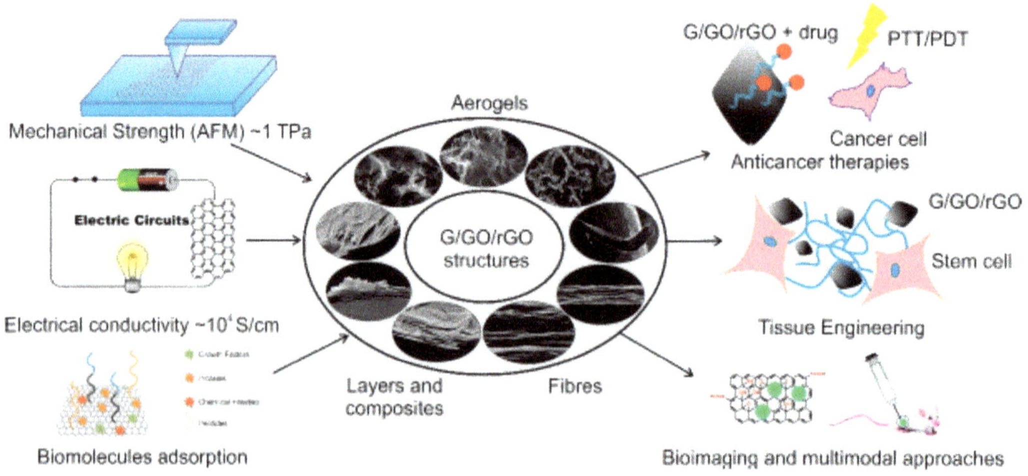

FIGURE 3.2 Biomedical applications of graphene and its derivatives (Tadyszak et al 2018).

3.3.2 Electronics Application

Interesting characteristics of graphene, like high electron mobility, high mechanical strength, transparency and flexibility, make it most sought-after for diverse applications in the field of electronics. It can be employed in building transducers for nanoelectromechanical accelerometers (Fan et al 2019). Also, graphene can be used in optical electronics, particularly in touchscreen devices, liquid crystal display and organic light-emitting diodes. This is possible because of its transparency attributes and high light-transmitting capacity of about 97.7%. Currently, touchscreen devices can be revolutionized by employing graphene and could be replaced by the use of indium tin oxide in the future. In addition, graphene can be utilized in transistors and integrated circuits. Replacing silicon with graphene resulted in a transistor with improved attributes better than the silicon counterpart. Research has also shown that blends of GO and titanium oxide can be used to create a flexible memory with high performance capacity. These flexible memories can be used in foldable electronic devices.

3.3.3 Batteries

Graphene is a good conductor of heat and electricity, which makes it a viable material for battery system. The overall performance of a battery can be improved by integrating graphene material into the system. They exhibit excellent properties such as higher capacity, high charging rate, light weight, flexibility and high temperature range. Also, a hybrid combination of graphene and vanadium oxide results in a material that can be used to achieve battery improvement. Lithium iron phosphate battery can be improved using graphene material at the cathode, which upgrades the performance in terms of fast charging rate, lightweight and great capacity when compared to the conventional counterpart. However, despite the numerous research inputs, graphene is still at the laboratory level and is yet to be commercialized (Cai et al 2017). Kim and friends researched on all-graphene battery where both the cathode and anode materials are made of graphene which has similarity with batteries and supercapacitors in performance and operation. Result from this studies show that it is a promising energy storage system (Kim et al 2014).

3.3.4 Solar Cell Applications

Integrating graphene in solar cell devices has improved the efficiency and reduces the solar reflectance. There are several researches documented in this area showing a diverse integration outcome of graphene-based solar cells such as graphene-silicon solar cells, graphene-polymer solar cells, dye-sensitized solar cells, graphene/quantum dot solar cells, graphene-tandem solar cells, graphene-perovskite solar cells, graphene-organic solar cells and graphene bulk-heterojunction solar cells (Feng et al 2011; Huang et al 2012a, b; Miao et al 2012; Ye and Dai 2012; Li et al 2014; Guo et al 2015; Li et al 2015a, b; Singh and Nalwa 2015;), in which the number of graphene layers and the effects of doping graphene materials can potentially improve the nature of the solar cell device. Doping is of two kinds: p-type and n-type. The impact of doping on graphene depends on the graphene derivatives and the doping process, like ball milling, thermal annealing, in situ doping, solid and liquid and gaseous chemical doping. Integrating graphene in dye-sensitized solar cells gives an output with enhanced light scattering at photoanode, effective dye molecule dispersal and improved efficiency. Also, research has shown that graphene-hybridized quantum dot offers a better efficiency in comparison with carbon nanotubes. Studies show that graphene can be utilized in heterojunction solar cells in diverse areas such as acceptor layers, donor layers, buffer layers, active layers and electrodes. This is possible because of the high conductivity, transparency and flexibility attributes of graphene.

3.3.5 Water Treatment

Drinkable water is one of the sustainable development goals in many countries due to the rising water scarcity resulting from fast population growth and pollution issues. Clean water is a basis for several human activities. Water treatment is imperative to purify the available water to get clean water that has all the contaminants safely below the maximum permissible limits. Several techniques and materials have been documented in the past research for water treatment such as reverse osmosis, nanofiltration, ultrafiltration, microfiltration, chlorination, sterilization by boiling and desalination. Graphene with its unique attributes (such as large surface area, low or no cytotoxicity, tunable chemical properties and delocalized Π-electron) serves as a membrane for water filtration process where they can chemically and physically remove pollutants from water via adsorption and photocatalysis. Graphene membranes possess better water vapour retention flux, high salt rejection rates and antifouling capability. These features make it a good membrane for use in water desalination process. Graphene-based filters are extensively researched and they have shown to be cheaper, faster and more eco-friendly. A graphene film has been used for water purification, pervaporation and desalination (Seo et al 2008; Maliyekkal et al 2013; Shanin and Mady 2017; Kaijie et al 2018; Yang et al 2018; Wang 2019).

3.3.6 Composites and Coating Applications

Generation of composites and coatings using graphene in combination with other materials is set to improve performance with increasing application possibilities in many industries as shown in Figure 3.3. Coating is usually employed in providing protection and surface quality for a substrate. Documented research work shows that graphene can be coated onto a plunger-in-needle microsyringe for a solid-phase microextraction device as a sorbent material towards UV filters (Zhang and Lee 2012), coating graphene on melamine sponge via dip coating shows high adsorbent capacity (Dong et al 2011), coating. Reduced graphene oxide nanoribbons (rGONRs) on Si/SiO_2 substrates

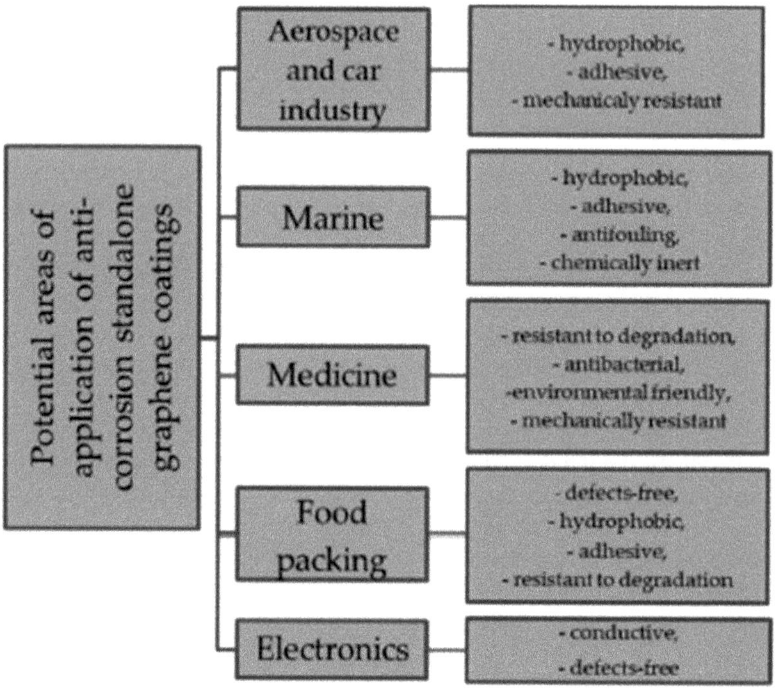

FIGURE 3.3 Potential applications of grapheme-based coating (Ollik and Lieder 2020).

by spray coating gave rise to a potential biosensor (Shanin and Mady 2017). Also, there are several studies on graphene use as an anticorrosion layer deposited using different methods to enhance the protective properties of its substrates against corrosion (Yu et al 2012; Chang et al 2014; Madhan Kumar et al 2015; Rajabi et al 2015; Wu et al 2015; Sai Pavan and Ramanan 2016; Cui et al 2017; Pourhashem et al 2017; Ahmed et al 2018; Krishnan et al 2018; Asgar et al 2019; Ollik et al 2019; Soler-Crespo et al 2019; Rekha and Srivastava 2020;). For instance, graphene is usually integrated with zinc-rich coating to improve its corrosion resistance. rGO as documented was observed to exhibit high corrosion protection in a zinc-rich coating (Liu et al 2018; Mohammadi and Roohi 2018; Cao et al 2019; Wang 2019).

3.3.7 Conductive Ink

This ink has the ability to print a working circuit that can conduct electricity. They are usually made of conductive materials, a thermoplastic polyvinylbuyral terpolymer binder and a glycol ether solvent. These conductive materials allow the movement of electric current either by electron or ions as the charge carrier. Numerous forms of conductive materials both electronic and ionic conductors exist like conductive carbon, conductive polymer, metallic conductor and polymer electrolytes. Graphene-based conductive ink has gained wide research attention. They are used in printed and flexible electronics as demonstrated in wireless and IoT applications by Pan et al (2018) as shown in Figure 3.4.

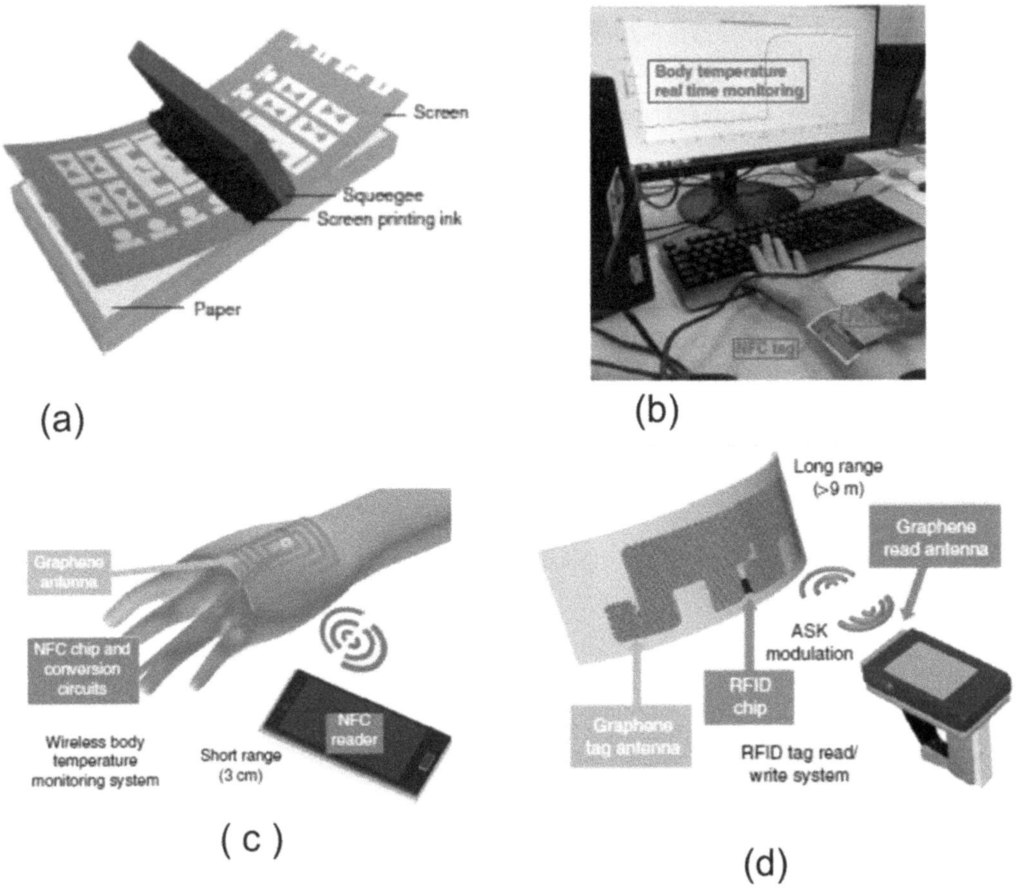

FIGURE 3.4 Printed graphene antennas and IoT applications (Pan et al 2018).

Other applications of graphene include its use in plasmonics and metamaterials, lubricating oil, radio wave absorption, nanoantennas, sound transducers, coolant additive, structural materials and aviation. Research shows that graphene and its derivatives and nanocomposites are good raw materials in tribology and lubrication because of their anti-friction, anti-wear and self-lubricating attributes (Jianlin and Shaonan 2019).

3.4 CONCLUSION

Graphene is becoming the bedrock of industrial transformation resulting from its numerous beneficial characteristics. Every field is gradually experiencing the miraculous impact of graphene. Conversion of biological waste such as animal waste is an added advantage in terms of pollution control and revenue generation from bio-graphene-based products. Also, animal waste is a renewable carbon source which implies less harm and cost. The diverse applications of graphene have drawn several research attentions; however, more investigation is needed. There are less research outputs in the utilization of animal waste in graphene extraction in comparison to lignocellulosic waste.

REFERENCES

Ahmed, M.K.; Shahid, M.; Khan, Z.A.; Ammar, A.U.; Saboor, A.; Khalid, A.; Hayat, A.; Saeed, A.; Koohgilani, M. Electrochemical comparison of SAN/PANI/FLG and ZnO/GO coated cast iron subject to corrosive environments. Materials 2018, 11, 2239.

Akhavan, O.; Bijanzad, K.; Mirsepah A. Synthesis of graphene from natural and carbonaceous waste. RSC Adv. 2014, 4, 20441–20448.

Asgar, H.; Deen, K.M.; Rahman, Z.U.; Shah, U.H.; Raza, M.A.; Haider, W. Functionalized graphene oxide coating on Ti6Al4V alloy for improved biocompatibility and corrosion resistance. Mater. Sci. Eng. C 2019, 94, 920–928.

Bhuyan, M.S.A.; Nizam Uddin, M.; Islam, M.; Bipasha, F.A.; Hossain, S.S. International Nano Letters volume 6, 65–83 (2016)

Biswas, C.; Lee, Y.H. Graphene versus carbon nanotubes in electronic devices. Adv. Funct. Mater. 2011, 21, 3798.

Cai, X.; Lai, L.; Shen, Z.; Lin, J. Graphene and graphene based composites as li-ion battery electrode materials and thin application in full cells. J. Mater. Chem. A 2017, 5, 15423–15446.

Cao, X.; Huang, F.; Huang, C.; Liu, J.; Cheng, Y.F. Preparation of graphene nanoplate added zinc-rich epoxy coatings for enhanced sacrificial anode-based corrosion protection. Corros. Sci. 2019, 159, 108120.

Chang, K.C.; Ji, W.F.; Li, C.W.; Chang, C.H.; Peng, Y.Y.; Yeh, J.M.; Liu, W.R. The effect of varying carboxylic-group content in reduced graphene oxides on the anticorrosive properties of PMMA/reduced graphene oxide composites. Express Polym. Lett. 2014, 8, 908–919.

Chen, F.; Yang, J.; Bai, T.; Long, B.; Zhou, X. Facile synthesis of few-layer graphene from biomass waste and its application in lithium ion batteries. J. Electroanal. Chem. 2016, 768, 18–26, https://doi.org/10.1016/j.jelechem.2016.02.035

Chen, Y.-S.; Yeager, D.; Emelianov, S.Y. Chapter 9—Photoacoustic Imaging for Cancer Diagnosis and Therapy Guidance. In Cancer Theranostics; Chen, X., Wong, S., Eds.; Academic Press: Oxford, UK, 2014; pp. 139–158.

Cui, C.; Lim, A.T.O.; Huang, J. A cautionary note on graphene anti-corrosion coatings. Nat. Nanotechnol. 2017, 12, 834–835.

Dong, X.; Long, Q.; Wang, J.; Chan-Park, M.B.; Huang, Y.; Huang, W., et al. A graphene nanoribbon network and its biosensing application. Nanoscale 2011, 3, 5156–5160.

Evans, C.L.; Xie, X.S. Coherent anti-stokes Raman scattering microscopy: chemical imaging for biology and medicine. Annu. Rev. Anal. Chem. 2008, 1, 883–909.

Fan X., Forsberg F., Smith A.D., Schroder S., Wagner S., et al. Graphene ribbons with suspended masses as transducers in ultra-small nanoelectromechanical accelerometers. Nat. Electron. 2019, 2, 394–404.

Feleppa, E.J.; Mamou, J.; Porter, C.R.; Machi, J. Quantitative ultrasound in cancer imaging. Semin. Oncol. 2011, 38, 136–150.

Feng, T.; Xie, D.; Lin, Y.; Xang, Y.; Ren, T. Graphene based Schottky junction solar cells on patterned silicon-pillar-array substrate. App. Phys. Lett. 2011, 99, 233505.

Goswami, S.; Banerjee, P.; Datta, S.; Mukhopadhayay, A.; Das, P. Graphene oxide nanoplatelets synthesized with carbonized agro-waste biomass as green precursor and its application for the treatment of dye rich wastewater. Process Saf. Environ. 2017, 106, 163–172, https://doi.org/10.1016/j.psep.2017.01.003

Guo, X.; Lu, G.; Chen, J. Graphene-based materials for photoanodes in dye-sensitized solar cells. Front. Energy Res. 2015, 3, 50.

Guo, W.; Qiu, J.; Liu, J.; Liu, H. Graphene microfiber as a scaffold for regulation of neural stem cells differentiation. Sci. Rep. 2017, 7, 5678.

Hao, P.; Zha, Z.; Leng, Y.; Tian, J. Graphene-based nitrogen self-doped hierarchical porous carbon aerogels derived from chitosan for high performance supercapacitors. Nano Energy 2015, 15, 9–23.

Huang, X.; Xiaoying, Q.; Boey, F.; Zhang, H. Graphene based composites. Chem. Soc. Rev. 2012a, 41, 666–686.

Huang, J.; Zong, C.; Shen, H.; Liu, M.; Chen, B.; Ren, B.; Zhang, Z. Mechanism of cellular uptake of graphene oxide studied by surface-enhanced Raman spectroscopy. Small 2012b, 8, 2577–2584.

Huff, T.B.; Cheng, J.X. In vivo coherent anti-Stokes Raman scattering imaging of sciatic nerve tissue. J. Microsc. 2007, 225, 175–182. Nanomaterials 2018, 8, 944 19 of 20.

Jakubovic, R.; Ramjist, J.; Gupta, S.; Guha, D.; Sahgal, A.; Foster, F.S.; Yang, V.X.D. High-frequency micro-ultrasound imaging and optical topographic imaging for spinal surgery: initial experiences. Ultrasound Med. Biol. 2018, 44, 2379–2387.

Jang, S.C.; Kang, S.M.; Lee, J.Y.; Oh, S.Y.; Vilian, A.T.E.; Lee, I.; Han, Y.K.; Park, J.H.; Cho, W.S.; Roh, C.; et al. Nano-graphene oxide composite for in vivo imaging. Int. J. Nanomed. 2018, 13, 221–234.

Jianlin, S.; Shaonan, D. Application of graphene derivatives and their nanocomposites in tribology and lubrication: a review. RSC Adv. 2019, 9, 40642–40661.

Kaijie, Y.; Jun, W.; Xiaoxiao, C.; Qiang, Z.; Abdul, G.; Baoliang, C. Application of graphene-based materials in water purification: from nanoscale to specific devices. Environ. Sci. Nano. 2018, 5, 1264–1297.

Kim, H.; Park, K.; Hong, J.; Kang-show, K. All-graphene-battery: bridging the gap between supercapacitor and lithium ion batteries. Sci. Rep. 2014, 4, 5278.

Krishna, M.C.; Devasahayam, N.; Cook, J.A.; Subramanian, S.; Kuppusamy, P.; Mitchell, J.B. Electron paramagnetic resonance for small animal imaging applications. ILAR J. 2001, 42, 209–218.

Krishnan, M.A.; Aneja, K.S.; Shaikh, A.; Bohm, S.; Sarkar, K.; Bohm, H.L.M.; Raja, V.S. Graphene-based anticorrosive coatings for copper. RSC Adv. 2018, 8, 499–500.

Li, P.; Chen, C.; Zhang, J.; Li, S.; Sun, B.; Bao, Q. Graphene based transparent electrodes for hybrid solar cells. Front. Mater. 2014, 1, 26.

Li, X.; Chen, W.; Zhang, S.; Wu, Z.; Wang, P.; Xu, Z.; Chen, H.; Yin, W.; Zhong, H.; Lin, S. 18.5% efficient graphene/GaAs van der Waals heterostructure solar cell. Nano Energy 2015b, 16, 310–319.

Li, X.; Zhang, S.; Wang, P.; Zhong, H.; Wu, Z.; Chen, H.; Liu, C.; Lin, S. High performance solar cells based on graphene-GaAs heterostructures. Nano Energy 2015a, 16, 310.

Li, N.; Zhang, Q.; Gao, S.; Song, Q.; Huang, R.; Wang, L.; Liu, L.; Dai, J.; Tang, M.; Cheng, G. Three-dimensional graphene foam as a biocompatible and conductive scaffold for neural stem cells. Sci. Rep. 2013, 3, 1640.

Li, Z.; Liu, Z.; Sun, H.; Gao, C. Superstructured assembly of nanocarbons: fullerenes, nanotubes, and graphene. Chem. Rev. 2015, 115, 7046–7117.

Liu, J.; Liu, T.; Guo, Z.; Guo, N.; Lei, Y.; Chang, X.; Yin, Y. Promoting barrier performance and cathodic protection of zinc-rich epoxy primer via single-layer graphene. Polymers 2018, 10, 591.

Madhan Kumar, A.; Suresh Babu, R.; Obot, I.B.; Gasem, Z.M. Fabrication of nitrogen doped graphene oxide coatings: experimental and theoretical approach for surface protection. RSC Adv. 2015, 5, 19264–19272.

Maliyekkal, S.M.; Sreeprasad, T.S.; Krishnan, D.; Kouser, S.; Mishra, A.K.; Waghmare, U.V.; Pradeep, T. Graphene: a reusable substrate for unprecedented adsorption of pesticides. Small 2013, 9, 273–283.

Marcano, D.C.; Kosynkin, D.V.; Berlin, J.M.; Sinitskii, A.; Sun, Z.; Slesarev, A.; Alemany, L.B.; Lu, W.; Tour, J.M. Improved synthesis of graphene oxide. ACS Nano. 2010, 4, 4806–4814, https://doi.org/10.1021/nn1006368.

Matsumoto, K.I.; Subramanian, S.; Murugesan, R.; Mitchell, J.B.; Krishna, M.C. Spatially resolved biologic information from in vivo EPRI, OMRI, and MRI. Antioxid. Redox Signal. 2007, 9, 1125–1141.

Miao, X., Tongay, S., Petterson, M., Berke, K., Rinzler, A., Appleton, B., Hebard, A. High efficiency graphene solar cells by chemical doping. Nano Lett. 2012, 12(6), 2745–2750.

Mohammadi, S.; Roohi, H. Influence of functionalized multi-layer graphene on adhesion improvement and corrosion resistance performance of zinc-rich epoxy primer. Corros. Eng. Sci. Technol. 2018, 53, 422–430.
Moo, J.G.S.; Khezri, B.; Webster, R.D; Pumera, M. Graphene oxides prepared by hummers', hofmann's, and staudenmaier's methods: dramatic influences on heavy-metal-ion adsorption. ChemPhysChem 2014, 15, 2922–2929, https://doi.org/10.1002/cphc.201402279
Ollik, K.; Rybarczyk, M.; Karczewski, J.; Lieder, M. Fabrication of anti-corrosion nitrogen doped graphene oxide coatings by electrophoretic deposition. Appl. Surf. Sci. 2019, 499, 143914.
Pan, K.; Fan, Y.; Leng, T.; Li, J.; Xin, Z.; Zhang, J.; Hao, L.; Gallop, J.; Novoselov, K. S.; Hu, Z. Sustainable production of highly conductive multilayer graphene ink for wireless connectivity and IoT applications. Nat. Commun. 2018, 9, 5197.
Patel SC, Lee S, Lalwani G, Suhrland C, Chowdhury SM, Sitharaman B. Graphene-based platforms for cancer therapeutics. Ther Deliv. 2016;7(2):101-116. doi:10.4155/tde.15.93
Pourhashem, S.; Vaezi, M.R.; Rashidi, A.; Bagherzadeh, M.R. Exploring corrosion protection properties of solvent based epoxy-graphene oxide nanocomposite coatings on mild steel. Corros. Sci. 2017, 115, 78–92.
Purkait, T.; Singh, G.; Singh, M.; Kumar, D; Dey, R.S. Large area few-layer graphene with scalable preparation from waste biomass for high-performance supercapacitor. Sci. Rep. 2017, 7, 1–14, https://doi.org/10.1038/s41598-017-15463-w.
Rajabi, M.; Rashed, G.R.; Zaarei, D. Assessment of graphene oxide/epoxy nanocomposite as corrosion resistance coating on carbon steel. Corros. Eng. Sci. Technol. 2015, 50, 509–516.
Rekha, M.Y.; Srivastava, C. High corrosion resistance of metal-graphene oxide-metal multilayer coatings. Philos. Mag. 2020, 100, 18–31.
Ruan, G.; Sun, Z.; Peng, Z; Tour, J.M. Growth of graphene from food, insects and waste. ACS Nano. 2011, 5(9), 7601–7607.
Safian, M.T.; Haron, U.S.; Ibrahim, M.N.M. Graphene from biomass waste. Bioresources 2020, 15 (4), 9756–9785.
Sai Pavan, A.S.; Ramanan, S.R. A study on corrosion resistant graphene films on low alloy steel. Appl. Nanosci. 2016, 6, 1175–1181.
Seo, D.H.; Pineda, S.; Woo, Y.C.; Xie, M.; Murdock, A.T.; Ang, E.Y.M.; ....Ostrikov, K. Anti-fouling graphene-based membrane for effective water desalination. Nat. Commun. 2008, 9(1).
Shams, S.S.; Zhang, L.S.; Hu, R.; Zhang, R; Zhu, J. Synthesis of graphene from biomass: a green chemistry approach. Mater. Lett. 2015, 161, 476–479, https://doi.org/10.1016/j.matlet.2015.09.022
Shanin, H.; Mady, E. Graphene membrane for water desalination. NPG Asia Mater. 2017, 9, e427.
Shi, J.; Wang, Y.; Du, W.; Hou, Z. Synthesis of graphene encapsulated Fe_3C in carbon nanotubes from biomass and its catalysis application. Carbon, 2016, 99, 330–337, https://doi.org/10.1016/j.carbon.2015.12.049
Shi, S.; Yang, K.; Hong, H.; Valdovinos, H.F.; Nayak, T.R.; Zhang, Y.; Theuer, C.P.; Barnhart, T.E.; Liu, Z.; Cai, W. Tumor vasculature targeting and imaging in living mice with reduced graphene oxide. Biomaterials 2013, 34, 3002–3009.
Silpa, P.S.; Sivmangai, N.M. One step synthesis of graphene. Inorg. Nano-Met. Chem. 2019, 1–7, https://doi.org/10.1080/24701556.2019.1661470
Singh, E.; Nalwa, H. Graphene-based bulk-heterojunction solar cells: a review. J. Nanosci. Nanotechnol. 2015, 15, 6237–6278.
Singh, P.; Bahadur, J.; Pal, K. One-step one chemical synthesis process of graphene from rice husk for energy storage applications. Graphene 2017, 6, 61–71, https://doi.org/10.4236/graphene.2017.63005
Soler-Crespo, R.A.; Mao, L.; Wen, J.; Nguyen, H.T.; Zhang, X.; Wei, X.; Huang, J.; Nguyen, S.B.T.; Espinosa, H.D. Atomically thin polymer layer enhances toughness of graphene oxide monolayers. Matter 2019, 1, 369–388.
Somanathan, T.; Prasad, K.; Ostrikov, K.K.; Saravanan, A.; Krishna, V.M. Graphene oxide synthesis from agro waste. Nanomaterials 2015, 5, 826–834, https://doi.org/10.3390/nano5020826
Tadyszak, K.; Wychowaniec, J. K.; Litowczenko, J. Biomedical applications of graphene-based structures. Nanomaterials 2018, 8, 944; https://doi.org/10.3390/nano8110944
Wang, S. A mechanistic study of corrosion of graphene and low zinc- rich epoxy coatings on carbon steel in salt environment. Int. J. Electrochem. Sci. 2019, 14, 9671–9681.
Wang, Y.; Huang, R.; Liang, G.; Zhang, Z.; Zhang, P.; Yu, S.; Kong, J. MRI-visualized, dual-targeting, combined tumor therapy using magnetic graphene-based mesoporous silica. Small 2014, 10, 109–116.
Widiatmoko,P.; Sukmana, I.F.; Nurdin, I.; Prakoso, T.; Devianto,H. Increasing yield of graphene synthesis from oil palm empty fruit bunch via two-stages pyrolysis. IOP Conf. Ser.: Mater. Sci. Eng. 2019, 543, 012032. https://iopscience.iop.org/article/10.1088/1757-899X/543/1/012032/pdf

William, S.; Hummers, J.R.; Offeman, R.E. Preparation of graphitic oxide. J. Am. Chem. Soc. 1958, 80, 1339–1339, https://doi.org/10.1021/ja01539a017

Wu, F.; Liang, J.; Li, W. Electrochemical deposition of $Mg(OH)_2$/GO composite films for corrosion protection of magnesium alloys. J. Magnes. Alloy 2015, 3, 231–236.

Yang, G.; Li, L.; Lee, W.B.; Ng, M.C. Structure of graphene and its disorders: a review. Sci. Technol. Adv. Mater. 2018, 19(1), 613–648, https://doi.org/10.1080/14686996.2018.1494493

Ye, Y.; Dai, L. Graphene based Schottky junction solar cells. J. Mater. Chem. 2012, 22, 24224.

Yu, L.; Lim, Y.S.; Han, J.H.; Kim, K.; Kim, J.Y.; Choi, S.Y.; Shin, K. A graphene oxide oxygen barrier film deposited via a self-assembly coating method. Synth. Met. 2012, 162, 710–714.

Zhang, H.; Lee, H.K. Simultaneous determination of ultraviolet filters in aqueous samples by plunger-in-needle solid-phase microextraction with graphene-based sol-gel coating as sorbent coupled with gas chromatography-mass spectrometry. Anal. Chim. Acta 2012, 742, 67–73.

Zhu, Y.; Murali, S.; Cai, W.; Li, X.; Suk, J.W.; Potts, J.R.; Ruoff, R.S. Graphene and graphene oxide: synthesis, properties, and applications. Adv. Mater. 2010, 22, 3906–3924.

Ziyad, S.H. Bio-inspired/-functional colloidal core-shell polymeric-based nanosystems: technology promise in tissue engineering, bioimaging and nanomedicine. Polymers 2010, 2, 323–352.

Zuchowska, A.; Chudy, M.; Dybko, A.; Brzozka, Z. Graphene as a new material in anticancer therapy-in vitro studies. Sens. Actuators B Chem. 2017, 243, 152–165.

4 Graphene from Essential Oils

Telli Alia

CONTENTS

4.1 INTRODUCTION

Graphene is a type of crystalline allotrope of carbon, which is a two-dimensional (2D) monolayer of sp^2-hybridized carbon atoms (Ding et al., 2020; Madurani et al., 2020). It is a basic structure block for graphitic materials of all other dimensionalities such as fullerenes, nanotubes or graphite (Polichitti et al., 2010). Owing to its extraordinary properties such as electronic, thermal, optical and mechanical, graphene-based nanomaterials have attracted tremendous interest in recent years (Sadhukhan et al., 2019).

These unique properties make graphene suitable for a wide range of applications. In 2014, the worldwide market for graphene is recorded to have amounted to $9 million where the majority of these sales are being focused on semiconductors, electronics, battery energy and composites. Nonetheless, the market grew rapidly to ~$100 billion as the material makes further incursion into the silicon market for extremely integrated circuits and in improving the lithium batteries performance (Radadiya, 2015).

Different methods have been reported for the production of graphene, such as chemical vapor deposition (CVD), epitaxial growth, electric arc discharge and graphene oxide reduction in solution (Xu et al., 2010; Suresh et al., 2015). However, the conventional methods of synthesis have a major drawback because of using hazardous and expensive chemical reductants (Lei et al., 2011; Bo et al., 2014). Also, additional stabilizers are required to get stabilized single-layered graphene (Lei et al., 2011). The green synthesis of graphene and graphene oxide through natural molecules or from agricultural residues has newly attracted more attention to produce nontoxic nanomaterial with high bioavailability, biocompatibility and/or bioactivity (Barhoum et al., 2020).

Many natural molecules could be employed in the fabrication of nanostructured materials as stabilizers or as reductants like gallic acid (Li et al., 2013), hydrolysable tannin (Lei et al., 2011), dopamine (Xu et al., 2010), vitamin C (Fernández-Merino et al., 2010), reducing sugars (Zhu et al., 2010), glycine (Bose et al., 2012) and many other molecules and plant extracts. In addition, biomass waste has become a new source for fabricating graphene because of its carbon-rich structure and renewable nature (Safian et al., 2020). Essential oils (EOs) produced by different aromatic plants are widely utilized as antimicrobial, antioxidants in the areas of food processing, food packaging or as fragrances in cosmetic products fabrication. These metabolites constitute another alternative in the fabrication of graphene and its derivatives. In this chapter, the graphene produced from EOs will be discussed.

DOI: 10.1201/9781003169741-4

4.2 GRAPHENE: STRUCTURE, PROPERTIES, SYNTHESIS AND APPLICATIONS

Graphene is among the most studied and used carbon nanostructures. It is a sheet of a single layer (2D) of carbon atoms, tightly bound in a hexagonal structure aligned on the same plane with a bonds angle of 120°, giving the graphene its honeycomb form (Safian et al., 2020). It forms many allotropes: fullerene (0D), carbon nanotube (CNT) (1D) and stacked on top of each other to form graphite (3D), with an interplanar spacing of 0.335 nm (Polichitti et al., 2010; Kamel et al., 2019). Graphene oxide and reduced graphene oxide are the derivatives of graphene with functional groups like epoxide, diol, ether, ketone and hydroxyl groups (Lee et al., 2011).

Graphene is an advanced carbon functional material with inherent unique properties that make it suitable for a wide range of applications. Either the top-down or the bottom-up approaches can be used in the synthesis of graphene (Yan et al., 2020). The easiest approach of producing graphene is mechanical exfoliation, as shown by Geim and Novoselov utilizing the Scotch tape technique. However, this method is only applicable to small-area production (Ani et al., 2018). One of the most promising and broadly used methods is CVD, which consists on the transformation of an organic gas carbon or a solid solution carbon to graphene by the catalytic action of an appropriate metal. Nevertheless, the CVD graphene is generally polycrystalline and contains a significant amount of domain boundaries that limit intrinsic physical properties of graphene (Rodríguez-Manzo et al., 2011; Ago et al., 2012). It is possible to obtain graphene in a solid-state transformation of amorphous carbon in the attendance of a catalytically active transition metal (Rodríguez-Manzo et al., 2011). To control the layer number, one promising strategy is the use of the segregation technique in which alloy substrates are used due to their complementarity in the catalytic growth of graphene (Li et al., 2020). Supercritical fluids are also applied as reaction medium for the exfoliation, surface modification and reduction of GO to high-quality graphene (Rangappa et al., 2011).

The intrinsic properties like huge surface area, extraordinary mechanical strength and high carrier mobility are reasons justifying the considerable interest in graphene (Lei et al., 2011). Graphene is a semimetal, which behaves electronically as a zero-gap semiconductor (Polichitti et al., 2010). Graphene sheets are promising candidates as channel materials of electronic devices, since both electron and hole in them have extremely high carrier mobilities (200,000 cm^2/V s). Graphene has also high thermal conductivity (~5000 W/m K), high transparency (~97.7%), high adsorption capacity and electrical conductivity and high theoretical specific surface area (~2630 m^2/g) (Kondo et al., 2011; Li et al., 2016).

Graphene, graphene-based nanocomposites and polymers are broadly applied in different fields. They are used in the construction of photocatalysts applied in pollutant degradation, H_2 production and CO_2 reduction (Li et al., 2016). Graphene oxide and reduced graphene oxide are widely employed in batteries owing to their important surface area, low electrical resistance, low mass density and high cyclic stability (Lawal, 2019). The use of graphene and its derivatives in solar cells has been studied by several researchers (Singh and Nalwa, 2015; Aydın, 2018; Lim et al., 2018; Mahmoudi et al., 2018). Graphene is an excellent anticorrosion coating for metals because of its impermeability to all molecules and has high chemical stability (Lawal, 2019). It has been one of the most popular choices to develop the electrodes of a sensor. For biosensors, the fabrication of graphene-based nanocomposite modified with biomacromolecules (DNA, proteins, peptides, etc.) improves the biocompatibility and bio-recognition ability of these materials, therefore, could greatly enhance their biosensing performances on both selectivity and sensitivity (Wang et al., 2017). Its higher permeate flux, higher selectivity and improved stability through controlled pore size and shape make graphene membrane ideal to solve environmental problems (Mohan et al., 2018). Carbon-based nanomaterials are potentially applied in biomedical fields such as molecular imaging, cancer and gene therapy, drug delivery, biosensors and tissue-engineering applications (Polichitti et al., 2010; Croitoru et al., 2019).

4.3 ESSENTIAL OILS: COMPOSITION, PROPERTIES AND APPLICATIONS

Terpenes are widely distributed in nature constituting the largest class of secondary metabolites with more than 50,000 molecules isolated and identified until today. Plants, animals, microorganisms, algae, endophytes and marine life produce them. They are part of the nonsaponifiable isoprenic lipids. This class of lipids is characterized like the other lipids by its insolubility in water and its solubility in organic solvents. These molecules have a common metabolic origin that is isoprene (2-methyl-1,3-butadiene). This precursor is synthesized via the mevalonate route from acetyl-CoA. Another pathway has been demonstrated in several bacteria and a green alga (*Scenedesmus obliquus*) leading to the production of isoprene from 2-methyl-D-erythritol-4-phosphate. This class contains EOs, resins, steroids, vitamins and polymers (latex and rubber). This class is also characterized by structural diversity, which gives these molecules different biological activities. These molecules can be simple hydrocarbons, oxygenated, sulfurized or aromatic hydrocarbons. Depending on the number of associated isoprenic units, there are monoterpenes (C10), sesquiterpenes (C15), diterpenes (C20), sesterterpenes (C25), triterpenes (C30), tetraterpenes (C40) and polyterpenes (Cn) (Paduch et al., 2007; Marouf and Tremblin, 2009; Martins et al., 2017; Parveen et al., 2018).

EOs, known as volatile oils, are complex mixtures of volatile substances produced by many aromatic plants (Bagetta et al., 2015). EOs are a variable blend of mainly terpenoids, specifically monoterpenes and sesquiterpenes but diterpenes may as well be found, and various low-molecular-weight aliphatic hydrocarbons (linear, ramified, saturated and unsaturated), acids, alcohols, aldehydes, acyclic esters or lactones and exceptionally nitrogen- and sulfur-containing molecules, coumarins and homologues of phenylpropanoids (Dorman and Deans, 2000).

EOs are among the secondary metabolites involved in the adaptation of the plant to its environment, the defense against physical and biological attacks (abiotic and biotic stresses) and in communication with plants. These metabolites are synthetized by some plant families such as Cupressaceae, Pinaceae, Apiaceae, Asteraceae, Geraniaceae, Illiciaceae, Lamiaceae, Lauraceae, Myristicaceae, Myrtaceae, Oleaceae, Rosaceae, Santalaceae, Rutaceae, Poaceae and Zingiberaceae. The secreting parts of EOs are flowers (rose), flowering tops (lavender), leaves (lemongrass), bark (cinnamon), roots (iris), fruits (vanilla), bulbs (garlic), rhizomes (ginger) or seeds (nutmeg) (Boukhatem et al., 2019).

Different procedures are used to extract EOs from plants (flowers, leaves, roots, buds or bark) like extraction by organic solvents, steam distillation, hydrodistillation, cold pressing, solvent free-extraction and supercritical fluid extraction (Frutuoso et al., 2013; Boukhatem et al., 2019; Vasile et al., 2020).

Numerous studies have documented the biological activities of EOs obtained from different aromatic plants. The most studied activity is the antimicrobial capacity of EOs against various microbial strains (Nakamura et al., 1999; Dorman and Deans, 2000; Soliman et al., 2016; Kerdudo et al., 2017; Weli et al., 2019). The antiprotozoal activity of EOs has also been proved (Le et al., 2018). EOs have different bioactive properties as antioxidant (Khodaei et al., 2021; Peng et al., 2021), analgesic (Silva et al., 2003), anti-inflammatory (Silva et al., 2003; Darwish et al., 2020), immunomodulatory (Lang et al., 2019) and cytotoxic effect on tumor cell cultures (Bayala et al., 2018; Zhang et al., 2020) as well as in the treatment of different ailments (Walsh et al., 2011; Bouyahya et al., 2020; Prall et al., 2020; Unusan, 2020) obtained from various plant sources.

Due to their antimicrobial and antioxidant activities, EOs are recently gaining more importance and used, as safer substitutes for chemical preservatives, and integrated in food products aiming to increase their useful life as well as help to further the ways to establish food safety (Frutuoso et al., 2013; Chen et al., 2020). Currently, antimicrobial, antioxidant and eco-friendly packaging by adding EOs is one among active packaging studied and used in the food industry to prevent negative effects caused by microorganisms and oxidation (Varghese et al., 2020). Several works have shown the potential of EOs in the preservation of different food products. EOs are applied as an alternative to low toxicity, extracted from a natural source, and as an antimicrobial agent in the preservation of meat food systems (da Silva et al., 2021). The use of chitosan-EOs films or nanoemulsion improves the preservation of fruits in terms of fungal decay (Tovar et al., 2018). Despite their effectiveness, their volatility, oxidizability, aromatic odor and insolubility are the principal factors that hinder the EOs utilization in food industry. The encapsulated thyme EO by casein-maltodextrin showed high efficiency as a preservative of hamburger-like meat products (Radünz et al., 2020). Other nanomaterials like liposomes, cyclodextrin, silicon dioxide and nanoemulsion, along with packaging approaches as nanofibers and food wraps, have been shown as efficient encapsulation systems to safeguard EOs.

4.4 USE OF ESSENTIAL OILS IN THE SYNTHESIS OF GRAPHENE

To upgrade the practical applications of graphene-based materials, the most important consideration should be given to the investigation for the large-scale making of high-quality graphene with a simple and economic transformation route. Several attempts have been made to achieve the reduction of graphene through green approaches (Suresh et al., 2015). Some natural materials, molecules and living organisms constitute promising substitutes for toxic/explosive reductants (Bo et al., 2014).

The incorporation of EOs into the packaging material has been extensively studied. Graphene is one among the packaging studied (Demitri et al., 2015; Tovar et al., 2018; Giannakas, 2020; Syafiq et al., 2020). Additionally, graphene is employed as a capsule in the encapsulation of active molecules like EOs (Ali et al., 2019). Nevertheless, the use of EOs as precursors or as reducing agents in graphene synthesis is little studied and reported in literature. This may be due to the low yield of EOs and the no sustainability and no acceptability of the destruction of plants for nanoparticles formation (Barhoum et al., 2020).

Various natural, renewable and cheaper materials are utilized as carbon precursors in the synthesis of carbon-based nanomaterials. Plant hydrocarbons are among the natural carbon precursors recently have attracted considerable interest. There are limited reports available on the use of natural precursors such as camphor ($C_{10}H_{16}O$) (Kumar and Ando, 2003, 2007), eucalyptus oil ($C_{10}H_{18}O$) (Ghosh et al., 2007), turpentine oil ($C_{10}H_{16}$) (Awasthi et al., 2010) and vegetable triglyceride oils (castor oil, sesame oil, palm oil, *Jatropha* oil and neem oil) (Kumar et al., 2011; Thakur and Karak, 2014; Salifairus et al., 2018; Harun et al., 2020; Suresha et al., 2020) to produce carbon-based nanomaterials (nanotubes and graphene).

Camphor is a natural product deriving from the wood of the camphor laurel (*Cinnamomum camphora* L.) trees through steam distillation and purification by sublimination (Zuccarini and Soldani, 2009). Great-purity CNTs are synthesized by CVD of camphor, an eco-friendly molecule. The found yield after 1 hour of treatment of 3 g of camphor is about ~1.62 g multiwall nanotubes and a diameter of ~10 nm with an as-grown purity up to 88%; that is, camphor-to-CNT fabrication effectiveness is 50%. This is the highest efficiency ever obtained from any material by any method (Kumar and Ando, 2007).

Turpentine oil ($C_{10}H_{16}$) obtained from *Pinus merkusii* is a colorless, flammable liquid with a characteristic odor; a mixture consisting mostly of terpenes, principally the isomers of pinene (Young, 2001; Hudaya et al., 2016). Turpentine oil is an effective precursor in carbon-based nanomaterials. Single-walled CNTs were prepared by a catalytic decomposition of turpentine oil over well-dispersed metal particles supported on high silica Y-type zeolite at 850°C by spray pyrolysis method with a reaction time of 25 minutes. The fabrication of the bundles of aligned carbon nanotubes (ACNTs) is done by a spray pyrolysis of turpentine oil (inexpensive precursor) and ferrocene mixture at 800°C. The obtained results showed that ACNTs are densely packed (~70–130 μm in length). The as-grown multi-walled CNTs are greatly graphitized. Their outer diameters are between ~15 and 40 nm (Awasthi et al., 2010).

The antibacterial composite film, prepared from polyvinyl alcohol, graphene oxide and oregano EOs, was studied by Lin et al. (2019). With the increase of oregano EOs content, the thicknesses were increased, the mechanical properties were improved, and the antibacterial activities were increased (Lin et al., 2019).

4.5 CONCLUSION

Currently, high attention has turned to graphene due to its exceptional properties (extreme mechanical strength, thermal conductivity, 2D films and peculiar electronic characteristics). The combination of biomolecules with graphene-based materials offers a promising way to fabricate novel graphene-biomolecule hybrid nanomaterials with unique functions in biology, medicine, nanotechnology and materials science. In this chapter, we try to demonstrate the use of EOs in the synthesis of graphene as precursors or as reducing and stabilizer agents, but there has been little research interesting in the application of EOs in the synthesis processing of graphene.

REFERENCES

Ago H., Ogawa Y., Tsuji M., et al., 2012. Catalytic growth of graphene: toward large-area single-crystalline graphene. The Journal of Physical Chemistry Letters, 3(16), 2228–2236. doi:10.1021/jz3007029

Ali M., Meaney S.P., Abedin M.J., et al., 2019. Graphene oxide-silica hybrid capsules for sustained fragrance release. Journal of Colloid and Interface Science, 552, 528–539. doi:10.1016/j.jcis.2019.05.061

Ani M.H., Kamarudin M.A., Ramlan A.H., et al., 2018. A critical review on the contribution of chemical and physical factors toward the nucleation and growth of large area graphene. Journal of Materials Science, 53, 7095–7111. doi:10.1007/s10853-018-1994-0

Awasthi K., Kumar R., Tiwari R.S. & Srivastava O.N., 2010. Large-scale synthesis of bundles of aligned carbon nanotubes using a natural precursor: turpentine oil. Journal of Experimental Nanoscience, 5(6), 498–508. doi:10.1080/17458081003664159

Aydın C., 2018. Synthesis of SnO_2:rGO nanocomposites by the microwave-assisted hydrothermal method and change of the morphology, structural, optical and electrical properties. Journal of Alloys and Compounds. doi: 10.1016/j.jallcom.2018.08.298

Bagetta G., Cosentino M. & Sakurada T. Aromatherapy: basic mechanisms and evidence-based clinical use. CRC Press Taylor & Francis Group (2015). Florida.

Barhoum A., Jeevanandam J., Rastogi A., et al., 2020. Plant celluloses, hemicelluloses, lignins, and volatile oils for the synthesis of nanoparticles and nanostructured materials. Nanoscale. doi:10.1039/D0NR04795C

Bayala B., Bassole I.H.N., Maqdasy S., et al., 2018. *Cymbopogon citratus* and *Cymbopogon giganteus* essential oils have cytotoxic effects on tumor cell cultures. Identification of citral as a new putative anti-proliferative molecule. Biochimie. doi:10.1016/j.biochi.2018.02.013

Bo Z., Shuai X., Shun M., et al., 2014. Green preparation of reduced graphene oxide for sensing and energy storage applications. Scientific Reports, 4, 4684. doi:10.1038/srep04684

Bose S., Kuila T., Mishra A.K., et al., 2012. Dual role of glycine as a chemical functionalizer and reducing agent in the preparation of graphene: an environmentally friendly method. Journal of Materials Chemistry, 22, 9696–9703. doi:10.1039/C2JM00011C

Boukhatem M.N., Ferhat A. & Kameli A., 2019. Méthodes d'extraction et de distillation des huiles essentielles: revue de littérature. Revue Agrobiologia, 9(2), 1653–1659. www.agrobiologia.net

Bouyahya A., Lagrouh F., El Omari N., et al., 2020. Essential oils of *Mentha viridis* rich phenolic compounds show important antioxidant, antidiabetic, dermatoprotective, antidermatophyte and antibacterial properties. Biocatalysis and Agricultural Biotechnology. doi:10.1016/j.bcab.2019.101471

Chen K., Zhang M., Bhandari B., Mujumdar A.S., 2020. Edible flower essential oils: a review of chemical compositions, bioactivities, safety and applications in food preservation. Food Research International. doi:10.1016/j.foodres.2020.109809

Croitoru A., Oprea O., Nicoara A., et al., 2019. Multifunctional platforms based on graphene oxide and natural products. Medicina, 55, 230. doi:10.3390/medicina55060230.

da Silva B.D., Campos Bernardes P., Fontes Pinheiro P., et al., 2021. Chemical composition, extraction sources and action mechanisms of essential oils: natural preservative and limitations of use in meat products. Meat Science, 176, 108463. doi:10.1016/j.meatsci.2021.108463

Darwish R.S., Hammoda H.M., Ghareeb D.A., et al., 2020. Efficacy-directed discrimination of the essential oils of three *Juniperus* species based on their in-vitro antimicrobial and anti-inflammatory activities. Journal of Ethnopharmacology, 259, 112971. doi:10.1016/j.jep.2020.112971

Demitri C., Tarantino A.S., Moscatello A., et al., 2015. Graphene reinforced chitosan-cinnamaldehyde derivatives films: antifungal activity and mechanical properties. First Workshop on Nanotechnology in Instrumentation and Measurement (NANOFIM). IEEE: N° 17974635. doi:10.1109/NANOFIM.2015.8425334

Ding Z., Yuan T., Wen J., et al., 2020. Green synthesis of chemical converted graphene sheets derived from pulping black liquor, Carbon, 158, 690–697. doi:10.1016/j.carbon.2019.11.041

Dorman H.J.D. & Deans S.G., 2000. Antimicrobial agents from plants: antibacterial activity of plant volatile oils. Journal of Applied Microbiology, 88, 308–316. https://citeseerx.ist.psu.edu/viewdoc/download?doi=10.1.1.838.5779&rep=rep1&type=pdf

Fernández-Merino M.J., Guardia L., Paredes J.I., et al., 2010. Vitamin C is an ideal substitute for hydrazine in the reduction of graphene oxide suspensions. The Journal of Physical Chemistry C, 114(14), 6426–6432. doi:10.1021/jp100603h

Frutuoso A.E., do Nascimento N.T., Gomes de Lemos T.L., et al., 2013. Óleos essenciais aplicados em alimentos: uma revisão. Revista Brasileira de Pesquisa em Alimentos, Campo Mourão (PR), 4(2), 69–81. doi:10.14685/rebrapa.v4i2.134

Ghosh P., Afre R.A., Soga T. & Jimbo T., 2007. A simple method of producing single-walled carbon nanotubes from a natural precursor: eucalyptus oil. Materials Letters, 61, 3768–3770. doi:10.1016/j.matlet.2006.12.030

Giannakas A., 2020. Na-montmorillonite vs. organically modified montmorillonite as essential oil nanocarriers for melt-extruded low-density poly-ethylene nanocomposite active packaging films with a controllable and long-life antioxidant activity. Nanomaterials, 10, 1027. doi:10.3390/n&no10061027

Harun N.H., Zainal Abidin Z., Abdullah A.H. & Othamna R., 2020. Sustainable *Jatropha* oil-based membrane with graphene oxide for potential application in Cu(II) ion removal from aqueous solution. Processes, 8, 230. doi:10.3390/pr8020230

Hudaya T., Widjaja O., Rionardi A., et al., 2016. Synthesis of biokerosene through electrochemical hydrogenation of terpene hydrocarbons from turpentine oil. Journal of Engineering and Technological Sciences, 48(6), 655–664. doi:10.5614/j.eng.technol.sci.2016.48.6.2

Kamel S., El-Sakhawy M., Anis B. & Tohamy H.-A.S., 2019. Graphene's structure, synthesis and characterization; A brief review. Egyptian Journal of Chemistry, 62(2), 593–608. doi:10.21608/ejchem.2019.15173.1919

Kerdudo A., Ellong E.N., Burger P., et al., 2017. Chemical composition, antimicrobial and insecticidal activities of flowers essential oils of *Alpinia zerumbet* (Pers.) L. from Martinique Island. Chemistry & Biodiversity, 14(4), e1600344. doi:10.1002/cbdv.201600344

Khodaei N., Nguyen M.M., Mdimagh A., et al., 2021. Compositional diversity and antioxidant properties of essential oils: predictive models. LWT, 138, 110684. doi:10.1016/j.lwt.2020.110684

Kondo H., Hori M. & Hiramatsu M. Nucleation and vertical growth of nano-graphene sheets. In Cong J.R. editor. Graphene: synthesis, characterization, properties and applications. Intech (2011):21–36. Croatia. www.intechweb.org

Kumar M. & Ando Y., 2003. A simple method of producing aligned carbon nanotubes from an unconventional precursor – camphor. Chemical Physics Letters, 374, 521–526. doi:10.1016/s0009-2614(03)00742-5

Kumar M. & Ando Y., 2007. Carbon nanotubes from camphor: an environment-friendly nanotechnology. Journal of Physics: Conference Series, 61, 129. doi:10.1088/1742-6596/61/1/129

Kumar R., Tiwari R.S. & Srivastava O.N., 2011. Scalable synthesis of aligned carbon nanotubes bundles using green natural precursor: neem oil. Nanoscale Research Letters, 6, 92. http://www.nanoscalereslett.com/content/6/1/92

Lang M., Ferron P.-J., Bursztyka J., et al., 2019. Evaluation of immunomodulatory activities of essential oils by high content analysis. Journal of Biotechnology. doi:10.1016/j.jbiotec.2019.07.010

Lawal A.T., 2019. Graphene-based nano composites and their applications. A review. Biosensors and Bioelectronics. doi:10.1016/j.bios.2019.111384

Le T.B., Beaufay C., Bonneau N., et al., 2018. Anti-protozoal activity of essential oils and their constituents against Leishmania, Plasmodium and Trypanosoma. Phytochimie, 18(1), 1–33. http://hdl.handle.net/2078.1/199166

Lee S., Lee K.K. & Lim E. Synthesis of aqueous dispersion of graphenes via reduction of graphite oxide in the solution of conductive polymer. In Cong J.R. editor. Graphene: synthesis, characterization, properties and applications. Intech (2011): 37–44. Croatia. www.intechweb.org

Lei Y., Tang Z., Liao R., Guo B., 2011. Hydrolysable tannin as environmentally friendly reducer and stabilizer for graphene oxide. Green Chemistry, 13(7), 1655–1658. doi:10.1039/c1gc15081b

Li J., Xiao G.Y., Chen C.B., et al., 2013. Superior dispersions of reduced graphene oxide synthesized by using gallic acid as a reductant and stabilizer. Journal of Materials Chemistry A, 1, 1481–1487. doi:10.1039/C2TA00638C

Li X., Yu J., Wageh S., et al., 2016. Graphene in photocatalysis: a review. Small. doi:10.1002/smll.201600382

Li Y., Sun L., Liu H., et al., 2020. Rational design of binary alloys for catalytic growth of graphene via chemical vapor deposition. Catalysts, 10, 1305. doi:10.3390/catal10111305

Lim E.L., Yap C.C., Hj Jumali M.H., et al., 2018. A mini review: can graphene be a novel material for Perovskite solar cell applications? Nano-Micro Letters, 10, 27. doi:10.1007/s40820-017-0182-0

Lin D., Wu Z., Huang Y., et al., 2019. Physical, mechanical, structural and antimicrobial properties of polyvinyl alcohol/oregano essential oil/graphene oxide composite films. Journal of Polymers and the Environment. doi:10.1007/s10924-019-01627-4

Maduranil K.A., Supraptol S., Machrital N. I., et al., 2020. Progress in graphene synthesis and its application: history, challenge and the future outlook for research and industry. ECS Journal of Solid State Science and Technology, 9, 093013. doi:10.1149/2162-8777/abbb6f

Mahmoudi T., Wang Y. & Hahn Y.-B., 2018. Graphene and its derivatives for solar cells applications. Nano Energy, 47, 51–65. doi:10.1016/j.nanoen.2018.02.047

Marouf A. & Tremblin G., 2009. Abrégé de Biochimie Appliquée. Ed. EDP Sciences, Grenoble, France, 484 pages.

Martins M.A.R., Carvalho P.J., Palma A.M., Domańska U., Coutinho J.A.P. & Pinho S.P., 2017. Selecting critical properties of terpenes and terpenoids through group-contribution methods and equations of state. Industrial & Engineering Chemistry Research, 56, 35, 9895–9905. doi:10.1021/acs.iecr.7b022.47

Mohan V.B., Lau K.-T., Hui D., et al., 2018. Graphene-based materials and their composites: a review on production, applications and product limitations. Composites Part B: Engineering, 142, 200–220. doi:10.1016/j.compositesb.2018.01.013

Nakamura C.V., Ueda-Nakamura T., Bando E., et al., 1999. Antibacterial activity of *Ocimum gratissimum* L. essential oil. Memórias do Instituto Oswaldo Cruz, Rio de Janeiro, 94(5), 675–678. https://www.scielo.br/pdf/mioc/v94n5/3781.pdf

Paduch R., Kandefer-Szerszeñ M., Trytek M. & Fiedurek J., 2007. Terpenes: substances useful in human healthcare. Archivum Immunologiae et Therapiae Experimentalis, 55, 315–327. doi:10.1007/s00005-007-0039-1

Parveen T., Sharma N. & Sharma K., 2018. Assay of botanical oils against important phytopathogenic fungi. World Journal of Pharmaceutical Research, 7(11), 1326–1347. https://wjpr.s3.ap-south-1.amazonaws.com/article_issue/1527939824.pdf

Peng X., Feng C., Wang X., et al., 2021. Chemical composition and antioxidant activity of essential oils from barks of *Pinus pumila* using microwave-assisted hydrodistillation after screw extrusion treatment. Industrial Crops and Products, 166, 113489. doi:10.1016/j.indcrop.2021.113489

Polichitti T., Miglietta M.L. & Di Francia G., 2010. Overview on graphene: properties, fabrication and applications. Chimica Oggi – Chemistry Today, 28(6), 6–9. https://www.researchgate.net/publication/230634280_Overview_on_graphene_Properties_fabrication_and_applications

Prall S., Bowles E. J., Bennett K., et al., 2020. Effects of essential oils on symptoms and course (duration and severity) of viral respiratory infections in humans: a rapid review. Advances in Integrative Medicine. doi:10.1016/j.aimed.2020.07.005

Radadiya T.M., 2015. A properties of graphene. European Journal of Material Sciences, 2(1), 6–18. www.eajournals.org

Radünz M., Hackbort H.C.D.S., Camargo T.M., et al., 2020. Antimicrobial potential of spray drying encapsulated thyme (*Thymus vulgaris*) essential oil in the conservation of hamburger-like meat products. International Journal of Food Microbiology, 330, 108696. doi:10.1016/j.ijfoodmicro.2020.108696

Rangappa D., Jang J.-H. & Honma I. Supercritical fluid processing of graphene and graphene oxide. In Cong J.R. editor. Graphene: synthesis, characterization, properties and applications. Intech (2011): 46–58. Croatia. www.intechweb.org

Rodríguez-Manzo J.A., Pham-Huu C. & Banhart F., 2011. Graphene growth by a metal-catalyzed solid-state transformation of amorphous carbon. ACSNANO, 5(2), 1529–1534. doi:10.1021/nn103456z

Sadhukhan S., Ghosh T.K., Roy I., et al., 2019. Green synthesis of cadmium oxide decorated reduced graphene oxide nanocomposites and its electrical and antibacterial properties. Materials Science & Engineering C, 99, 696–709. doi:10.1016/j.msec.2019.01.128

Safian M.T., Haron U.S. & Mohamad Ibrahim M.N., 2020. A review on bio-based graphene derived from biomass wastes. BioRes, 15(4), 9756–9785. doi:10.15376/biores.15.4.Safian

Salifairus M.J., Soga T., ALrokayan S.A.H., et al., 2018. The synthesis of graphene from oil palm at different annealing time of nickel substrate via thermal chemical vapor deposition. AIP Conference Proceedings, 1963, 020015. doi:10.1063/1.5036861

Silva J., Abebe W., Sousa S.M., et al., 2003. Analgesic and anti-inflammatory effects of essential oils of *Eucalyptus*. Journal of Ethnopharmacology, 89(2–3), 277–283. doi:10.1016/j.jep.2003.09.007

Singh E. & Nalwa H.S., 2015. Graphene-based dye-sensitized solar cells: a review. Science of Advanced Materials, 7(10), 1863–1912. doi:10.1166/sam.2015.2438

Soliman F.M., Fathy M.M., Salama M.M., et al., 2016. Comparative study of the volatile oil content and antimicrobial activity of *Psidium guajava* L. and *Psidium cattleianum* Sabine leaves. Bulletin of Faculty of Pharmacy, Cairo University, 54, 219–225. doi:10.1016/j.bfopcu.2016.06.003

Suresh D., Udayabhanu, Pavan Kumar M.A., et al., 2015. Cinnamon supported facile green reduction of graphene oxide, its dye elimination and antioxidant activities. Materials Letters. doi:10.1016/j.matlet.2015.03.035i

Suresha B., Hemanth G., Rakesh A. & Adarsh K.M., 2020. Tribological behaviour of neem oil with and without graphene nanoplatelets using four-ball tester. Advances in Tribology, 2020, Article ID 1984931. doi:10.1155/2020/1984931

Syafiq R.M.O., Sapuan S.M. & Zuhri M.R.M., 2020. Effect of cinnamon essential oil on morphological, flammability and thermal properties of nanocellulose fibre–reinforced starch biopolymer composites. Nanotechnology Reviews, 9, 1147–1159. doi:10.1515/ntrev-2020-0087

Thakur S. & Karak N., 2014. Ultratough, ductile, castor oil-based, hyperbranched, polyurethane nanocomposite using functionalized reduced graphene oxide. ACS Sustainable Chemistry & Engineering, 2(5), 1195–1202. doi:10.1021/sc500165d

Tovar C.D.G., Chaves-Lopez C., Serio A., et al., 2018. Chitosan coatings enriched with essential oils: effects on fungal decay of fruits and mechanisms of action. Trends in Food Science & Technology. doi:10.1016/j.tifs.2018.05.019

Unusan N., 2020. Essential oils and microbiota: implications for diet and weight control. Trends in Food Science & Technology. doi:10.1016/j.tifs.2020.07.014

Varghese S.A., Siengchin S. & Parameswaranpillai J., 2020. Essential oils as antimicrobial agents in biopolymer-based food packaging – a comprehensive review, Food Bioscience, 38, 100785. doi:10.1016/j.fbio.2020.100785

Vasile B.S., Birca A.C., Musat M.C. & Holban A.M., 2020. Wound dressings coated with silver nanoparticles and essential oils for the management of wound infections. Materials, 13(7), 1682. doi:10.3390/ma13071682

Walsh M.E., Reis D., Jones T., 2011. Integrating complementary and alternative medicine: use of essential oils in hypertension management. Journal of Vascular Nursing, 29(2): 87–88. doi:10.1016/j.jvn.2011.01.001

Wang L., Zhang Y., Wu A., et al., 2017. Designed graphene-peptide nanocomposites for biosensor applications: a review. Analytica Chimica Acta. doi:10.1016/j.aca.2017.06.054

Weli A., Al-Kaabi A., Al-Sabahi J., et al., 2019. Chemical composition and biological activities of the essential oils of *Psidium guajava* leaf. Journal of King Saud University – Science, 31, 993–998. doi:10.1016/j.jksus.2018.07.021.

Xu L.Q., Yang W.J., Neoh K.-G., et al., 2010. Dopamine-induced reduction and functionalization of graphene oxide nanosheets. Macromolecules, 43(20), 8336–8339. doi:10.1021/ma101526k

Yan Y., Nashath F.Z., Chen S., et al., 2020. Synthesis of graphene: potential carbon precursors and approaches. Nanotechnology Reviews, 9, 1284–1314. doi:https://doi.org/10.1515/ntrev-2020-0100

Young J.A., 2001. Turpentine. Journal of Chemical Education, 78(11), 1459. doi:10.1021/ed078p1459

Zhang Y., Xin C., Cheng C. & Wang Z., 2020. Antitumor activity of nanoemulsion based on essential oil of *Pinus koraiensis* pinecones in MGC-803 tumor-bearing nude mice. Arabian Journal of Chemistry, 13:8226–8238. doi:10.1016/j.arabjc.2020.09.058

Zhu C., Guo S., Fang Y., Dong S., 2010. Reducing sugar: new functional molecules for the green synthesis of graphene nanosheets. ACS Nano, 4(4), 2429–2437. doi:10.1021/nn1002387

Zuccarini P. & Soldani G., 2009. Camphor: benefits and risks of a widely used natural product. Acta Biologica Szegediensis, 53(2), 77–82. http://www.sci.u-szeged.hu/ABS

5 Synthesis of Graphene from Biowastes

Nadia Akram, Muhammad Shahbaz, Khalid Mahmood Zia, and Fozia Anjum

CONTENTS

5.1 STRUCTURE OF GRAPHENE

Graphene is a miraculous material that possesses two-dimensional (2D) skeletons with a hexagonal architecture of a single monomolecular layer of sp^2-hybridized carbon atoms as shown in Figure 5.1. It exhibits excellent properties in various dimensions for diverse applications such as electronic, thermodynamic, and mechanical properties [1, 2]. Just as the emergence of plastics in the twentieth century, the graphene is considered the miracle of the twenty-first century [3, 4]. Actually, the sheets of graphene are very thin in texture organized as a single-atom sheet in thickness. This monolayer sheet is coordinated with adjacent atoms producing high surface area. It shows great compound dependability that depicts its mechanical behavior as well. In the year 2007, the group of Noble laureates Prof. Andre K. Geim and Prof. Konstantin Novoselov uncovered the gigantic capability of graphene in different applications by recognizing its electrical property. They likewise remarked "Graphene as quickly emerging materials not too far off of dense materials physical science". This is undeniably right as graphene was changing in all areas from energy, climate to well-being of all [3]. Its striking properties highlight its applications in different areas. Optical properties and its adaptability provide significant highlights in a number of electronic appliances [5, 6]. Currently, it is a smart option of various forms of conductors. Graphene has significant influence on optic gadgets, for example, terahertz gadgets, optical recurrence converters, and adaptable brilliant windows. Also, the change in Fermi level can be used for the optical preservation. Likewise, graphene is as of late archived as a diverse nanomaterial for state-of-the-art applications [7, 8]. The graphene layers are used for the cleaning and purification of water [9].

DOI: 10.1201/9781003169741-5

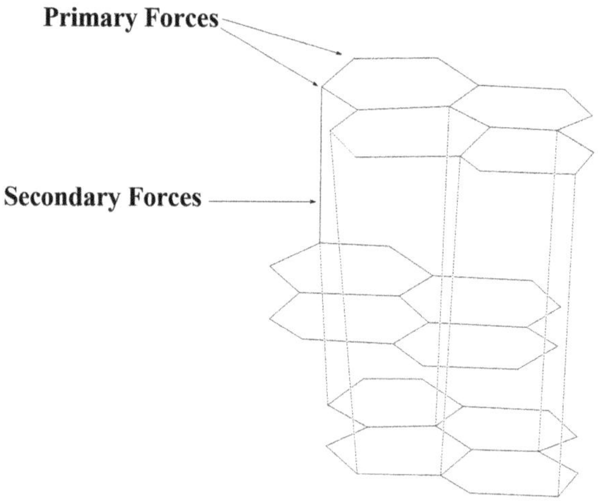

FIGURE 5.1 Graphene sheet structure.

Accordingly, it has been improving the general effectiveness of the impetus. In this manner, graphene is being investigated for super-quick semiconductors, super capacitors, memory gadgets, adaptable showcases, light-emanating diodes, batteries, sun-oriented cells, power modules, water treatment, and organic applications. In each try, the scientists can get a decent result [2].

Few companies of the United States that started to make graphene in large volume five years ago are Angstron Materials, Vorbeck Materials, and XG Sciences lab. Afterward, tens of grapheme-producing companies have gained progress all over the world, which made graphene sheets as well as graphene films of high quality on large scale. Nowadays, the graphene production industry in China is expanding very rapidly, and its annual making capacity of graphene films and sheets is 110,000 m^2 and 400 t, respectively. So, the materials world research is targeting on the mass production, characterization, and applications in real world of ultrathin films of carbon and the graphene, which is the thinnest element of carbon. Graphene has been a modern material for modern chemistry and physics because of its attractive properties. Since its first separation, scientific development with its production method and other applications has been progressing day by day, recommending that graphene can grow the industry because of its superlative properties. It can be added into polymers and inorganic systems to increase the thermal and mechanical strength as well as electrical conductivity. Many companies are starting the production of graphene for various purposes. Graphene is looking to exceed £150 million by 2022 in global market. But in real, the success of graphene at commercial level is a little bit difficult task. It is common that the material that shows the better properties than others finds the place in market. Considering the history of carbon fiber development, its industrialization was largely confined because of the lack of carbon fiber products with great properties and performance to satisfy the requirements that are application dependent. At that time, the uses of carbon fibers were only within a few places, including brassie golf clubs and fishing poles. In recent decades, the progress in the production of high-grade carbon fibers has confirmed expanded commercial applications in automotive, civil engineering, military, and aerospace industries. So the best properties of graphene, in improving the strength of material and synthesis, may prove graphene as a wide commercial material.

5.2 PRODUCTION OF GRAPHENE

The exfoliation technique was initially used for the first fabrication of single-layer graphene in the form of nanosheets. The method is known as Scotch-tape method that utilizes bulk graphite by epitaxial chemical vapor deposition (CVD) [10]. Although all these methods never produce graphene on industrial scale, there is always the need for the discovery of alternative methods. Hence, substitute approaches are obligatory for the mega-scale production of graphene. One of the predominant and exciting methods for the exfoliation of graphene on mega scale is the use of active oxidation chemical reaction to yield graphene oxide (GO). This produces the graphene with nonconductive hydrophilic characteristics [11]. Graphite and organic sources are considered most important for the production of graphene. There are abundant deposition and other methodologies in practices to produce graphene. These techniques include liquid exfoliation for the processing of graphite crystal [2, 12]. Some of the investigators [2, 13] have developed the comprehensive approaches to describe these techniques. These are common methods for the preparation of graphene. Each technique is marked with strengths and weaknesses. CVD utilizes carbon from various sources of hydrocarbons. It requires sufficient processing parameters; however, it does not produce adequate yield. Additionally, the hydrogen expelled in the surroundings shows propensity of ozone reaction, ultimately destroying the layer. Another technique is mini cleavage strategy, which is a straightforward cycle, here sticky films are utilized to deliver graphene, despite the fact that it creates a great nature of graphene, yet it's anything but an undoable technique for large-scale manufacturing. The liquid exfoliation method is used for the processing of graphite crystal by chemical reduction. The method utilizes graphite in the form of a precursor to produce graphene, which is recognized as Hummers' method. This technique releases poisonous gases. Mega-scale graphene synthesis is very challenging based on the suitable approach; there are substantial alterations in yield and price. Improvement of sheet structure and uniformity is a critical task. Ultimately, graphene obtained from various means barring graphite has gotten incredible consideration. The chapter focuses on the preparation of graphene using various precursors instead of conventional precursors [2, 13].

5.3 PREPARATION OF GRAPHENE FROM WASTE MATERIAL

The waste production and accumulation are inevitable; however, the utilization of waste materials can be performed efficiently. Moreover, transforming and recycling of surplus materials is always valuable. A number of biomaterials can be used for the productions of graphene as shown in Figure 5.2. Biowastes actually provide very unusual precursors. All the developed approaches have their own advantages and disadvantages, yet all techniques are novel. Graphene, environmentally and financially created from biowastes, is utilized in various applications. Consequently, the chase of raw biomaterials for the production of graphene has been a challenge with sufficient return in progress. Subsequently, this empowers the analysts to consider over different sources which are yet to be revealed. Any prerequisite for manageable assets is unavoidable; hence, inexhaustible biowaste provides reasonable methodology for graphene antecedent. Also, coal was initially utilized as a raw source; however, it is nothing but an inexhaustible asset. At this point we included it due to the creative and reasonable methodology associated with the creation of graphene quantum specks [2].

5.3.1 GRAPHENE FROM GLUCOSE

Glucose is a sustainable source of carbon that can be used to develop graphene due to an abundance of carbon in the main structure and ecological variances. It is also an important material for analysts to develop graphene [2, 14, 15]. Ferric chloride was used by researchers during the formation of glucose. As per their revealed strategy, in the initial step, the reaction was carried out in between glucose and $FeCl_3$ that were broken up into H_2O. The reaction was carried out at 80°C. Later, calcination was performed at 700°C in order to prompt the arrangement of graphene along with FeO. At last, the removal of Fe was carried out using HCl. $FeCl_3$ was assumed an imperative part in delivering excellent graphene on the grounds that it went about as format and impetus. From the previous technique, the large-scale manufacturing of graphene from glucose is an adequate method of creation. This basic technique can be received for graphene planning from any biomass. Xin-Hao Li and collaborators [2, 16] also integrated graphene by utilizing glucose. In their methodology,

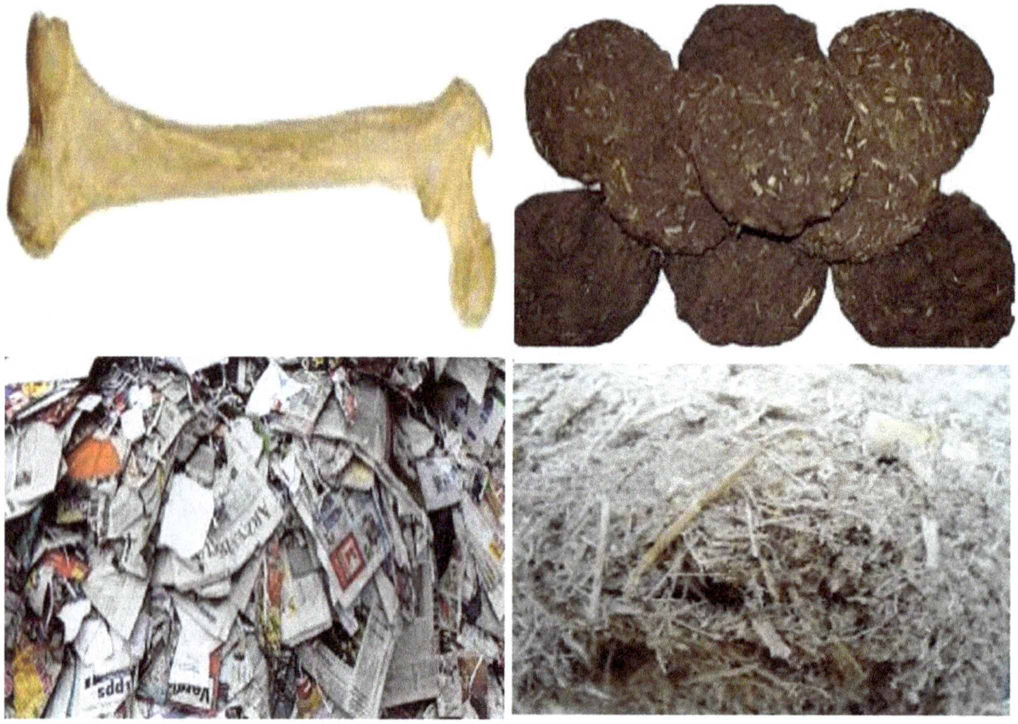

FIGURE 5.2 Synthesis of graphene from biowastes.

they used dicyandiamide (DCDA) along with nitrogen using in situ arrangement. In this manner, glass-like graphene was produced. The technique provided an impression to acquire graphene using nitrogen contents. It is an altered Hummers' technique for the production of thick graphene sheets with no decontamination and detachments of sheets. All the graphene sheets were neatly arranged as well. The researchers asserted the previous technique for graphene planning as an effortless methodology for wide applications [2, 16].

5.3.2 Graphene from Food, Insect, and Wastes

Another group of investigators [2, 17] developed a methodology for graphene production from crude resources, including chocolate, grass, plastics, and cockroaches. They used CVD method for the development of graphene using Cu foil. The extra Cu was removed later to obtain pure graphene. The X-ray photoelectron spectroscopy (XPS) information was used to uncover the carbon structure.

5.3.3 Graphene from Chitosan

Among the various available resources of carbon, chitosan is explored as a biomaterial. The basic source is obtained from the shells. This material is a rich source of carbon; hence, a group of investigators developed methods to produce graphene from this material, particularly using the pyrolysis technique. Additionally, the group used a photocatalyst to produce hydrogen under daylight [2, 18–20].

5.3.4 Graphene from Recycling Wastes Precursors

A number of precursors have been explored by the researchers for the development of graphene by the use of recycling waste materials [21, 22]. The recycling waste is a huge industry that generates a huge pile of waste. In view of prior evaluations, worldwide urban communities have delivered approximately billions of squanders annually [23]. Squanders transformation altogether changes because of the sort and creation of waste forerunners. For example, change of biomass relies upon different sources. Recently, the transformation of battery squanders into graphene has passed on a fundamental job for cutting-edge gadgets. Also, an excessive use of tire squanders into graphene has analyzed for cost-accommodating and adaptable creation. Furthermore, monolayered, bi-layered, and diverse graphene that ran among 3 and down technique have been effectively used for graphene combination. Moreover, different methodologies have been intended for the combination of graphene subordinates. Prior investigations have shown ordinary strategies which have commanded gigantic notice of academic local area. For instance, the union of graphene 2D-heterostructures was found for diverse product formation [21].

5.3.5 Graphene from Domestic Wastes

The domestic squander is also a huge raw material that contains a major component of carbon, which varies with the available domestic squanders containing a huge source of carbon as a raw material. This may include the carbon from vegetation, paper, plastic, plants, or any other available source. Researchers have developed various forms of graphene from these wastes, including the waste from insects using CVD technology. Panahi-Kalamuei and his gathering concocted reusing waste bread extra to create graphene subsidiaries by aqueous blend. The gathering acquired a gel by blending bread squander with water at 70–80°C. The gel was exposed to aqueous treatment in Teflon-lined steel autoclaved at 180°C under various states of pH and response time. At that point, the suspension was sifted to recuperate the hasten shape. Essentially, Akhavan et al. used different, common, and modern waste materials for the combination of graphene and GO. These squanders included wood, bagasse, leaves, and organic product squanders, powder residue gathered from fumes of diesel motor and waste papers [21, 24, 25].

5.3.6 Graphene from Waste Newspapers

The paper industry is a huge industry that utilizes the paper on daily basis; likewise its compositions, the huge waste is also produced on the same frequency. As the main ingredient of paper is cellulose, it clearly indicates the rich source of carbon present in the form of paper. The cellulose from this paper industry was converted into allotropes of carbon in the form of nanospheres [26]. Some other investigators [27] revealed the production of graphene from leaves of tea. Here, the graphene was produced by the modified Hummer's method. Another group of researchers [28] revealed another technique using the waste shells of coconut. Here, the shells were converted into allotropes of carbon developing graphene as well. They also used modified Hummer's methodology. Different techniques were utilized to affirm the fruitful combination of GO. Long et al. [29] detailed the planning of thickly permeable carbon structures imitating graphene like materials from some parasites. The pulp industry has also provided the biowaste for the production of graphene [30], while the agrowaste was also used for the development of graphene sheets [31, 32].

5.3.7 Graphene from Bread Squander

Scientists utilized food squander (bread squander) for the preparation of multilayered graphene using hydrothermal strategy. Various methods are used to utilize reuse squander produced from waste. The formation of fertilizers from food squander doesn't produce any additional value; however, it can be a good source to produce single-layered and multilayered graphene. During the processing in the first step, bread squander was ground and afterward suspension was formed using water [33]. These days, the development of graphene comprised two comparable sub–cross sections of carbon that fortified along with σ bonds. Traditionally, graphene is shaped using a mechanical procedure. However, various strategies are adopted these days to prepare graphene [34].

At present, bio-squander materials have been an extraordinary hotspot for the creation of carbon. Furthermore, bio-squander materials are plentiful and need for reusing. Also, the administration of waste biomass is testing particularly in created nations. In this way, it is proposed to change over the material into carbonaceous materials that can be utilized in colossal applications. Subsequently, specialists have invested a great deal of energy to incorporate graphene sheets utilizing eco-accommodating biomass forerunners, including squander corn shell, egg shell, and gelatin. Graphene is also obtained from the biowaste of glucose products, known as sugar atom ($C_6H_{12}O_6$), with bountiful and inexhaustible carbon sources. As of late, analysts have begun to utilize glucose as a beginning material to plan graphene. Zhang and collaborators [35] acquired graphene through a series of chemical processes initiating specialist combinations.

5.3.8 Graphene from Rice-Husks

A huge quantity of rice husks is delivered yearly on our planet. Because of gigantic mass, it gets consideration attention as a beginning material to produce good quality biomass. The extraction of graphene ordinarily involves chemical reactions. The potassium hydroxide can extraordinarily help in this extraction process. This can induce porous structure converting the carbon into its nanotubes or filaments, consequently improving various properties. In addition, KOH, likewise, aids the arrangement of high virtue graphene materials containing steady and clean edges. Singh and collaborators revealed the composites developed from graphene and rice husk that produced considerable toughness at high temperature which was verified by various techniques such as transmission electron microscopy (TEM), by utilizing a similar technique [34, 36].

5.3.9 Graphene from Corn Stalk Core (CSC)

The fibers of cellulose can be extracted from corn stalk. It can be utilized in the production of feasible fillers. Other than this scope, it is important to develop crude materials as well. Corn stalk produces sufficient horticultural squanders. The composition is diverse, comprising all the main ingredients such as cellulose, hemicellulose, and lignin. The core is normally permeable. Liu and colleagues have incorporated CSC-GO by means of get-together system (freeze-dried cycle) where GO is initially set up by utilizing changed Hummer's technique [34].

5.3.10 Graphene from Plastic Waste

The waste materials are abundant and a variety can be observed in waste materials. Plastic is one of them covering the earth in the form of flasks, baggage, and sheets, and all of these forms can result in a serious threat to resource. This is a huge source of environmental pollution. Alternatively, a number of methods are also available to transform these surplus materials into carbon-based materials. A group of researchers, Essawy and collaborators, have observed that graphene can be extracted from polyethylene terephthalate (PET). PET is widely available plastic that generates a huge quantity of waste materials available from pharmaceutical and food packaging industry. The waste of PET is eradicated using ignition, chemical, and feed stock to produce gaseous and liquid products, and carbon-enriched materials. The PET is marked with a high concentration of carbon and minimal contents impurities produced by minerals. Hence, this material has the great tendency to be converted into its raw materials of carbon [34, 37]. Recently, numerous opportunities and challenges have been encountered for the synthesis and characterization of graphene (G) and GO in several areas of interdisciplinary science [38–40].

5.4 GRAPHENE FROM SUCROSE (TABLE SUGAR)

Sucrose is a disaccharide that goes under hydrolysis and gives monosaccharide glucose and fructose. Sucrose, which is also known as sugar, more correctly α-D-glucopyranosyl-D-fructofuranoside, has been predicted to have the world's highest making of any unmixed, single, natural, and organic chemical. Sugar beet and sugarcane are two main crops that are growing for sugar production nowadays. Sugar cane now adds up to 75% of the world's synthesis of sugar.

Sucrose ($C_{12}H_{22}O_{11}$) is found in two crystalline shapes, the common stable sucrose A with melting point 184–185°C from water, and the other is metastable sucrose B with melting point 169–170°C, which is obtained by recrystallization by methanol. The advantages of sucrose are as follows:

1. It is pure
2. It has high sweetness degree
3. It is cheap
4. Solubility in water is easy
5. It has color
6. It can be handled easily

The heating process converts the sucrose into caramel. It results in the loss of H_2O and volatile organic components, developing a polymeric material. The process of caramel-melt insertion in the layers of silicates produces hybrid nanocomposites. The process is performed using microwave irradiation on caramel-clay. It results in the production of graphene-clay nanocomposites. The process is carried out at 750°C in the absence of oxygen. Here, the sucrose is a precursor of carbon just like other available sources such as acrylonitrile. The important aspect is the development of interlayer distance in between carbon-clay materials. The distance produced in between layers is approximately 0.4 nm. It is the same interlayer distance as observed in graphene. The processing was carried out by thermal alteration of caramel-clay nanocomposites into the graphene-clay.

5.5 GRAPHENE FROM METHANE

Methane is an odorless gas with a troposphere concentration of 1.8 parts per million in dry air. After water vapors and CO_2, it is the 3rd most important greenhouse gas. Nowadays, its concentration is 2.5 times more than the 0.7 parts per million noticed in ice cores from the period of 1000–1750 AD. Over the past 800,000 years, its concentration is higher than ever seen in ice core record. Atmospheric increase in methane since 1750 tells an emission of gases of 340 ± 50 teragram methane/year. This is round about two-thirds of total emissions now days. Emission from fossil fuel and agriculture together make 230 teragram/CH4/year. The donation of natural methane from wetland is 174 teragram/year. This amount is more sensitive to water table height and temperature. Emissions of wetland thus react to wetting and global warming. Natural sources are termites, wetlands sources, hydrates, oceans, wild animals, geological sources, and wildfires sources.

Methane can react with copper at very high temperatures to make graphene. Simply, copper is heated to about 1,000°C and takes it under the methane gas. From the carbon atoms in the methane gas, graphene layers will form on the surface of copper by carbons atoms.

5.6 CONCLUSION

The various innovative routes of the graphene preparation have been developed successfully and a lot more are under progress. The brilliant idea of developing the graphene from biowaste is a successful way to utilize the resources in an appropriate manner. However, still researchers are required to explore the authentic ways to develop this material to meet the demand in wide industrial applications. Some of the uncovered areas the defect distribution of graphene should also be addressed while producing it from various resources. While all the mentioned approaches are simple, clear, effective, and economic, there is always a window for developing the materials in accurate manners that should be explored for graphene in coming years, making its convenient availability for diverse applications.

REFERENCES

1. N. Akram, M. Saeed, M. Usman, M. Mansha, F. Anjum, K.M. Zia, I. Mahmood, N. Mumtaz, W. G. Khan. Influence of graphene oxide contents on mechanical behavior of polyurethane composites fabricated with different diisocyanates. Polymers 13 (2021) 444.
2. N. Raghavana, S. Thangavela, G. Venugopalb. A short review on preparation of graphene from waste and bioprecursors. Appl. Mater. Today 7 (2017) 246–254.
3. A.K. Geim, K.S. Novoselov, The rise of graphene. Nat. Mater. 6 (2007) 183–191, http://dx.doi.org/10.1038/nmat1849
4. P. Avouris, C. Dimitrakopoulos, Graphene: synthesis and applications. Mater. Today 15 (2012) 86–97, http://dx.doi.org/10.1016/S1369-7021(12)70044-5
5. C. Soldano, A. Mahmood, E. Dujardin, Production, properties and potential of graphene. Carbon (N. Y.) 48 (2010) 2127–2150, http://dx.doi.org/10.1016/j.carbon.2010.01.058
6. B. Marinho, M. Ghislandi, E. Tkalya, C.E. Koning, G. de With, Electrical conductivity of compacts of graphene, multi-wall carbon nanotubes, carbon black, and graphite powder. Powder Technol. 221 (2012) 351–358, http://dx.doi.org/10.1016/j.powtec.2012.01.024
7. F. Bonaccorso, Z. Sun, T. Hasan, A.C. Ferrari, Graphene photonics and optoelectronics. Nat. Photonics 4 (2010) 611–622, http://dx.doi.org/10.1038/nphoton.2010.186
8. R.K. Joshi, S. Alwarappan, M. Yoshimura, V. Sahajwalla, Y. Nishina, Graphene oxide: the new membrane material. Appl. Mater. Today 1 (2015) 1–12.
9. F. Schwierz, Graphene transistors. Nat. Nanotechnol. 5 (2010) 487–496, http://dx.doi.org/10.1038/nnano.2010.89
10. T. Guo, X. Chen, L. Su, C. Li, X. Huang, X.-Z. Tang, Stretched graphene nanosheets formed the 'obstacle walls' in melamine sponge towards effective electromagnetic interference shielding applications. Mater. Des. 182 (2019) 108029.
11. S. Gurunathan, J. Woong Han, V. Eppakayala, J. Kim, Green synthesis of graphene and its cytotoxic effects in human breast cancer cells. Int. J. Nanomed. 1015 (2013) 1015–1027.
12. J.N. Coleman, Liquid exfoliation of defect-free graphene. Acc. Chem. Res. 46 (2013) 14–22, http://dx.doi.org/10.1021/ar300009f
13. W. Ren, H.-M. Cheng, The global growth of graphene. Nat. Nanotechnol. 9 (2014) 726–730, http://dx.doi.org/10.1038/nnano.2014.229
14. F. Wang, L. Liu, W.J. Li, Graphene-based glucose sensors: a brief review. IEEE Trans. Nanobiosci. 14 (2015) 818–834, http://dx.doi.org/10.1109/TN.B.2015.2475338
15. B. Zhang, J. Song, G. Yang, B. Han, K.S. Novoselov, A.K. Geim, S.V. Morozov, D. Jiang, Y. Zhang, S.V. Dubonos, I.V. Grigorieva, A.A. Firsov, K.S. Novoselov, et al., Large-scale production of high-quality graphene using glucose and ferric chloride. Chem. Sci. 5 (2014) 4656–4660, http://dx.doi.org/10.1039/C4SC01950D
16. X.-H. Li, S. Kurasch, U. Kaiser, M. Antonietti, Synthesis of monolayer-patched graphene from glucose. Angew. Chem. Int. Ed. 51 (2012) 9689–9692, http://dx.doi.org/10.1002/anie.201203207
17. G. Ruan, Z. Sun, Z. Peng, J.M. Tour, Growth of graphene from food, insects, and waste. ACS Nano 5 (2011) 7601–7607, http://dx.doi.org/10.1021/nn202625c
18. A. Primo, P. Atienzar, E. Sanchez, J.M. Delgado, H. García, A.K. Geim, K.S. Novoselov, K.S. Novoselov, A.K. Geim, S.V. Morozov, D. Jiang, Y. Zhang, S.V. Dubonos, I.V. Grigorieva, et al., From biomass wastes to large-area, high-quality, N-doped graphene: catalyst-free carbonization of chitosan coatings on arbitrary substrates. Chem. Commun. 48 (2012) 9254, http://dx.doi.org/10.1039/c2cc34978g
19. Z. Terzopoulou, G.Z. Kyzas, D.N. Bikiaris, Recent advances in nanocomposite materials of graphene derivatives with polysaccharides. Materials 8 (2015) 652–683.
20. P. Hao, Z. Zhao, Y. Leng, J. Tian, Y. Sang, R.I. Boughton, C.P. Wong, H. Liu, B. Yang, Graphene-based nitrogen self-doped hierarchical porous carbon aerogels derived from chitosan for high performance supercapacitors. NanoEnergy 15 (2015) 9–23, http://dx.doi.org/10.1016/j.nanoen.2015.02.035

21. R. Ikram, B.M. Jan, W. Ahmad, Advances in synthesis of graphene derivatives using industrial wastes precursors; prospects and challenges. J. Mater. Res. Technol. 9 (2020) (6) 15924e1–5951. https://doi.org/10.1016/j.jmrt.2020.11.043
22. Z. Fang, Y. Gao, N. Bolan, S.M. Shaheen, S. Xu, X. Wu, et al. Conversion of biological solid waste to graphene-containing biochar for water remediation: a critical review. Chem. Eng. J. 390 (2020) 124611.
23. V. Forti, C.P. Balde, R. Kuehr, G. Bel. The global e-waste monitor: quantities, flows and the circular economy potential. United Nations University/United Nations Institute for Training and Research, International Telecommunication Union, and International Solid Waste Association (2020). Bonn, Geneva and Rotterdam.
24. G. Ruan, Z. Sun, Z. Peng, J.M. Tour. Growth of graphene from food, insects, and waste. ACS Nano 5 (2011) (9) 7601e7.
25. M. Panahi-Kalamuei, O. Amiri, M. Salavati-Niasari. Green hydrothermal synthesis of high quality single and few layers graphene sheets by bread waste as precursor. J. Mater. Res. Technol. 9 (2020) (3) 2679e90.
26. K.H. Adolfsson, S. Hassanzadeh, M. Hakkarainen. Valorization of cellulose and waste paper to graphene oxide quantum dots. RSC Adv. 5 (2015) (34) 26550e8.
27. M.A. Faiz, C.C. Azurahanim, Y. Yazid, A. Suriani, M.S.N. Ain. Preparation and characterization of graphene oxide from tea waste and it's photocatalytic application of TiO_2/graphene nanocomposite. Mater. Res. Express 7 (2020) (1) 015613.
28. E.H. Sujiono, D. Zabrian, M.Y. Dahlan, B.D. Amin, J. Agus. - Heliyon. Graphene oxide based coconut shell waste: synthesis by modified Hummers method and characterization. Heliyon 6 (2020) (8) e04568.
29. C. Long, X. Chen, L. Jiang, L. Zhi, Z. Fan. Porous layer-stacking carbon derived from in-built template in biomass for high volumetric performance supercapacitors. Nanomater. Energy 12 (2015) 141e51.
30. Z. Ding, T. Yuan, J. Wen, X. Cao, S. Sun, L.-P. Xiao, et al. Green synthesis of chemical converted graphene sheets derived from pulping black liquor. Carbon 158 (2020) 690e7.
31. A. Hashmi, A.K. Singh, B. Jain, A. Singh. Muffle atmosphere promoted fabrication of graphene oxide nanoparticle by agricultural waste. Fullerenes Nanotubes Carbon Nanostruct. 28 (2020) (8) 627e36.
32. S. Kang, K.M. Kim, K. Jung, Y. Son, S. Mhin, J.H. Ryu, et al. Graphene oxide quantum dots derived from coal for bioimaging: facile and green approach. Sci. Rep. 9 (2019) (1) 4101.
33. M. Panahi-Kalamueia, O. Amirib, M. Salavati-Niasari, Green hydrothermal synthesis of high quality single and few layers graphene sheets by bread waste as precursor. J. Mater. Res. Technol. 9 (2020) (3) 2679–2690, https://doi.org/10.1016/j.jmrt.2020.01.001
34. N.F. Tajul Arifin, N. Yusof, A.F. Ismail, J. Jaafar, F. Aziz, W.N. Wan Salleh, Graphene from waste and bioprecursors synthesis method and its application: a review. Malays. J. Fundam. Appl. Sci. 16 (2020) (3) 342–350.
35. B. Zhang, J. Song, G. Yang, B. Han. Large-scale production of high-quality graphene using glucose and ferric chloride. Chem. Sci. 5 (2014) 4656–4660.
36. M. Priyanka, M.P. Saravanakumar. A short review on preparation and application of carbon foam. IOP Conf. Ser. Mater. Sci. Eng. 263 (2017) 246–254.
37. L. Cui, X. Wang, N. Chen, B. Ji, L. Qu. Trash to treasure: converting plastic waste into a useful graphene foil. Nanoscale 9 (2017) 9089–9094.
38. R. Ikrama, B.M. Jana, W. Ahmad. An overview of industrial scalable production of graphene oxide and analytical approaches for synthesis and characterization. J. Mater. Res. Technol. 9 (2020) (5) 11587–11610, https://doi.org/10.1016/j.jmrt.2020.08.050
39. G. Cui, Z. Bi, R. Zhang, J. Liu, X. Yu, Z. Li. A comprehensive review on graphene-based anti-corrosive coatings. Chem. Eng. J. 373 (2019) 104–121.
40. K. Chojnacka, M. Mikulewicz. Green analytical methods of metals determination in biosorption studies. TrAC Trends Anal. Chem.116 (2019) 254–265.

6 Graphene from Rice Husk

Hosam M. Saleh and Amal I. Hassan

CONTENTS

6.1 INTRODUCTION

Carbon is one of the most substantial elements in the periodic table. It exists in a diversity of allotropes, including the traditional ones like diamond and graphite and later, fluorine and carbon nanotubes (CNTs) [1]. The two-dimensional (2D) carbon chemical formula (graphene) is one of the most recently found formulas, and it has piqued the interest of scientists in recent years. Graphene is the thinnest compound on Earth, its thickness is about the thickness of one atom, and it is a very light material (the weight of a square meter of it is about 0.77 mg), and the best types of it are 100–300 times stronger than steel [2].

Graphene is a flat, one-atom-thick sheet of carbon sp^2 atoms packaged in a wave-glass structure. For graphite, carbon nanotube, and fullerenes, it is the fundamental structural element. Samples of graphene are available in Si/SiO_2 substratum wafers in nanoflakes. With a space of approx. 0.34 nm, each layer is monoatomically thin, but multilayer flakes can be manufactured [3].

In 2004, a group of scientists from the University of Manchester in the United Kingdom discovered graphene, which ushered in a global revolution in scientific study. Graphene is a cross-linked carbon sheet made up of sp^2-hybridized [4]. Graphene's most notable features of thermal and electrical conductivity are because of its hexagonal lattice structure [5]. Chemically, graphene comprises a single element, carbon isotope C-12, which has four free electrons, which facilitates formatting bonds with other atoms. Graphene takes part in many chemical reactions as a reactant, an electron donor, or an oxidizing agent (electron acceptor). It directly results from graphene's electron structure, which gives it an electron affinity and an ionization potential of about 4.6 eV [6]. Because of its 2D crystals, it has novel electronic and mechanical properties, high electronic mobility, structural flexibility, and the ability to be tuned from p-type to n-type by applying gate voltage [7].

In many countries, rice is the main crop and the most commonly produced husk that is used to burn, which contributes to environmental pollution. Therefore, RH ash (RHA) has created a major concern because it is harder to discard. RH is the rice grain's outer covering. When the rice is treated, its cover is removed, and its waste contains 15–20% silica, besides cellulose and lignin. Cellulose, lignin, and ash are all eliminated during burning [8]. In 2017, RH production worldwide was estimated to be around 759.6 million tons, providing an enormous amount of essentially free raw material if it can be put to good scientific use [9].

6.2 THE CHEMICAL COMPOSITION AND PROPERTIES OF RH

Because of geographical conditions, type, weather, soil chemistry, and paddy crop manures, the chemical ingredients are found to differ. The RH represents the covering that surrounds the rice grain before it is put out as food for human consumption. The factories put these husks out; they are

DOI: 10.1201/9781003169741-6

TABLE 6.1
Chemical Composition of RH

Compound	Value (%)
Silicon dioxide SiO_2	20
Cellulose	30
Lignin	25
CaO	0.89
MgO	0.83
SO_2	0.41
Fe_2O	0.12
Al_2O_3	0.20
Blaine fineness (cm^2/g)	4238
Specific gravity	2.30
Color	Whitish gray

thrown out for disposing of, causing environmental pollution. Chemically, RH comprises 20% silica, 40–30% cellulose, 25–30% lignin, 10–15% ash, and 5–10% moisture. Organic matter and water make up 74%, and oxides (Al_2O_3 + Fe_2O_3 + CaO + MgO) constitute 4% (Table 6.1). The content of RH of organic compounds after extracting silica from it, in which the organic part is composed of cellulose, lignin, and hemicelluloses. Hemicelluloses are a type of carbohydrate that has branched and shorter chains than other polysaccharides [10]. They are especially significant in the manufacture of biofuels because they can make up to 35% of the biomass weight [11]. Lignin is an oxygenated polymer of *p*-propylphenol units that comprises *p*-coumaryl, coniferyl, and sinapyl alcohols. In recent years, lignin modifications to make functional carbon materials for batteries or supercapacitors have piqued interest [12]. After the treatment with biotechnology, RHs can be used on animal feed, with urea, ammonia, certain acids, and bases to enhance the protein content and increase the digestion factor [13]. In addition, it is used as a ground for spreading barley seeds, so when the seedlings grow, it becomes fodder for feeding animals [14]. It is used in the cellulosic industries and compressed panels because it contains fibers at rates close to what involves tree wood, making it an alternative material in obtaining these fibers [15], which constitute the basis for an enormous number of industries based on cellulose and lignin. The husk is fuel for biomass that needs to be processed thermally, such as combustion, pyrolysis, or gasification [16]. Heat, electricity, syngas, and biofuels are the energy generated through thermal processing. The capacity of husk to produce power from biomass is based on small-scale gasification that includes heating RHs to high temperatures, which leads to a decomposition of materials into a fuel-gas mixture [17]. The gases are consumed to generate heat or steam that activates and generates electricity from turbines. The husk is used as fuel and steam in the rice mills to provoke the rice parboiling process [18].

6.3 CHARACTERISTICS AND GRAPHENE SYNTHESIS METHODS

Initial investigation on graphene showed that its characteristics exceed the mass graphite from which it was obtained. The mobility of both holes and electrons is equivalent to that of bulk metal, such as copper [19]. The conductivity of thermal products is so high that the ballistic conductivity without the impedance of the material itself is taken into account, and the opacity of single layers is over 2% at nearly all levels (Figure 6.1). These unique characteristics enable graphene to be used in several applications, including electrochemical condensers, anticorrosion coatings, composites, transparent film conduction, thermal pastes, and sensing and bi-sensing [20].

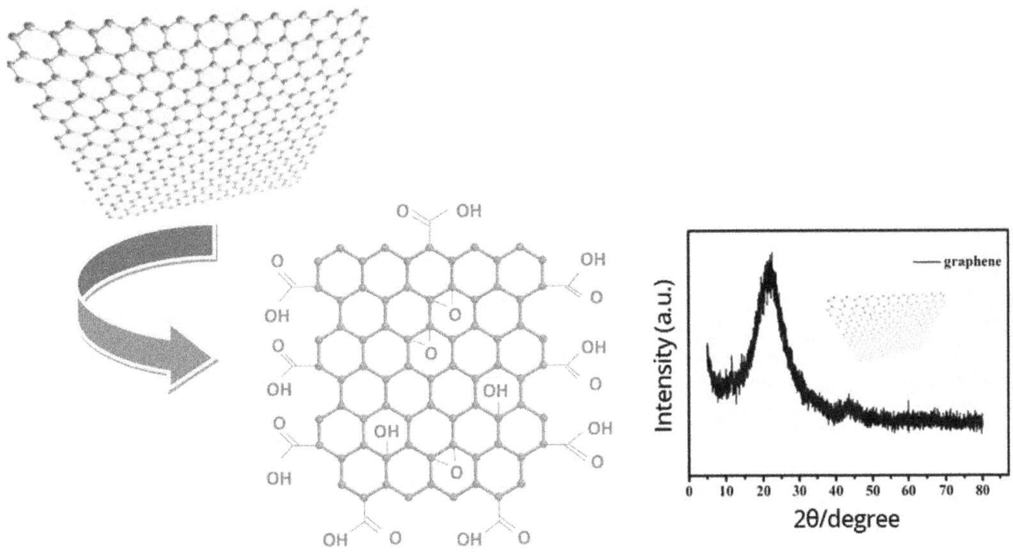

FIGURE 6.1 Schematic of graphene.

Most graphene synthesis approaches involve many complicated procedures or high temperature, or costly advanced devices. A cost-effective and straightforward graph synthesis could therefore be highly valuable for different purposes [21].

The graphene synthesis can be classified as a top-down and bottom-up technique in two basic categories. The initial material in the top-down approach comprises graphite and is intended to make graphene and exfoliate sheets using solid-state, liquid conditions, or electrochemical exfoliation [22]. The graphite oxide exfoliation to graphene oxide (GO) followed by chemical and heat reduction is another technique in this area [23]. Chemical vapor deposition (CVD) or epitaxial growth of graphene made of molecular precursor blocks bases the bottom-up process on the production [24]. By deleting all oxygenated functions and restoring the π-network, a direct reduction of the GO has garnered the most attention to producing enormous amounts of graphene [25]. However, both endothermic and exothermic processes require reduction steps, depending on the different functional groups. Epoxide group reduction in GO can be achieved using hydrazine as a reducing agent. Heating is necessary for dehydroxylation and decarboxylation. Thermal dehydroxylation and deep oxidation of hydrazones react to heat and thermal treatment. The decreased GO in physical characteristics and structure is more comparable to graphene. The degree of restoration via β-network is determined by reductions, which lead to graphics with various qualities [26].

Depending on the production process, the technique shows the resulting graphene's structure, morphology, and attributes such as number of layers, defect level, electrical and thermal conduciveness, solubility, hydrophilic, or hydrophobicity. The interlayer strength of the van der Waals chemical method may be reduced by the graphite oxidation using strong oxidizing agents, such as concentration of sulfuric acid (H_2SO_4), sodium nitrate ($NaNO_3$), and potassium permanganate ($KMnO_4$), which has achieved a similar level of oxidation [27].

Process parameters manually affect graphene quality like the gas flow, catalyst, precursor, temperature, growth time, and gas pressures in the CVD process (such as layer number, grain size, band distance, and the doping impact) [28]. The graphene-up technique shows the carbon precursor atoms formed on the substratum surface directly, which results in graphene crystal planes. Graphene is produced on one-crystal silicon carbide (SiC) using thermal vacuum graphics for the epitaxial thermal growth of graphene using the substrate approach. At around 1300°C SiC, the silicon atoms are sublimated because their surface is reorganized and graphitized with carbon-enriched materials [29]. The CVD technology involves the chemical reaction of gaseous precursor carbon molecules spread on a high-temperature metal substratum (900–1500°C) [30]. When the substratum is refreshed, carbon solubility is lowered on the substratum, and the carbon precipitates to one to several layer graphene layers on the substratum [30]. To decompose, diffuse, and segregate, carbon atoms need extremely high temperatures (above 2500°C without catalyst), but it must be reduced to roughly 1000°C. The metallic catalyst is one or more binary alloys of transition metals, which can dissolve carbon and afterward separate to generate polycrystalline graphene, with uncompleted orbitals (e.g., Co, Cu, or Ni).

GO has hydroxyl, carbonyl, and epoxy groups on the basal plate, besides carboxyl groups on the edges, which operate as a 2D material for graphene synthesis and their combination products (Figure 6.2).

Graphene has the advantage of being compatible with planning technologies used in the semiconductor sector. Like CNTs, graphene is highly mobile, μ exhibits over 15,000 cm^2/V, and can be either metallic or semiconducting. If the graphene nanoribbons are adapted to less than 100 mm, a bandgap can open because of the electron containment [31]. Graphene electronic states depend on the border of the nanoribbon and its breadth. The transmission channel in a metal oxide semiconductor (MOS) is one of the most appreciable uses of graphene [32].

A green synthesis approach for producing graphene from precursors of biomass, such as chitosan, alfalfa, sugar, green tea, RH, and other foodstuffs, has been introduced recently. An abundant sort of agricultural waste is the RH, a prerequisite result of rice milling [33]. The organic ingredients constitute about 80% of the RH composition SiO_2 and are around 20% of the inorganic silicon dioxide [34]. RH is used for producing biofuels, energy generation, fuel boilers, and the bulking agent for animal manure composting. Silicon and C-based products, such as silica, silicon (Si), silicon nitride (Si_3N_4), silicon tetrachloride, SiC, zeolites, activated carbon, and graphene, are typically prepared with RH and RHA. As a result, the use of RH in creating graphene will pave the way for a range of applications [35].

Manufacturing RH graphene, employing KOH and carbon as protective barriers against oxidation, was initially described by Ref. [36]. Singh and others then revised the synthesis procedure to substitute RH for black carbon [37]. RH graphene is manufactured by activating RHA with KOH and washed, centrifugated, sounded, and vacuum-filtering with distilled water to yield graphic materials [38]. RH was washed many times to remove huge silica and other impurities as feasible in a standard synthesis method. The RHA was made by combusting RH into the air after it had been washed. An amount of 3 mg of RHA was combined with 15 g of KOH and ground for 15 minutes. In a porcelain crucible, a mixture of RH and KOH was compacted. The crucible was then wrapped

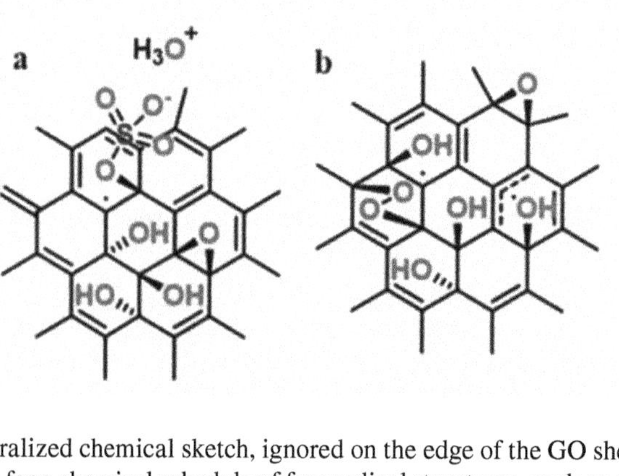

FIGURE 6.2 (a) GO's generalized chemical sketch, ignored on the edge of the GO sheet flaws or functional groups. (b) GO is responsible for a chemical schedule of free radical structures, such as carbon centers, groups of allyl, and cytotoxic endoperoxide groups.

in ceramic wool and placed inside a massive graphite crucible. A sufficient amount of sacrificial RHA was applied to the top of the graphite crucible to form a barrier against the oxidation of the sample inside the porcelain crucible.

Recently, to make GO-RH biochar (GO-RHB) composites, 10-g RH biomass was mixed with 0.5-g GO-RHB, and the mixture was then agitated for 2 hours at 450°C to ensure thorough mixing. The mixture of GO-RHB composite was then sonicated for 3 hours to adsorb the GO suspension onto the RHB surface [39]. The blended suspension was then placed in an oven and dried for 12 hours at 100°C. Finally, the materials were placed in a tube furnace with N_2 at 300°C for 2 hours. An amount of 5 g of raw RH was pulverized and cooked in a Carbolite furnace for 2 hours at 400°C in the air before the pretreatment process. The solid residue, known as RHA, was collected, weighed, and the yield was 47%. The RHA was then impregnated with KOH powder in a 1:2 impregnation ratio at 800°C for 2 hours in the air for activation.

6.4 RH NANOCOMPOSITE PREPARATION

A one-step in situ approach based on the electrostatic force between positive SiO_2-NH_2 and negative GO dispersion was developed for producing an RHA-SiO_2/GO nanocomposite. To function the surface by binding the monolayer-NH_2 nanoparticles with Aminopropyltriethoxysilane (APS), the RHA-SiO_2 was initially altered. For 30 minutes, a combination with 10-g RHA-SiO_2 powder was agitated for a good dispersion of RHA-SiO_2 in a 150-ml solution of ethanol [40]. The Langmuir isotherm model determined maximum adsorption capacity for Ni^{2+} for nanocomposites of 256.4 mg/g. The adsorption data were also well matched for the kinetic equation pseudo-secondary. Tien et al. showed that RHA/SiO_2/GO removes Ni^{2+} ions from the water solution as a highly efficient adsorbent. About 335,016 m^2/g was measured in the high specific surface area of RHA-SiO_2/GO. The density functional theory approach is used to measure the pore size distribution. There was a total pore volume of 0.747 cm^3/g measured [40].

RHs were sonicated, rinsed, and cleaned with distilled water to remove undesirable impurities. RH was then dried for 24 hours at 65°C. The dried RH was mechanically crushed into powder using a commercial blender machine and sieved using a laboratory test sieve of 20-m ferrocene, and $Fe(C_5H_5)_2$, as a catalyst, was dissolved in ethanol and sonicated at room temperature for 20 minutes [41]. An amount of 120-mg dried RH powder and 80-mg $Fe(C_5H_5)_2$ solution were combined and applied to a 2.5 cm × 2.5 cm aluminum sheet as a substrate. After that, a 40 mm (D) × 10 mm handmade metal casing was applied to the aluminum sheet (H) [41].

Brown-RH is used to make graphene nanosheets (GNs), and their electrochemical energy-storage performance is investigated regarding its use as an electrode material. Despite its simplicity, the production approach yields crystalline ultrathin GNs with a low defect density, and this is most likely because of the optimization of each fabrication stage. The GNs have a crumpled-silk-veil-wave- and sheet-like structure with a large surface area and porosity. With edges and several folded regions, the GNs had a crumpled-silk-veil-wave- and sheet-like structure. In an aqueous 1-M Na_2SO_4 electrolyte, the performance of ultrathin GN electrode's electrochemical supercapacitors was investigated [41, 42].

The nanosheets of graphene show an ultrathin crumpled, silk-veil-like structure, with a B1225-m^2 high surface area and high porosity. The electrode GNs exhibit a specified capacity of 115 F/g in 0.5 mA/cm^2 with a power density of 36.8 W h/kg, with an optimum cyclic stability of 88% over 2000 cycles, at a power density of 323 W/kg [43]. Its synergistic effects, the enhanced ion diffusion, and the better electric conductivity of the inherently large electrical surface area showed the positive electric energy storage performance of the biomaterial formed from the GNs electric electrode [44].

RH has a high silica content (20 wt%) and is found as waste in rice milling. Silica is a graphene contaminant and must be removed before conversion, and an alkaline solution is required before treatment [45]. The product is a smoother carbohydrate, appropriate for creating pristine graphene, which is utilized to extract amorphous carbon from RH [36]. The KOH promotes carbon porosity, which makes it easier to remove contaminants. The amount of KOH affects the morphology of the graphene, because KOH opens the structure of the carbon, and it increases the particular graphene surface area [46].

Nanosilica precursors could make sophisticated materials include carbon/silica composites, photocatalysts, hydrogen production, and CO_2 capture materials, and metal ion removal adsorbents in the future [47]. Various ways can make nanosilica with a porous RH composition. Halim et al. [48] investigated the synthesis of sodium hydroxide–dissolved xerogels using RH as a raw material [48]. Thermal methods include the use of furnace muffles, fixed bed furnace, the reactor of fluidized bed, and other thermal methods comprising sloping step-grate oven, and cyclone oven [49]. The thermal technology has a lot of drawbacks: it takes much time to react, produces hot spots, and lacks free flow of air. Nanosilica is obtained via an electric/muffle furnace from agricultural waste in a laboratory size. The downside is delayed reaction time and reduced output rate while employing this technique. The best HR nanosilica extraction was examined by Patil et al. [50] using thermal treatment at different temperatures with electric ovens for 6 hours at 700°C. Amorphous silica RH succeeded for 4 hours in an electronic laboratory muffle furnace at 600°C in a rubber, stainless steel reactor [50]. The electric furnace can boost, in particular, the purity of silica ingredients derived from burning.

Narayanan [51] has described a reduced GO nanocomposite production of new RHA and the catalytic use thereof in the reaction of Biginelli. An RHA and GO mixture was utilized to generate a homogeneous composite in a hydrothermal way. In addition, the spectrum tests X-ray diffraction (XRD) and Fourier transform infrared spectroscopy (FTIR) showed that nanocomposites have partially reduced GO while being treated. X-ray photoelectron spectroscopy analysis shows the connection between silica particles of the RH and graphene sheets by Si–O–C bonding. RH was scrubbed with mud and other contaminants multiple times. Then it was dried and calcined for 3 hours at 650°C to get RHA under sunshine. An amount of 2.5 g of RHA was added to 250 ml of distilled water at 1300 rpm and heated to 80°C to synthesize 10% of GO-loaded RHA. Then a drop of GO dispersion was added (0.25-g GrO in 500-ml water) and agitated by 5 hours. The material is next transferred to an autoclave and stored for the day at 120°C. The RHA-G nanocomposite had been filtered and deionized water washed many times. The material was dried and treated with heat at 200°C for 3 hours. The RHA-G10 is the nanocomposite. The same approach was taken for RHA-G5 and RHA-G15 synthesis, where 5 and 15 denote the GO to RHA weight [51].

Mishra et al. [52] examined silicon purification by reducing RHA calcium. The RHA was blended in stoichiometric measurements at 500°C and calcium completely. The temperature of the mixture was then lowered to 720°C. The lower product was leached to 99.9% silicone with HNO_3 and HF. Several sequential acids (HF, H_2SO_4, and HCl mixes) leaching therapies followed for obtaining silicone purity (99.999%) by decreasing RHA and magnesium at a temperature of 800°C. MgO, Mg_2Si, and unreacted Mg can be removed as by-products using acid therapy.

Larbi [53] showed that a maximum silicon yield could be attained by using an RHA pellet with 5 wt% of Mg content above the stoichiometric requirements and 900°C heating under flowing argon. Magnesiothermal reactions have also produced silicon nanocrystals. The reduction of SiO_2 to Si has been quite popular since magnesiothermic reduction has allowed the three-dimensional Si replicas to be formed with the parent SiO_2 diatom [54]. The synthesis of silicone/carbon (SiO_2/C) composites and the SiC materials, which are of considerable use in the applications of lithium-ion batteries (LIBs) or non-metallic oxide ceramics, has been often employed to decrease magnesiothermically different types of silica/carbon (SiO_2/C) [55]. In magnesiothermic reduction SiO_2 and Mg_2Si were assumed to be intermediate Si [56]. Based on these facts, Si's kinetic product can be generated when the transformation is completed from SiO_2 to Si before the intermediate silicone reaches carbon. That means that the spread of Si to carbon is not sufficiently prompt, so Si can be produced. Shen et al. [57] presented an integrated MA technique for manufacturing RH nano-Si. Thermal decomposition and magnesiothermic reduction to create nano-SiO_2 have initially turned RH into nano-SiO_2. There are diverse advances to this novel

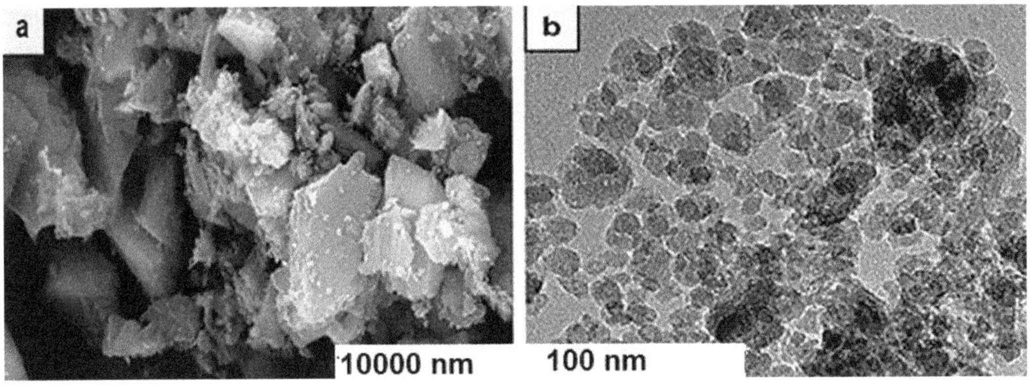

FIGURE 6.3 (a) SEM and (b) TEM of the silica from rice husk.

method: (1) the recuperated silicone inherits the unique and intrinsic nanostructure of the RH silica to provide outstanding battery performance through pulverizing reduction; (2) the RH is an abundant and durable source of silica; (3) it is easier, energy-efficient, and easier to scale up the entire process; and (4) it does not use expensive pre-silica and reagents. This green procedure uses only HCl after electrolysis and transforms it into Cl_2. In addition, silicone nanoparticles generated from RH silica were produced by Wong et al. [58] by the magnesiothermic reduction method. The anode of Si-graphene as an anode material for LIBs was subsequently exploited to manufacture a binder-free Si-graphene composite. The initial capacity of 1000 mA h/g with a high current density of 1000 mA/g was obtained using Si-graphene composite.

We prepared nano-silica from RH in our labs in 2017 [59]. In the experiment, we collected RHs from Egyptian farms. El Nasr pharmaceuticals Abu Zaabal, Egypt, was supplied with 34% of hydrochloric acid. A 150-W grinding stainless steel bladder grinder first repeatedly ground the RH, then refluxed with a hydrochloric acid solution (17%) with a 1:10 w/v ratio during a 1-hour time. Afterward, the treated RHs were filtered with distilled water, washed multiple times, and dried overnight. The dried RHs were subsequently calcinated for 2 hours at 650°C and suddenly cooled. The use of the PHILIPS® MPD X'PERT diffractometer with the geometry of Bragg-Brentano analyzed the resulting white silica powder by employing JOEL JSM-6510 LA energy distribution technology and XRD technique. The X-ray tube was operated at 40 kV and 30 mA, the angle of divergence slit was fixed at 0.5°C, the receiving slit was at 0.1°, and the step scan size 0.03, and the time of scan was 2 seconds. Adsorption-desorption isotherm from purified N_2, at 77 K, was carried out using Nova 2000, Quanta Chrome (commercial BET unit) to eliminate any moisture adsorbed on silica surfaces and pores at a residual pressure of 10–5 Torr at 150°C overnight. The Barrett-Joyner-Halenda (BJH) method was used in the calculation of the distribution of the pores [59]. An exterior RH silica morphology has been seen using an FEI Quanta 250 FEG scanning electron microscope, while a JEOL JEM-1230 microscope was used for testing its particle's form and size using transmission electron microscopy (TEM) (Figure 6.3a and b).

6.5 CONCLUSION

Agriculture debris is a worldwide concern, and regulations have been developed to recycle agricultural waste. Rice growing is worldwide and is one of the principal sources of food, in particular, in Asia. Rice consumption increases dramatically every year, and it has also increased agricultural waste by meeting these demands. The debris produced is mostly RH; because of their high calorific value, RH has been proving a significant role as a fuel. Graphene and its variants are new materials that still have to peak in the realm of technology. Although there has been much debate over this structure, challenges remain to be addressed to achieve the aim of materials, which is readily available. Graphene technology is restricted to the manufacturing of graphene. Thus, even though graphite is manufactured from inexpensive and plentifully available materials, graphite production is expensive.

In future applications, we can use RH silica, such as surface-functional mesoporous silica, hybrid mesoporous silica development, fluorescent silica synthesis, CNTs, mesoporous silica, and silica-supported nanoparticles.

REFERENCES

1. Naz A, Kausar A, Siddiq M, Choudhary MA (2016) Comparative review on structure, properties, fabrication techniques, and relevance of polymer nanocomposites reinforced with carbon nanotube and graphite fillers. Polym Plast Technol Eng 55:171–198.
2. Qiu B, Xing M, Zhang J (2018) Recent advances in three-dimensional graphene based materials for catalysis applications. Chem Soc Rev 47:2165–2216.
3. Radadiya TM (2015) A properties of graphene. Eur J Mater Sci 2:6–18.
4. Yadav R, Subhash A, Chemmenchery N, Kandasubramanian B (2018) Graphene and graphene oxide for fuel cell technology. Ind Eng Chem Res 57:9333–9350.
5. Wang J, Ma F, Sun M (2017) Graphene, hexagonal boron nitride, and their heterostructures: properties and applications. RSC Adv 7:16801–16822.
6. Muhich CL (2014) Metal oxide catalysts for renewable energy generation and green chemistry purposes. Chem Biol Eng Grad Theses Diss.
7. Koppens FHL, Mueller T, Avouris P, et al. (2014) Photodetectors based on graphene, other two-dimensional materials and hybrid systems. Nat Nanotechnol 9:780–793.
8. Nanayakkara MPA, Pabasara WGA, Samarasekara AMPB, et al. (2017) Synthesis and characterization of cellulose from locally available rice straw. In: 2017 Moratuwa Engineering Research Conference (MERCon). IEEE, pp 176–181.
9. Lo F-C, Lee M-G, Lo S-L (2021) Effect of coal ash and rice husk ash partial replacement in ordinary Portland cement on pervious concrete. Constr Build Mater 286:122947.
10. Requena R, Jiménez-Quero A, Vargas M, et al. (2019) Integral fractionation of rice husks into bioactive arabinoxylans, cellulose nanocrystals, and silica particles. ACS Sustain Chem Eng 7:6275–6286.
11. Zabed HM, Akter S, Yun J, et al. (2019) Recent advances in biological pretreatment of microalgae and lignocellulosic biomass for biofuel production. Renew Sustain Energy Rev 105:105–128.
12. Wang F, Ouyang D, Zhou Z, et al. (2021) Lignocellulosic biomass as sustainable feedstock and materials for power generation and energy storage. J Energy Chem 57:247–280.
13. Van Soest PJ (2006) Rice straw, the role of silica and treatments to improve quality. Anim Feed Sci Technol 130:137–171.
14. Athinarayanan J, Periasamy VS, Alhazmi M, et al. (2015) Synthesis of biogenic silica nanoparticles from rice husks for biomedical applications. Ceram Int 41:275–281.
15. Owodunni AA, Lamaming J, Hashim R, et al. (2020) Adhesive application on particleboard from natural fibers: a review. Polym Compos 41:4448–4460.
16. Ríos-Badrán IM, Luzardo-Ocampo I, García-Trejo JF, et al. (2020) Production and characterization of fuel pellets from rice husk and wheat straw. Renew Energy 145:500–507.
17. Situmorang YA, Zhao Z, Yoshida A, et al. (2020) Small-scale biomass gasification systems for power generation (< 200 kW class): a review. Renew Sustain Energy Rev 117:109486.
18. Chakrovorty RS, Roy R, Forhad HM, et al. (2020) Modification of conventional rice parboiling boiler to enhance efficiency and achieve sustainability in the rice parboiling industries of Bangladesh. Process Saf Environ Prot 139:114–123.
19. Bhimanapati GR, Lin Z, Meunier V, et al. (2015) Recent advances in two-dimensional materials beyond graphene. ACS Nano 9:11509–11539.
20. Othman NH, Ismail MC, Mustapha M, et al. (2019) Graphene-based polymer nanocomposites as barrier coatings for corrosion protection. Prog Org Coatings 135:82–99.
21. Whitener Jr KE, Sheehan PE (2014) Graphene synthesis. Diam Relat Mater 46:25–34.
22. Ciesielski A, Samorì P (2016) Supramolecular approaches to graphene: from self-assembly to molecule-assisted liquid-phase exfoliation. Adv Mater 28:6030–6051.
23. Acik M, Chabal YJ (2013) A review on thermal exfoliation of graphene oxide. J Mater Sci Res 2:101.
24. Wang X-Y, Narita A, Müllen K (2017) Precision synthesis versus bulk-scale fabrication of graphenes. Nat Rev Chem 2:1–10.

25. Mahmoudi T, Wang Y, Hahn Y-B (2018) Graphene and its derivatives for solar cells application. Nano Energy 47:51–65.
26. Sun Z, Fan Q, Zhang M, et al. (2019) Supercritical fluid-facilitated exfoliation and processing of 2D materials. Adv Sci 6:1901084.
27. Smith AT, LaChance AM, Zeng S, et al. (2019) Synthesis, properties, and applications of graphene oxide/reduced graphene oxide and their nanocomposites. Nano Mater Sci 1:31–47.
28. Han Z, Kimouche A, Kalita D, et al. (2014) Homogeneous optical and electronic properties of graphene due to the suppression of multilayer patches during CVD on copper foils. Adv Funct Mater 24:964–970.
29. Chen X, Zhang L, Chen S (2015) Large area CVD growth of graphene. Synth Met 210:95–108.
30. Ismail MS, Yusof N, Yusop MZM, et al. (2019) Synthesis and characterization of graphene derived from rice husks. Malays J Fundam Appl Sci 15:1–6.
31. Overbeck J, Barin GB, Daniels C, et al. (2019) A universal length-dependent vibrational mode in graphene nanoribbons. ACS Nano 13:13083–13091.
32. Tiwari SK, Mishra RK, Ha SK, Huczko A (2018) Evolution of graphene oxide and graphene: from imagination to industrialization. ChemNanoMat 4:598–620.
33. Ovais M, Khalil AT, Ayaz M, Ahmad I (2020) Metal oxide nanoparticles and plants. In: Phytonanotechnology: Challenges Prospect; 123–141. Elsevier. https://doi.org/10.1016/B978-0-12-822348-2.00007-3
34. Todkar BS, Deorukhkar OA, Deshmukh SM (2016) Extraction of silica from rice husk. Int J Eng Res Dev 12:69–74.
35. Perea-Moreno M-A, Manzano-Agugliaro F, Hernandez-Escobedo Q, Perea-Moreno A-J (2020) Sustainable thermal energy generation at universities by using loquat seeds as biofuel. Sustainability 12:2093.
36. Muramatsu H, Kim YA, Yang K, et al. (2014) Rice husk-derived graphene with nano-sized domains and clean edges. Small 10:2766–2770. Wiley-VCH Verlag GmbH & Co. KGaA, Weinheim.
37. Singh C, Tiwari S, Singh JS (2017) Impact of rice husk biochar on nitrogen mineralization and methanotrophs community dynamics in paddy soil. Int J Pure Appl Biosci 5:428–435.
38. Lakshmi SD, Avti PK, Hegde G (2018) Activated carbon nanoparticles from biowaste as new generation antimicrobial agents: a review. Nano-Structures & Nano-Objects 16:306–321.
39. Liou T-H, Wang P-Y (2020) Utilization of rice husk wastes in synthesis of graphene oxide-based carbonaceous nanocomposites. Waste Manag 108:51–61.
40. Luyen NT, Linh HX, Huy TQ (2020) Preparation of rice husk biochar-based magnetic nanocomposite for effective removal of crystal violet. J Electron Mater 49:1142–1149.
41. Allafchian A, Mousavi ZS, Hosseini SS (2019) Application of cress seed musilage magnetic nanocomposites for removal of methylene blue dye from water. Int J Biol Macromol 136:199–208.
42. Sekar S, Aqueel Ahmed AT, Kim DY, Lee S (2020) One-pot synthesized biomass C-Si nanocomposites as an anodic material for high-performance sodium-ion battery. Nanomaterials 10:1728.
43. Ali SH, Emran MY, Gomaa H (2021) Rice husk-derived nanomaterials for potential applications. In: Waste Recycling Technologies for Nanomaterials Manufacturing; pp. 541–588. Springer, Cham.
44. Wang M, Yang J, Liu S, et al. (2020) Nitrogen-doped hierarchically porous carbon nanosheets derived from polymer/graphene oxide hydrogels for high-performance supercapacitors. J Colloid Interface Sci 560:69–76.
45. Nguyen H, Moghadam MJ, Moayedi H (2019) Agricultural wastes preparation, management, and applications in civil engineering: a review. J Mater Cycles Waste Manag 21:1039–1051.
46. Yin C, Tao C, Cai F, et al. (2016) Effects of activation temperature on the deoxygenation, specific surface area and supercapacitor performance of graphene. Carbon N Y 109:558–565.
47. Salman M, Jahan S, Kanwal S, Mansoor F (2019) Recent advances in the application of silica nanostructures for highly improved water treatment: a review. Environ Sci Pollut Res 26:21065–21084.
48. Halim ZAA, Yajid MAM, Hamdan H (2020) Effects of solvent exchange period and heat treatment on physical and chemical properties of rice husk derived silica aerogels. Silicon 13(1):1–7.
49. Chintala V (2018) Production, upgradation and utilization of solar assisted pyrolysis fuels from biomass – a technical review. Renew Sustain Energy Rev 90:120–130.
50. Patil R, Dongre R, Meshram J (2014) Preparation of silica powder from rice husk. J Appl Chem 27:26–29.
51. Narayanan DP, Sankaran S, Narayanan BN (2019) Novel rice husk ash-reduced graphene oxide nanocomposite catalysts for solvent free Biginelli reaction with a statistical approach for the optimization of reaction parameters. Mater Chem Phys 222:63–74.
52. Mishra P, Chakraverty A, Banerjee HD (1985) Production and purification of silicon by calcium reduction of rice-husk white ash. J Mater Sci 20:4387–4391.

53. Larbi KK (2010) Synthesis of high purity silicon from rice husks. (Doctoral dissertation). A thesis submitted in conformity with the requirements for the degree of Master of Applied Science, University of Toronto. P.70
54. Bao Z, Weatherspoon MR, Shian S, et al. (2007) Chemical reduction of three-dimensional silica microassemblies into microporous silicon replicas. Nature 446:172–175.
55. Ahn J, Kim HS, Pyo J, et al. (2016) Variation in crystalline phases: controlling the selectivity between silicon and silicon carbide via magnesiothermic reduction using silica/carbon composites. Chem Mater 28:1526–1536.
56. Kim W-S, Hwa Y, Shin J-H, et al. (2014) Scalable synthesis of silicon nanosheets from sand as an anode for Li-ion batteries. Nanoscale 6:4297–4302.
57. Shen Y (2017) Rice husk silica-derived nanomaterials for battery applications: a literature review. J Agric Food Chem 65:995–1004.
58. Wong DP, Suriyaprabha R, Yuvakumar R, et al. (2014) Binder-free rice husk-based silicon–graphene composite as energy efficient Li-ion battery anodes. J Mater Chem A 2:13437–13441.
59. Al-Adham EK, Hassan AI, Shebl A, Hazem MM (2018) Evaluation of the therapeutic effects of rice husk nanosilica combined with platelet-derived growth factor in hepatic veno-occlusive disease. Biochem Cell Biol 96:682–694.

7 Synthesis of Graphene from Vegetable Waste

R Imran Jafri, Adona Vallattu Soman, Athul Satya, Sourav Melethethil Surendran, and Akshaya S Nair

CONTENTS

7.1 INTRODUCTION

Carbon is a unique and omnipresent material present in nature. It is a non-metallic tetravalent element capable of forming complicated networks, which is the basis of existence of life. Carbon and its allotropes in zero-dimension (fullerene), one-dimension (nanotubes), two-dimensional (graphene), and three-dimension (diamond and graphite) exhibit different chemical, physical, and electronic properties due to the way carbon atoms are attached to one another (Katsnelson 2007). Graphite is one of the oldest known pure forms of carbon. Graphite poses a hexagonal layered structure consisting of six carbons attached in the form of a ring (Reich and Thomsen 2004), and Figure 7.1 depicts graphene's layered structure.

A group of physicists from Manchester University led by Andre Geim and Kostya Novoselov paved the way to the discovery of graphene in 2004 from graphite (Novoselov et al. 2005). Graphene, a wonder material of 21st century, possesses massive strength, high electrical conductivity, transparency, and flexibility (Edwards and Coleman 2013). This innate ability of graphene makes them a suitable material in the bio-medicine field, optical-based device, water treatment applications, fuel

DOI: 10.1201/9781003169741-7

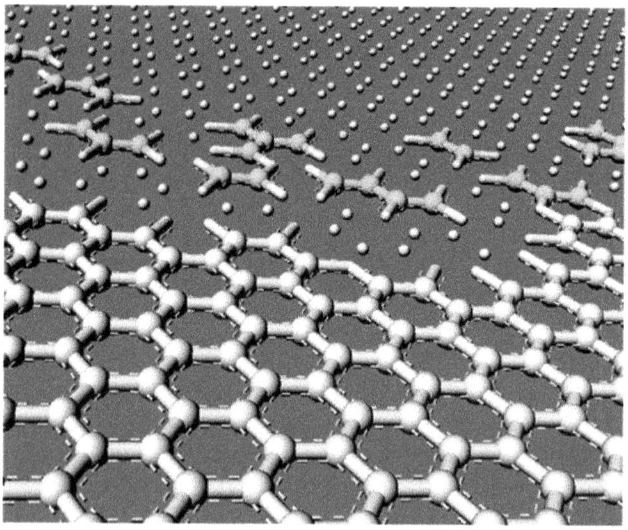

FIGURE 7.1 Hexagonal layered structure of graphene. (Reprinted with permission from Sun et al. (2011). Copyright (2011) American Chemical Society.)

cells, super capacitors, batteries, etc. (Raghavan et al. 2017). This includes applications in the field of bio-medicine, optical-based device, water treatment applications, fuel cells, super capacitors, batteries, etc. (Raghavan et al. 2017). Graphene contains carbon atoms with sp^2 hybridization arranged in a hexagonal honeycomb lattice with p-orbitals above and below the plane of the sheet which are partially filled (Edwards and Coleman 2013). Graphene can be monoatomic or single-layered (one graphitic layer), bilayer (two graphitic layer), tri-layer (three graphitic layer), few-layer graphene (more than 5 layer up to 10 graphitic layer), multilayered graphene or thick graphene or nanocrystalline thin graphite (20–30 layers of graphene) (Bhuyan et al. 2016) as shown in Figure 7.2. There is zero bandgap between the conduction band and the valance band in

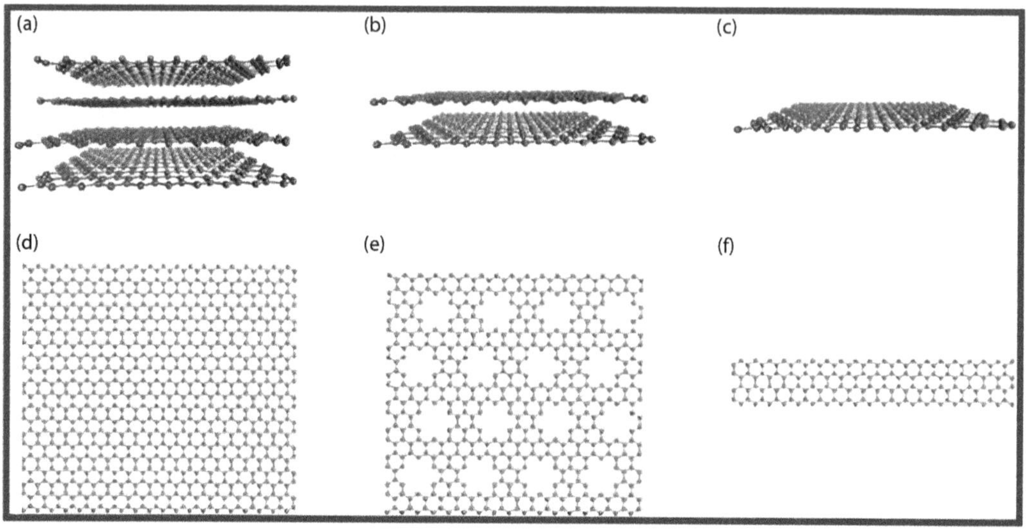

FIGURE 7.2 Different structures of graphene. For more information refer "Graphene chemistry: synthesis and manipulation" (Sun et al. 2011). (Reprinted with permission from Sun et al. (2011). Copyright (2011) American Chemical Society.)

monolayer graphene due to the half-filled p-orbitals, which permits the free electrons. The adjacent graphene layers in layered graphene have weak van der Waals interactions due the p bonds present between them (Yang et al. 2018).

In the late 1970s, thin graphite layers were synthesized by the precipitation of carbon on the transition metal surface (Bhuyan et al. 2016). Graphene was first synthesized in 2004 using a recognized method known as micromechanical cleavage by Novoselov et al. (2005). Using this mechanical exfoliation method, they extracted single-layer graphene flakes. A scotch tape or adhesive tape was used to exfoliate the graphite using micromechanical cleavage technique. The graphite layers were cleaved using this adhesive tape, and thus, graphene was obtained. Micromechanical cleavage is a top-down method. The repeated cleavage of graphite using this method produces monolayer, bilayer, and few-layer graphene (Edwards and Coleman 2013; Novoselov et al. 2005). Graphene is the thinnest and toughest material ever created. The interlayer spacing of graphene with different layers varies such as the interlayer spacing for turbostratic graphene (few-layer graphene with no discernible stacking order) is larger than 0.342 nm, whereas crystalline graphene has an interlayer spacing of 0.335 nm (Edwards and Coleman 2013). Since graphene has a hexagonal lattice, if we are considering a monolayer graphene then it will be having an armchair or zigzag-shaped edges. Due to this different edge shapes, it will have different magnetic and electronic properties. Along with appreciable electrical and thermal properties, it possesses antibacterial properties (Arvidsson et al. 2018). Graphene also has an unusual quantum hall effect (Dong and Chen 2010). Various significant properties of graphene are recorded in Table 7.1.

As earlier mentioned, the remarkable mechanical properties of graphene arise from the stability of sp^2 bonds found in the graphene. Due to this stability of sp^2 bonds, graphene can oppose different types of in-plane deformations. It has a young modulus ~1 TPa. The 2p orbitals which form the π state bands make graphene extremely stiff (Papageorgiou et al. 2017). Since the single-layer graphene is a zero-gap semiconductor or a semimetal, it poses high electrical conductivity and very high electron mobility of around 15,000 $cm^2/(V\ s)$ at room temperature. Due to these properties, graphene is used in nanoelectronics such as ballistic field effect transistor due to its large mean-free paths of charge carriers (Dong and Chen 2010). Moreover, the theoretical capacitance of single-layer graphene is ~21 $\mu F/cm^2$, and the corresponding specific capacitance is ~550 F/g (Ke and Wang 2016). The electrochemical surface area of the film of bilayer graphene flakes was significantly higher than that of the film of trilayer graphene flakes. The basic capacitance of the bilayer graphene film (373 F/g) was around 1.5 times higher than that of the trilayer graphene film (238 F/g) (Mir et al. 2019). Graphene as electrodes show specific energy density of 116 W h/kg while lithiation process, making it a potential candidate to be used as an electrode material in lithium batteries (Wang et al. 2009). The basic capacitance of graphene materials as electrodes for electrochemical double-layer capacitors (EDLCs) has been stated to be 135 F/g in aqueous KOH electrolyte and 117 F/g in aqueous H_2SO_4 electrolytes (Yan et al. 2010). The electrochemical response for different layers of graphene is different. Even though graphene is considered as a nontoxic material like carbon, under certain conditions, it behaves as a toxic or as a risk material (Randviir et al. 2014).

TABLE 7.1
Properties of Graphene

Properties	Value	Reference
Electron mobility	15,000 $cm^2/(V\ s)$	Dong and Chen (2010)
Optical transmittance	~97.7%	Bhuyan et al. (2016)
Thermal conductivity	~5000 W/(m k)	Edwards and Coleman (2013)
Young's modulus	~1 TPa	Edwards and Coleman (2013)
Surface area	2630 m^2/g	Bhuyan et al. (2016)

7.2 CONVENTIONAL GRAPHENE SYNTHESIS METHODS

Fabricating or extracting graphene of desired size, shape, and purity relies on its synthesis route (Bhuyan et al. 2016). Conventionally, graphene can be fabricated by two techniques: top-down and bottom-up. In top-down method, stacked-layered graphite is broken down to form a single graphene sheet, whereas in bottom-up method, graphene is synthesized from alternate carbon containing sources (Edwards and Coleman 2013).

7.2.1 Top-Down Method

The desired materials which are in the bulk form are broken down into nano-sized particles in the top-down method. There are serious restrictions in this method due to the surface imperfections in the product (Yadi et al. 2018). Some of the top-down approaches are given below.

7.2.1.1 Mechanical Exfoliation

It is referred as the first recognized method for the synthesis of graphene. Since graphite is layered structure of monolayered graphene, which are stacked together by van der Waals forces, it can be synthesized by applying force. Since the interlayer distance of graphene is 0.34 nm and the interlayer bond energy is 2 eV/nm^2, an external force of ~300 $nN/\mu m^2$ is needed for mechanical cleavage. Graphene sheets of different thickness can be synthesized by this method. For exfoliation, different agents can be used such as scotch tape, ultrasonication, electric field, etc. (Bhuyan et al. 2016).

7.2.1.2 ARC Discharge

It is a traditional method for synthesis of fullerene and carbon nanotubes (CNT), which is an extremely reliable method for graphene synthesis too (Lee et al. 2019). Synthesis of graphene from graphite electrode through arc discharge is similar to that of synthesis of carbon nanotube. The produced graphene looks like a black powder since it consists of nanoflakes and nanosheets. For pure graphene, arc discharge is usually performed at high pressures of hydrogen + helium. Unlike other methods, this method does not require substrate or catalyst. The arc discharge method can also be used to make doped graphene by incorporating nitrogen and boron-containing gases into discharge atmospheres alongside hydrogen or by using ammonia or boron-stuffed graphite electrodes (Jariwala et al. 2011).

7.2.1.3 Chemical Exfoliation

In this method, graphene exfoliation from graphitic layers takes place by adding solvents directly or by oxidation process; the oxygen-containing group increases the separation between the graphite layers, lowering the van der Waals forces between the layers. Reducing solvents such as hydrazine hydrate, *N*-methyl-2-pyrrolidone, methanesulfonic acid and ionic solvents are used to exfoliate graphene from graphite (Farjadian et al. 2020).

7.2.1.4 Unzipping CNT Method

Multi-walled CNTs (MWCNT) consist of multiple rolled layers (concentric tubes) of graphene. Unzipping is a technique that cut MWCNT crossway to form graphene (Rangel et al. 2009). The procedure is divided into two chemical steps: the initial of which includes multistage oxidation to unzip the CNT. Following that, a reduction step is conducted in order to obtain the final graphene nanosheets (GNS). During the oxidation process, the oxidant substance was optimized and adjusted with a longer curing period. The purpose of this modification is to reduce the oxygen-functional groups at the edges of graphene basal planes, reducing the material's electrical conductivity (Al-Tamimi et al. 2018). Cooper et al. (2012) immersed MWCNT in sulfuric acid and then treated with $KMnO_4$ to cut it longitudinally. The oxidized graphene nanoribbons were subsequently reduced by chemical processes. Due to the existence of the oxygen defect sites, the fabricated nanoribbons have poorer electronic properties (Cooper et al. 2012).

7.2.1.5 Electrochemical

A two-step process was used by Alanyalıoğlu et al. (2012) to synthesize graphene having thickness near to monolayer graphene. This method consists of electrochemical intercalation of sodium dodecyl sulfate (SDS) into graphite and electrochemical exfoliation of the SDS intercalated graphite electrode using electric current as oxidizing and reducing agents (RAs). The electric potential for the electrochemical exfoliation has a significant impact on the produced graphene sheets. For the synthesis of monolayer graphene, a strong interaction potential is needed. This method has advantage that there will not be a restacking of the produced individual graphene films due to the surfactant's adsorption in the surface of the graphene films (Alanyalıoğlu et al. 2012).

7.2.2 Bottom-Up Method

The bottom-up technique is another alternative of obtaining graphene. Major kinds of bottom-up methods are chemical vapor deposition (CVD), epitaxial method, and pyrolysis (Bhuyan et al. 2016).

7.2.2.1 Chemical Vapor Deposition

In this process, graphene is made by high-temperature pyrolysis of carbon-holding gases. This technique is commonly used to produce graphene films on transition metal substrates. This method is done through a surface catalyst or segregation method based on metal type. In the surface catalyst method, the growth is found at the surface and mainly known as self-limiting to monolayer graphene and in the segregation method, the amount of graphene layers created depends on several facts such as the quantity of C dissolved and the degree of cooling (Edwards and Coleman 2013). The first method of graphene synthesis by CVD is done using camphor as a precursor material (Bhuyan et al. 2016). Latest milestone in the development of graphene by CVD method is the replicability of great quality graphene on centimeter scale and efficient transition to several other substrates (Choi et al. 2010).

7.2.2.2 Epitaxial Method

The most lauded method among the synthesis of graphene is the epitaxial method. Epitaxial growth is the process of depositing a single crystalline layer on a single crystalline substrate to generate an epitaxial film. It produces high-crystalline graphene on single-crystalline substrates. Based on the substrate, there are two types of epitaxial methods, heteroepitaxial method, and homoepitaxial method. When a film is deposited on a substrate of the similar type, it is defined as a homoepitaxial layer; if the film and substrate are made of dissimilar components, it is defined as a heteroepitaxial layer. Epitaxial graphene growth has been illustrated as a much more suitable strategy for the processing and wide scale making of graphene for electronic applications (Bhuyan et al. 2016).

7.2.2.3 Spray Pyrolysis

Spray pyrolysis is a technique for depositing thin films by spraying the precursor onto a heated surface where the chemicals react to form chemical compounds. Spray pyrolysis of a mixture of iron pentacarbonyl ($Fe(CO)_5$) and pyridine(C_5H_5N), where carbon atoms from C_5H_5N are coated on iron nanoparticles that operate as a substrate, can yield greater quality GNS with a well-developed graphitic structure. The carbonyl and pyridine are mixed and pyrolyzed. The reactive solutions were fed into a quartz reactor from the upper part, while samples were collected constantly from the

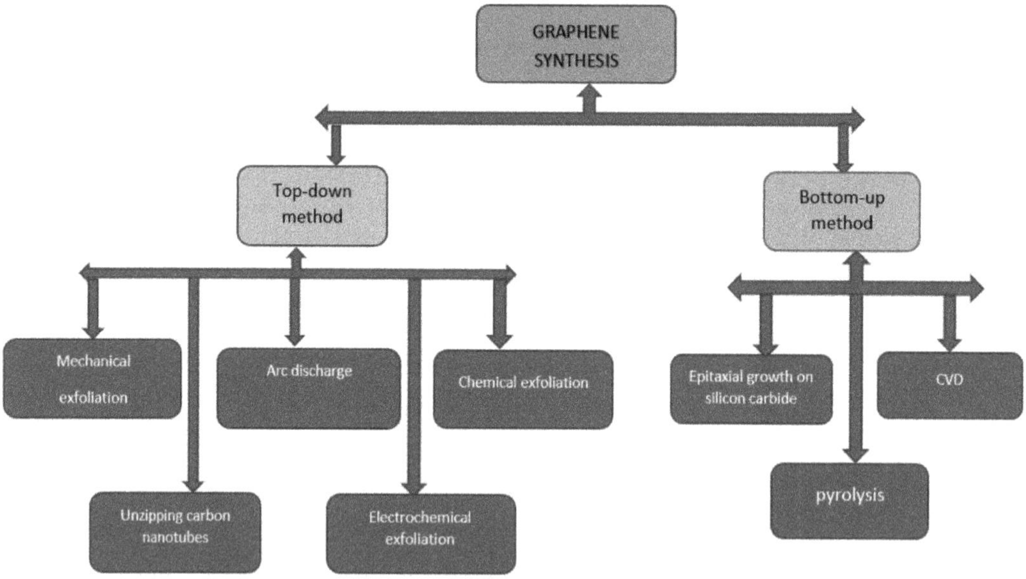

FIGURE 7.3 Flowchart depicting conventional synthesis method.

lower part. After eliminating the Fe substrates, GNS with multiple graphitic layers can be formed (Zou et al. 2015). Figure 7.3 depicts a diagram of the conventional synthesis procedure.

Graphene can be synthesized using various sources containing carbon as precursors. The need for a carbon precursor which is eco-friendly and easily available leads to the investigation of vegetable waste as a carbon precursor. The next section deals with the discussion of synthesis of graphene using various vegetable wastes as precursors.

7.3 SYNTHESIS OF GRAPHENE FROM VEGETABLE WASTE AS A PRECURSOR

Nowadays, synthesizing graphene in an inexpensive way from sustainable and biomass-related sources is in arena. The need for preparing graphene from renewable natural material is to reduce the cost, the amount of toxic materials released during graphene synthesis when conventional methods are employed (Kumar et al. 2016). Graphene can be synthesized from vegetable waste such as sesame oil, coconut shell, castor oil, and *Colocasia esculenta* leaves. Hydrocarbon precursors produced naturally are cheap, easily available, and less expensive (Kumar et al. 2016). The main aim behind using plant-based precursors as a carbon source is due to its sustainable features. A few among them have been exploited for graphene synthesis as shown in Figure 7.4. These plant-based precursors are having cheap raw materials, and they are renewable (Azmina et al. 2012).

7.3.1 Synthesis of GNS and Rectangular Aligned CNTs (RA-CNTs) Using Sesame Oil

GNS and RA-CNTs were synthesized using sesame oil as a carbon source by Kumar et al. (2014). Sesame oil is derived from sesame seeds which is used as an edible vegetable oil (Kumar et al. 2014). GNS and RA-CNTs were synthesized by spray pyrolysis method assisted by CVD. The experimental set up consists of a precursor containing a mixture of ferrocene in sesame oil at a concentration of 20 mg/ml. There was a quartz tube containing Si/SiO_2 substrate placed inside the reaction furnace. The whole experimental set up was conducted at a temperature range of 750°C–850°C. The deposition of RA-CNTs and GNSs takes place at the Si/SiO_2 substrate after a time period of 20 minutes on cooling. The temperature gradient plays an important role in determining whether the synthesized

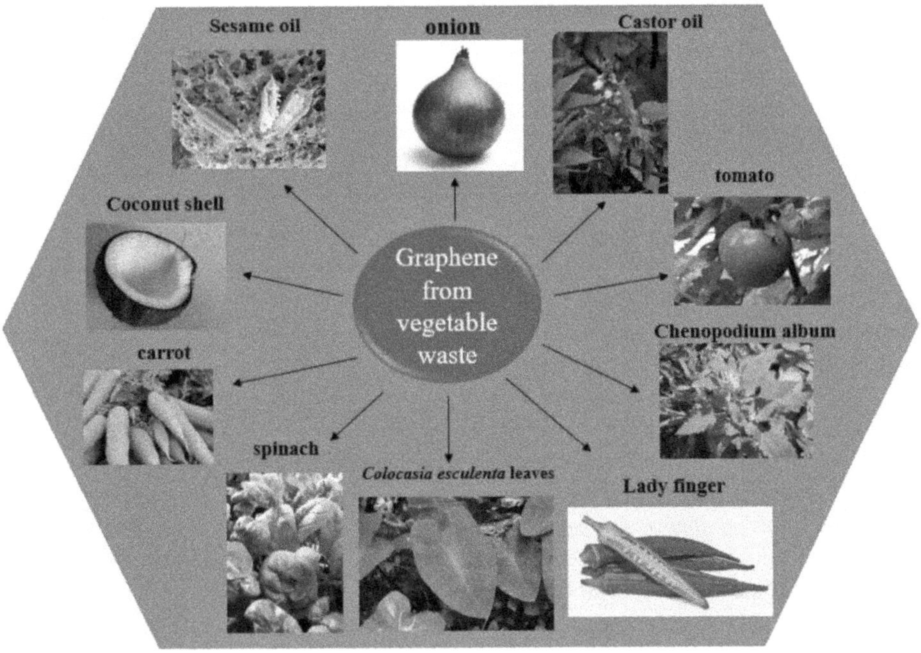

FIGURE 7.4 Depiction of various vegetable wastes explored for graphene synthesis.

material is RA-CNTs or graphene. Using sesame oil as a precursor helps in production of a high yield better quality of CNTs bundles (Kumar et al. 2014). The X-ray diffraction (XRD) indicated that the interlayer separation for both GNSs and RA-CNTs are almost similar, which is in a range of 0.35 nm. The results of the Raman spectrum done at 800°C by Kumar et al. (2014) indicated that both GNSs and RA-CNTs had fewer plane defects. Also, the results showed that both RA-CNTs and GNSs synthesized using sesame oil have well crystalline structure. The Fourier-transform infrared spectroscopy (FTIR) data revealed absence of oxygen containing functional group in GNSs. Also, RA-CNTs have both carboxyl C–O and aromatic C=C stretching vibrations. Thermogravimetric analysis results showed that a significant weight loss of ~20 wt% was observed at 750°C and 850°C and above 350°C for RA-CNTs due to amorphous carbon oxidation. Amorphous carbon was developed along with RA-CNTs. Also, the results indicated that GNSs has less thermal stability than that of RA-CNTs (Kumar et al. 2014).

7.3.2 Synthesis of Porous GNS (PGNS) Using Coconut Shell

PGNS can be synthesized using coconut shells as a carbon precursor through SAG (simultaneous activation-graphitization) route as done by Sun et al. (2013). In this method, $FeCl_3$, $ZnCl_3$, and coconut shell were used to synthesis PGNS. An amount of 9 g of $ZnCl_3$ and 3 g of coconut shell were mixed in 50 ml of 3 M ferric tri-chloride ($FeCl_3$) solution. The carbon precursor was obtained from this mixture by heating under 100°C in a standard oven. In this process, $FeCl_3$ was used as a catalyst that can form carbides on reacting with carbon atoms. Graphene was formed by the decomposition of these carbides (Xia et al. 2018). The XRD analysis of the sample done by Sun et al. (2013) showed that the resultant PGNS has a high crystallinity. The results of Raman spectroscopy showed that I_G/I_D value was 3.98 (relative intensity between the G-band and D-band) which suggests that PGNS have a high graphitization degree. The spectrum of PGNS obtained in X-ray photoelectron spectroscopy analysis showed no trace of impurities in PGNS (Sun et al. 2013). PGNSs are used to make supercapacitors because of their superior capacitive properties as mentioned by Sun et al. (2013).

Surface area (187 m^2/g) of PGNS materials was appropriate for the electrolyte penetration and ion diffusion. This provides an extraordinary capability for ion transport within electrode materials made of PGNSs. Also, PGNs show a specific capacitance of 268 F/g at 1 Ag^{-1}. These properties make PGNSs appropriate for making supercapacitors (Sun et al. 2013).

7.3.3 Synthesis of CNTs and Graphite Nanofibers (GNFs) Using Castor Oil

CNTs and GNFs were prepared using castor oil by Awasthi et al. (2011). CNTs were prepared using spray pyrolysis method and GNFs were prepared using Catalytic thermal decomposition by using hydrocarbon gas like C_2H_2. CNT-based nanocomposites are becoming more common viable substitute to traditional smart devices materials due to their greater sensitivity and sound good conductive peculiarities. Graphitic nanophase (crystallite size 3.6 nm) and nickel were confirmed by XRD analysis. The GNFs recorded d-spacing was ca. 0.342 nm, similar to that of ideal graphite d value of 0.335 nm. Internal structure analysis reveals the planar GNFs are obtained at 600°C and helical GNFs at 700°C. Brunauer-Emmett-Teller analysis study reveals GNFs carry large surface area (207 m^2/g). At room temperature conductivity is observed as 1.05×10^{-7} S/cm (Awasthi et al. 2011).

The demerits of using chemical RAs can be overcome by using green RAs. The plant extracts which are rich in amino acids, polysaccharides, alkaloids, vitamins, proteins, enzymes, etc. can effectively reduce graphene oxide (Rai et al. 2020).

7.3.4 Synthesis of Graphene from *Colocasia Esculenta* Leaves

Mohan and Balachandran (2020) were successfully able to synthesis graphene from *C. esculenta.* They synthesized graphene using modified Hummer's method. $SnCl_2{\cdot}2H_2O$ was used as a precursor for tin oxide. During the XRD analysis, they observed broad peaks which stipulate the smaller size of crystalline nanomaterial formed. The FTIR analysis proved the presence of oxygen rich surface states in the nanomaterial synthesized. Raman spectra analysis showed the existence of defect filled sp^2 carbon which elevates the properties of graphenic carbon. The synthesized graphene had high antibacterial potential. The tin oxide increases the surface-to-volume ratio which helps nanoparticle to resist the degradation. The reactive oxygen species generation can be activated by SnO_2 with release of metal ions and piercing through cell membrane to the death of cell. The synthesized graphene had a robust bactericidal potential over natural illumination. The minimum inhibitory concentration value of synthesized graphene was recorded as 125 μg/ml, which is reminiscent to the antibiotic drug cephalexin. From the antibacterial analysis, the synthesized graphene is effective in annihilating *Pseudomonas aeruginosa* (Mohan and Balachandran 2020).

The use of plant extracts as an RA for the production of graphene by reducing GO (graphite oxide) produces graphene that is rich in carbon content. Also, using plant extract as RA would reduce toxicity which arises when chemicals like hydrazine etc. are used (Ismail 2019).

7.4 VEGETABLE WASTE AS A REDUCING AGENT (RA) FOR THE PRODUCTION OF GRAPHENE

Vegetable waste from carrot, tomato, onion, spinach, *Chenopodium album*, and lady's finger can be used as an RA to synthesis graphene by reducing GO.

Carrot: Graphene can be synthesized in a bulk amount using reduced graphene oxide (RGO). The processing is done in order to reduce the oxygen content in GO, so that it can be used to produce graphene more economically. Vegetable waste such as carrot extract can be used to reduce GO. GO is usually prepared using Hummer's method. Carrot extracts are prepared to reduce the GO. RGO can be used to synthesize graphene easily by a less expensive method. The complete reduction of GO to RGO was affirmed by the analysis of UV-visible spectra of GO and carrot extract RGO (Ct-RGO) done by Vusa et al. (2014).The value of max from 230 nm (at t = 0 min) to 270 nm (at t = 60 min) confirmed the change of GO to graphene. The Raman spectroscopy analysis showed that the I_D/I_G ratio of Ct-RGO (1.198) was higher than the I_D/I_G ratio of GO (0.979). The higher value of I_D/I_G ratio of Ct-RGO was due to the formation of unpaired defects as a result of eliminating oxygen functionalities and due to sp^2 network restoration. The Ct-RGO synthesized by Vusa et al. (2014) was considered to be a bio-compatible product and scientists are trying to find different fields like biosensors, drug delivery, and bio-composites where this can be effectively used (Vusa et al. 2014).

Tomato: GO prepared by Hummer's method was used to synthesize graphene using tomato juice by Almaadeed and Patan (2017). GO is dissolved in a solvent (deionized water or distilled water). The RA that is used to reduce the suspension was tomato juice. The XRD analysis done by Almaadeed and Patan (2017) shows that in the case of GO, the XRD spectra showed a peak at 10.82°. After the addition of tomato juice, the peak at 10.82° vanished, but broad peak appeared around 22°–24°, indicating the presence of few-layers graphene. The Raman spectrum of graphene done by Almaadeed and Patan (2017) had both G and D bands at 1578 cm^{-1} and 1348.3 cm^{-1}. In addition to this, at 2700 cm^{-1}, a new band also appeared which is referred to as the 2D band. The I_{2D}/I_G less than 1 intensity ratio of graphene prepared by Almaadeed and Patan (2017) using tomato juice indicated that it's a few layers graphene (Almaadeed and Patan 2017).

Onion: In order to synthesis graphene from GO, onion (*Allium cepa*) extracts were used as an RA by Khanam and Hasan (2019). GO was prepared by modified Hummer's method. The characteristic studies of the graphene done using XRD indicated that GO showed a peak at 10.82°, and in the case of graphene, there was no peak at 10.82° (Khanam and Hasan 2019). This confirmed the complete reduction of GO to graphene. The Raman spectroscopy analysis of graphene confirmed that the G bands which occurs due to the scattering of E_{2g} phonons of sp^2 atoms appears at 1585 cm^{-1} and the D band appear at 1355 cm^{-1} (Khanam and Hasan 2019). The new peak at 2700 cm^{-1} of the graphene spectrum indicated the 2D band. The I_{2D}/I_G was found to be less than 1, which indicated the multilayered structure of graphene (Khanam and Hasan 2019).

Spinach: Due to the exciting properties, synthesis of graphene has become one among the breakthroughs during recent years. Spinach, a leafy vegetable belonging to *Amaranthus* family, was used to synthesize graphene by Suresh et al. (2015). GO was prepared using hummers method. The graphite powder diffraction peak appeared at 26°, which leads to interlayer distance of 0.33 nm. Due to the (0 0 2) plane, the diffraction peak of GO was recorded at 10°, with an interlayer distance of 0.76 nm. Then the produced GO was reduced by the introduction of spinach extract. Diffraction peak shifted from 10° to 26° due to formation of RGO from GO. The absorption peak of GO was obtained at 225 nm caused by π-π^* transition of aromatic C=C bond which shifted to 282 nm after reduction, when analyzed using UV-visible spectroscopy (Suresh et al. 2015).

***Chenopodium album*:** *C. album* is commonly well-known as "bathua". It is a green leaf with short life span because of its high moisture content. *C. album* contains high-grade vitamins, proteins, nutrients, and antioxidants. Umar et al. (2020) has shown that *C. album* can be utilized as an excellent reducing and stabilizing agent while preparing reduced graphene. The UV spectroscopy showed the absorbance peak at 263 nm. Conversion of GO to RGO is done by vitamin C and supplementary antioxidants existing in the *C. album*. In XRD analysis, a diffused peak band 2θ = 22.50° is observed. In FTIR analysis, a peak at 1570 cm^{-1} because of the elongation of C=C and a broad peak at 1730 cm^{-1} because of the C-C stretching was exhibited. The reduced graphene formed have antibacterial (Gram-positive and Gram-negative) and antibiofilm characteristics. Synthesized graphene has also showed anti-breast cancer activity (Umar et al. 2020).

Lady's finger: *Abelmoschus esculentus* (lady's finger) vegetable extract was employed as both a reducing and stabilizing agent for GO. The *Abelmoschus* extract was prepared by boiling 100 ml of deuterium-depleted water (DDW) and 20.0 g of lady's finger. The characterization studies were done using UV-visible spectroscopy, where the RGO formation was affirmed by the absorption

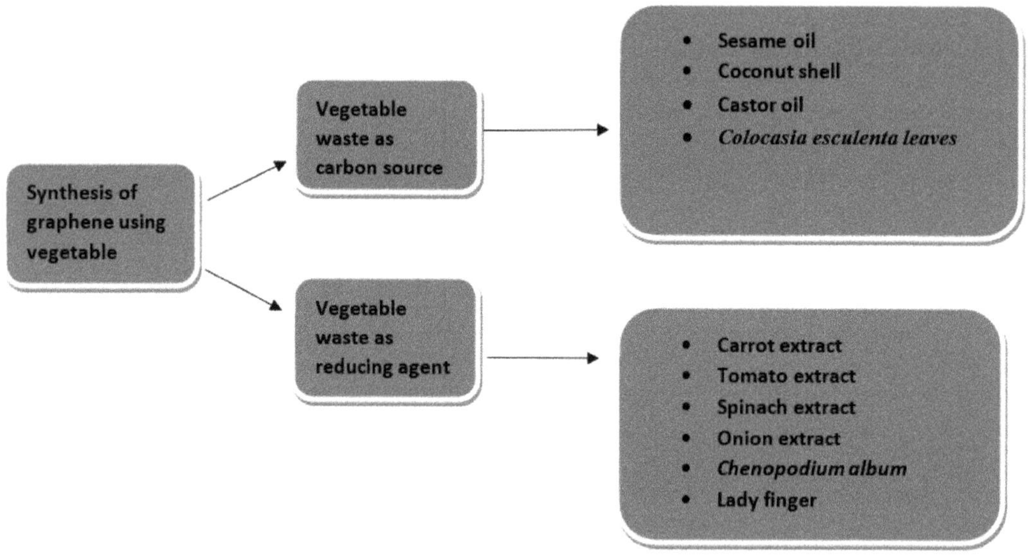

FIGURE 7.5 Classification of vegetable wastes based on its utility as carbon precursors or RA for graphene synthesis.

peak at 263 nm. Also, the vanishing hump in the UV-visible absorption spectra near 298 nm also confirmed the formation of RGO (Gnanaprakasam and Selvaraju 2014).

As evident from the literature mentioned in Sections 7.3 and 7.4, vegetable waste classified based on its use as a precursor or as a reducing agent is depicted in Figure 7.5.

7.5 CONCLUSION

Graphene is evolving into extremely proficient nanomaterials because of its extraordinary properties. It's a good conductor of heat and electricity, mechanically strong, and optically transparent. Yet graphene can be synthesized using many ways, the synthesis of graphene using natural precursors have lot more advantages. Graphene synthesized using vegetable waste is eco-friendly and cost effective. Natural carbon precursors such as vegetable waste are easily available. Recent literature reports suggest that vegetables such as sesame oil, coconut shells, castor oil, and *C. esculenta* leaves are the sources from which graphene had been easily synthesized. Also, some vegetables such as carrot, tomato, onion, spinach, *C. album* and lady's finger stood promising as an RA for graphene synthesis. In the case of GNFs prepared using castor oil, the GNFs recorded interplanar spacing ca. 0.342 nm, similar to that of ideal graphite interplanar spacing of 0.335 nm (Awasthi et al. 2011). XRD analysis of the graphene sample prepared using coconut shell by Sun et al. (2013) showed that the resultant PGNS had high crystallinity. The results of Raman spectroscopy showed that I_G/I_D value of graphene synthesized using coconut shell was 3.98 which suggests that PGNS synthesized using coconut shell have a high graphitization degree (Sun et al. 2013). Even though the research in this field is in infancy, graphene made out of natural precursors have the potential to overcome the existing barrier of expensive production cost and can also be used in electronic devices such as super capacitors and others effectively.

REFERENCES

Alanyalıoğlu, M., Segura, J. J., Oró-Solè, J. et al. 2012. The synthesis of graphene sheets with controlled thickness and order using surfactant-assisted electrochemical processes. *Carbon* 50:142–152. doi:10.1016/j.carbon.2011.07.064

Almaadeed, M. A. A., and Patan, N. K. 2017. Green synthesis of graphene by using tomato Juice. Qatar University. 1–5. https://patents.google.com/patent/US20170320739A1/en

Al-Tamimi, B. H., Farid, S. B. H., and Chyad, F. A. 2018. Modified unzipping technique to prepare graphene nano-sheets. *J. Phys. Conf. Ser.* 1003:012020. doi: 10.1088/1742-6596/1003/1/012020

Arvidsson, R., Boholm, M., Johansson, M. et al. 2018. Just carbon: Ideas about graphene risks by graphene researchers and innovation advisors. *Nanoethics* 12:199–210. doi:10.1007/s11569-018-0324-y

Awasthi, K., Kumar, R., Raghubanshi, H. et al. 2011. Synthesis of nano-carbon (nanotubes, nanofibres, graphene) materials. *Bull. Mater. Sci.* 34:607–614. doi:10.1007/s12034-011-0170-9

Azmina, M. S., Suriani, A. B., Salina, M. et al. 2012. Variety of bio-hydrocarbon precursors for the synthesis of carbon nanotubes. *Nano Hybrids* 2:43–63. doi:10.4028/www.scientific.net/nh.2.43

Bhuyan, M. S. A., Uddin, M. N., Islam, M. M. et al. 2016. Synthesis of graphene. *Int. Nano Lett.* 6:65–83. doi: 10.1007/s40089-015-0176-1

Choi, W., Lahiri, I., Seelaboyina, R. et al. 2010. Synthesis of graphene and its applications: A review. *Crit. Rev. Solid State* 35:52–71. doi:10.1080/10408430903505036

Cooper, D. R., D'Anjou, B., Ghattamaneni, N. et al. 2012. Experimental review of graphene. *Int. Scholarly Res. Not.* 2012:1–57. doi:10.5402/2012/501686

Dong, L.-X., and Chen, Q. 2010. Properties, synthesis, and characterization of graphene. *Front. Mater. Sci.* 4:45–51. doi:10.1007/s11706-010-0014-3

Edwards, R. S., and Coleman, K. S. 2013. Graphene synthesis: Relationship to applications. *Nanoscale* 1:38–51. doi:10.1039/c2nr32629a

Farjadian, F., Abbaspour, S., Sadatlu, M. A. A. et al. 2020. Recent developments in graphene and graphene oxide: Properties, synthesis, and modifications: A review. *ChemistrySelect* 5:10200–10219. doi:10.1002/slct.202002501

Gnanaprakasam, P., and Selvaraju, T. 2014. Green synthesis of self assembled silver nanowire decorated reduced graphene oxide for efficient nitroarene reduction. *RSC Adv.* 4:24158–24525. doi:10.1039/c4ra01798f

Ismail, Z. 2019 Green reduction of graphene oxide by plant extracts: A short review. *Ceram* 45:23857–23868. doi:10.1016/j.ceramint.2019.08.114

Jariwala, D., Srivastava, A., and Ajayan, P. M. 2011. Graphene synthesis and band gap opening. *J. Nanosci.* 11:6621–6641. doi:10.1166/jnn.2011.5001

Katsnelson, M. I. 2007. Graphene: Carbon in two dimensions. *Mater. Today* 10:20–27. doi:10.1016/S1369-7021(06)71788-6

Ke, Q., and Wang, J. 2016. Graphene-based materials for supercapacitor electrodes–A review. *J. Materiomics* 2:37–54. doi:10.1016/j.jmat.2016.01.001

Khanam, P., and Hasan, A. 2019. Biosynthesis and characterization of graphene by using non-toxic RA from *Allium cepa* extract: Anti-bacterial properties. *Int. J. Biol. Macromol.* 126:151–158. doi: 10.1016/j.ijbiomac.2018.12.213

Kumar, R., Singh, R. K., Kumar, P. et al. 2014. Clean and efficient synthesis of graphene nanosheets and rectangular aligned-carbon nanotubes bundles using green botanical hydrocarbon precursor: Sesame oil. *Sci. Adv. Mater.* 6:76–83. doi:10.1166/sam.2014.1682

Kumar, R., Singh, R. K., and Singh, D. P. 2016. Natural and waste hydrocarbon precursors for the synthesis of carbon-based nanomaterials: Graphene and CNTs. *Renew. Sustain. Energy Rev.* 58:976–1006. doi:10.1016/j.rser.2015.12.120

Lee, X. J., Hiew, B. Y. Z., Lai, K. C. et al. 2019. Review on graphene and its derivatives: Synthesis methods and potential industrial implementation. *J. Taiwan Inst. Chem. Eng.* 98:163–180. doi:10.1016/j.jtice.2018.10.028

Mir, A., Abhilesh, G. N., Tamgadge, R. M. et al. 2019. Capacitance of graphene films: Effect of the number of layers of the constituent graphene flakes. *J. Solid State Electrochem.* 23:2281–2290. doi:10.1007/s10008-019-04344-z

Mohan, A., and Balachandran, M. 2020. Extraction of graphene nanostructures from *Colocasia esculenta* and *Nelumbo nucifera* leaves and surface functionalization with tin oxide: Evaluation of their antibacterial properties. *Chem. Eur. J.* 26:8105–8114. doi:10.1002/chem.202000590

Novoselov, K. S., Geim, A. K., Morozov, S. V. et al. 2005. Two-dimensional gas of massless Dirac fermions in graphene. *Nature* 438:197–200. doi:10.1038/nature04233

Papageorgiou, D. G., Kinloch, I. A., and Young, R. J. 2017. Mechanical properties of graphene and graphene-based nanocomposites. *Prog. Mater. Sci.* 90:75–127. doi:10.1016/j.pmatsci.2017.07.004

Raghavan, N., Thangavel, S., and Venugopal, G. 2017. A short review on preparation of graphene from waste and bio precursors. *Appl. Mater. Today* 06:246–254. doi:10.1016/j.apmt.2017.04.005

Rai, S., Bhujel, R., Biswas J. et al. 2020. A green approach for synthesis of graphene. *AIP Conf. Proc.* 2273:050032. doi:10.1063/5.0024237

Randviir, E. P., Brownson, D. A. C., and Banks, C. E. 2014. A decade of graphene research: Production, applications and outlook.*Mater. Today Commun.* 17:426–432. doi:10.1016/j.mattod.2014.06.001

Rangel, N. L., Sotelo, J. C., and Seminario, J. et al. 2009. Mechanism of carbon nanotubes unzipping into graphene ribbons.*J. Chem. Phys.* 131:031105. doi:10.1063/1.3170926

Reich, S., and Thomsen, C. 2004. Raman spectroscopy of graphite. *Philos. Trans. A Math. Phys. Eng. Sci.* 362:2271–2288. doi:10.1098/rsta.2004.1454

Sun, Z., James, D. K., and Tour, J. M. 2011. Graphene chemistry: Synthesis and manipulation. *J. Phys. Chem. Lett.* 2:2425–2432. doi:10.1021/jz201000a

Sun, L., Tian, C., Li, M. et al. 2013. From coconut shell to porous graphene-like nanosheets for high-power supercapacitors. *J. Mater. Chem.* 1:6462–6470. doi:10.1039/c3ta10897j

Suresh, D., Nethravathi, P. C., Udayabhanu et al. 2015. Spinach assisted green reduction of graphene oxide and its antioxidant and dye absorption properties. *Ceram* 41:4810–4813. doi:10.1016/j.ceramint.2014.12.036

Umar, M. F., Ahmad, F., Saeed, H. et al. 2020. Bio-mediated synthesis of reduced graphene oxide nanoparticles from *Chenopodium album*: Their antimicrobial and anticancer activities. *Nanomaterials* 10:1096(1–14). doi:10.3390/nano10061096

Vusa, C. S. R., Berchmans, S., and Alwarappan, S. 2014. Facile and green synthesis of graphene. *RSC Adv.* 43:22470–22475. doi:10.1039/c4ra01718h

Wang, C., Li, D., Too, C. O., and Wallace, G. G. 2009. Electrochemical properties of graphene paper electrodes used in lithium batteries. *Chem. Mater.* 21:2604–2606. doi:10.1021/cm900764n

Xia, J., Zhang, N., Chong, S. et al. 2018. Three-dimensional porous graphene-like sheets synthesized from biocarbon via low-temperature graphitization for supercapacitor. *Green Chem.* 20:694–700. doi:10.1039/c7gc03426a

Yadi, M., Mostafavi, E., Saleh, B. et al. 2018. Current developments in green synthesis of metallic nanoparticles using plant extracts: A review. *Artif. Cell Nanomed. Biotechnol.* 46:1–8. doi:10.1080/21691401.2018.1492931

Yan, J., Wei, T., Shao, B. et al. 2010. Electrochemical properties of graphene nanosheet/carbon black composites as electrodes for supercapacitors. *Carbon* 48:1731–1737. doi:10.1016/j.carbon.2010.01.014

Yang, G., Li, L., Lee, W. B. et al. 2018. Structure of graphene and its disorders: A review. *Sci. Technol. Adv. Mater.* 1:613–648. doi:10.1080/14686996.2018.1494493

Zou, B., Wang, X. X., Huang, X. X. et al. 2015. Continuous synthesis of graphene sheets by spray pyrolysis and their use as catalysts for fuel cells. *ChemComm* 51:741–744. doi:10.1039/c4cc08197h

ABBREVIATIONS

Ct-RGO	carrot extract reduced graphene oxide
RA-CNT	rectangular aligned carbon nanotubes
GO	graphite oxide
GNS	graphene nanosheets
PGNS	porous graphene nanosheets
FTIR	Fourier-transform infrared spectroscopy
XRD	X-ray diffraction
RGO	reduced graphene oxide
GNFs	graphite nanofibers
RAs	reducing agents

8 Graphene Oxide from Natural Products and Its Applications in the Agriculture and Food Industry

Seyyed Sasan Mousavi and Akbar Karami

CONTENTS

8.1 INTRODUCTION

Graphene, a two-dimensional carbon framework, has presented good potential in different usages because of its special construction and properties (Geim, 2009; Geim and Novoselov, 2010). Graphene has been generated using a variety of methods, the most common of which are mechanical or ultrasonic exfoliation (Hernandez et al., 2008), epitaxial growth (Sutter et al., 2008), electric arc discharge (Sun et al., 2010), chemical intercalation (Malik et al., 2010), and thermal/chemical reduction of graphene oxide (GO) (Hong et al., 2013; Hu et al., 2013; Mao et al., 2012).

DOI: 10.1201/9781003169741-8

Among the above procedures, the chemical reduction of GO is known as a flexible and cost-effective process for producing graphene in large amounts. Unfortunately, a significant number of commonly used reduction factors, including hydrazine hydrate (HH) and sodium borohydride (Shin et al., 2009), are poisonous and/or explosive (Stankovich et al., 2007). As a result, ongoing efforts have been focused on the production and expansion of environmentally friendly reducing agents for GO reduction. Benjamin Brodie discovered graphite oxide in 1859 (Galande et al., 2014).

Graphite oxide comprises a few layers of graphene with oxygen-containing functional groups, including hydroxy, epoxy, and carboxylic acids. GO is a single graphene layer that results from complete exfoliation of graphite oxide (He et al., 1996). At first, GO was intensely examined as a precursor to large-scale graphene processing to produce a graphene-like substance known as reduced graphene oxide (rGO). The layered structure of graphite oxide is similar to that of graphite, but its interlayer gap is more and enables the atomic-thick structures to hydrophilize (Novoselov et al., 2004). The oxidation of graphite crystals produces GO, which is inexpensive and abundant (Stergiou et al., 2014). GO has a hexagonal carbon framework similar to graphene (Freudenberg and Neish, 1968). Apart from the easy synthesis process, these oxygenated groups have several benefits over graphene, such as higher solubility and the ability to surface functionalize, all of which have led to a slew of applications in nanocomposite materials (Smith et al., 2019). GO has two main aspects: (i) it could be formed through graphite by utilizing low-cost chemical processes with a good yield, and (ii) it is very hydrophilic and could make stable aqueous suspensions by easy and inexpensive methods (Ray, 2015).

Recently, researchers have found that GO, as a hot topic, has great features (Stankovich et al., 2007). According to the need of various fields, it could be utilized as catalytic active centers for covalent/non-covalent modification design. Furthermore, the presence of oxygen-containing groups expands the GO interlayer gap. Small molecules or polymer intercalations may be used to functionalize it. The functionalization of GO has made significant progress to date. It has been used in the fields of desalination, water treatment, drugs, oil-water isolation, solar panels, healthcare, etc. (Yu et al., 2020).

8.2 HISTORY OF GRAPHENE OXIDE

The residual carboxylic acid, epoxide, and hydroxide groups in GO obtained through oxidative exfoliation of graphite affect hydrophilicity and other structural and functional characteristics of the material. As a result, reducing these residual functional groups aids in the enhancement of different GO characteristics synthesized GO for the first time in 1859, several years before graphene was discovered (Geim, 2012; Shubha et al., 2017). He developed a technique for producing graphite oxide. The technique was made by oxidizing and exfoliating graphite and finally produced a significant quantity of single-layer GO. Regretfully, no one had heard of graphene at the time. It was only more than a century and a half after "The rise of graphene (Geim and Novoselov, 2010)" the old invention resembled as a quick and inexpensive method of making the novel important and promising material. Brodie, Staudenmaier, and Hoffman have defined a method of preparing GO that requires specific acids (nitric and/or sulfur) and potassium chlorate (Dideikin and Vul, 2019). You can see a summary of the history and different methods of preparing GO in Figure 8.1.

8.3 PREPARATION/SYNTHESIS OF GRAPHENE OXIDE

Graphite, which is used to make GO, is made up of crystals or granules and may come from both natural and synthetic resources. The popular source is natural graphite, which is utilized in a wide variety of chemically modified usages (Pendolino and Armata, 2017). Carbon layers in

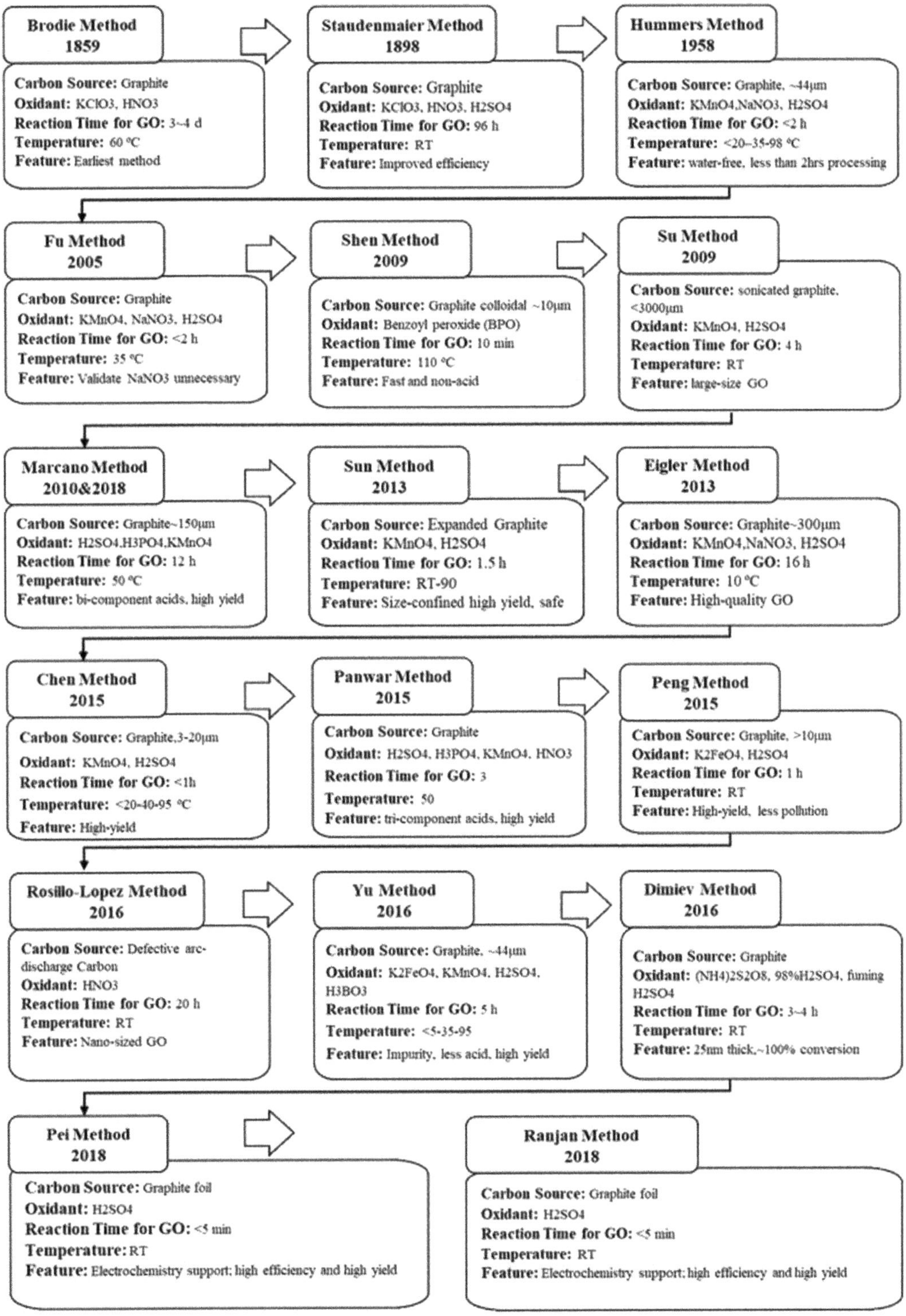

FIGURE 8.1 Summary of the history and different methods of preparing graphene oxide (Sun, 2019). RT: Room temperature.

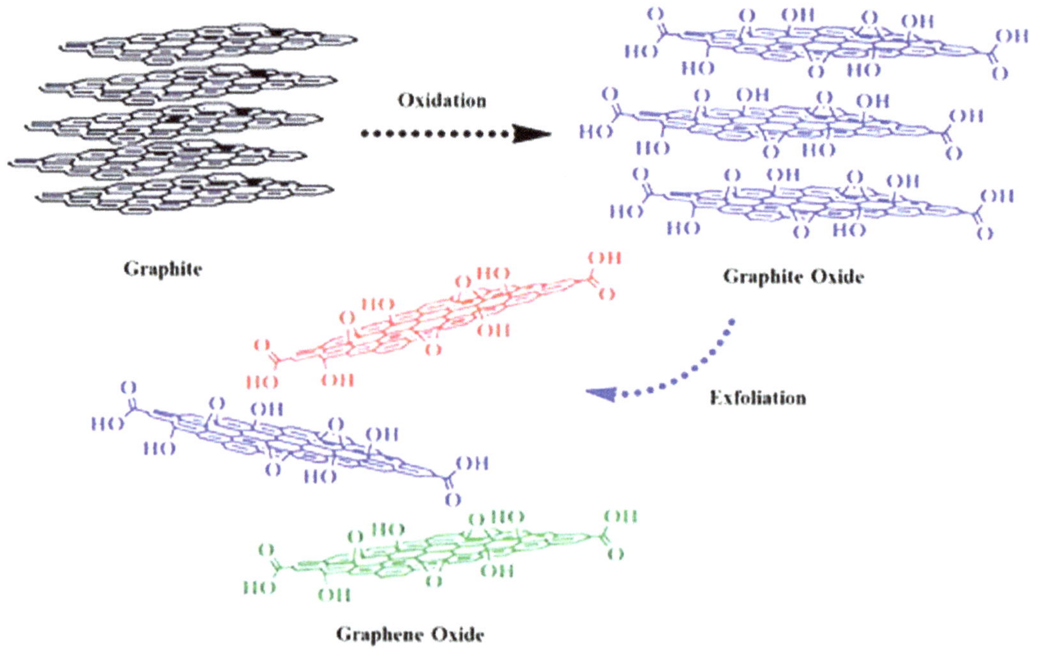

FIGURE 8.2 Preparation of graphene oxide (Yu et al., 2020).

multilayer GO are separated by functional groups bonded to each layer of carbon atoms. While graphene and GO are both two-dimensional carbon materials, their properties are significantly different. It does not absorb visible light, has very low electrical conductance in comparison with graphene, and shows notably greater chemical activity (Dideikin and Vul, 2019). Figure 8.2 shows the two major stages involved in preparing GO. First, graphite powder is oxidized to make graphite oxide, which could be easily spread in a polar solvent such as water, due to the presence of hydroxyl and epoxide groups around the basal planes of graphite oxide, as well as carbonyl and carboxyl groups at the edges. Second, in various solvents, the bulk graphite oxide may be exfoliated by sonication to form colloidal suspensions of monolayer, bilayer, or few-layer GO sheets (Wissler, 2006).

The selection of an appropriate oxidant to oxidize graphite is the key point in the preparation of GO. The production of GO can be classified into two categories: bottom-up methods in which simple carbon molecules are utilized to build pristine graphene and "top-down" procedures in which layers of graphene derivatives are extracted from a carbon source, commonly graphite (Chua and Pumera, 2014; Wang et al., 2013c). Dong introduced the most widely used approach in 2012, in which citric acid is used to acquire GO (Dong et al., 2012). Many other compounds, such as glucose (Shehab et al., 2017), mango leaves (Kumawat et al., 2017), limeade (Suvarnaphaet et al., 2016), coffee grounds (Wang et al., 2016), or honey (Mahesh et al., 2016), have been proposed as potential GO precursors. The features of obtained GO will vary depending on the synthesis method and raw material used. Consequently, the choice of the method has to be compatible with the features and application that are planned. For realizing graphene derivatives, top-down methods that first generate GO and/or rGO are more common, particularly for use in nanocomposite materials (Poniatowska et al., 2019). Most GO has been synthesized in recent years by chemical oxidation and exfoliation of pristine graphite using the Brodie, Staudenmaier, or Hummers methods, or variants of these procedures.

Brodie discovered that the oxidizing mixture ($KClO_4$ + fuming HNO_3) could only form GO with graphitizable carbons that have graphitic structure regions (Brodie, 1859). Researchers stated the formation of GO while graphite was heated with H_2SO_4, HNO_3, and $KClO_4$ (Staudenmaier, 1898). Later, Hummers and Offeman developed a simple procedure for preparing GO by H_2SO_4 and $KMnO_4$ (Hummers and Offeman, 1958). The above three methods and their preparation processes are depicted in Figure 8.1. Exfoliation is a crucial step in the production of GO, and it is normally done in a liquid solution using the ultrasonic process with various solvents (Cui et al., 2011; Paredes et al., 2008). The typical procedures involve exposure of graphite or graphite oxide powders to particular solvents and then exposing these solutions to sonication. Recently, other methods were also developed for simultaneously exfoliation and reduction of GO to obtain graphene nanosheets (GNs), such as thermal treatment (Cao et al., 2012), the chemical method (Zhang et al., 2010b), plasma (Cardinali et al., 2011), and microwave (Zhu et al., 2010).

8.4 GREEN SYNTHESIS OF GRAPHENE OXIDE

Chemically reducing GO is a common synthesis technique among the many available. Reductant agents including hydrazine or sodium borohydride are often used in chemical reduction (Ren et al., 2010; Shin et al., 2009; Stankovich et al., 2007). The use of the method is limited by its high prices, the toxicity of reducing agents (Wang et al., 2012), and the permanent accumulation of GO. Different green synthesis procedures and bio-reduction processes, in which plant extracts or microorganisms serve as eco-friendly reductants, are being investigated to address these difficulties (Mhamane et al., 2011; Nasrollahzadeh et al., 2015; Salas et al., 2010). Microorganisms need aseptic conditions to develop, which makes their maintenance more difficult (Roy and Das, 2015). The utilization of herb extracts to reduce GO decreases the usage of harmful reductant chemicals, rendering the procedure both environmentally and economically friendly. As a slower reaction rate, they also aid in the regulation of crystal formation (Singh et al., 2010).

Various biodegradable substances, including chitosan, polyethylene oxide, and polyvinyl alcohol, can be used to modify the surface of GO. There are a lot of marginal hydrophilic groups in GO. Al-Marri et al. (2015) made a highly reduced graphene (HRG)-Ag nanocomposite (NP) with *Pulicaria glutiinosa* plant extract, which served as a reductant and a ligand for the binding of silver NPs to HRG layers (Al-Marri et al., 2015). Scientists have shown the bactericidal, antitumor, and fungicidal impacts of secondary metabolites isolated from various herbs toward different microbes, which allows one to continue exploring the biological potentials of medicinal compounds for utilizing in the pharmacy industry. *Citrus sinensis* (orange peel) (Namasivayam et al., 1996), taro (*Colocasia esculenta*), and Ceylon ironwood (*Mesua ferrea* Linn) leaf extracts (Thakur and Karak, 2012), and grape extract (Upadhyay et al., 2015) are just a few examples.

Plant extracts are a good alternative to dangerous chemicals for reducing GO because of their local distribution, cost-effectiveness, and eco-friendliness. Although a variety of herbs had been utilized in the reduction procedure, their use is limited due to global unavailability and ineffective reduction capacity (Bhattacharya et al., 2017). Herb extracts could serve as reductant and capping factors. Iravani et al. examined some plant extracts for metal nanoparticles, including *Pelargonium graveolens* (Shankar et al., 2003), *Azadirachta indica* (Shankar et al., 2004), *Tamarindus indica* (Ankamwar et al., 2005), *Cinnamomum zeylanicum* Blume bark extract (Sathishkumar et al., 2009), etc., working on environmentally friendly ways to make metal nanoparticles and use them in electrochemical sensing functions in continuous research interests (Emmanuel et al., 2014; Karuppiah et al., 2015; Li et al., 2012). Summary of process setting in GO reduction by plant extracts is shown in Table 8.1.

TABLE 8.1
Summary of Process Setting in Graphene Oxide Reduction by Plant Extracts

No	Reducing Agent	Type of Extract	Method	Time	Temperature (°C)	GO Concentration	Reference
1	*Allium cepa*	Hydroalcoholic	Stirring	6 h	RT	50 mg/ml	Khanam and Hasan (2019)
2	*Allium sativum*	Hydroalcoholic	Heated	3 h	100	5 mg	Srivastava et al. (2014)
3	*Aloe vera*	Ethanolic	Refluxing	8 h	80	6 mmol	Ramanathan et al. (2017)
4	*Artemisia annua*	Ethanolic	Sonication and stirring	24 h	95	1 mg/ml	Hou et al. (2018)
5	*Artemisia vulgaris*	Aqueous	Refluxing	12 h	90	0.5 mg/ml	Chettri et al. (2016)
6	*Cinnamomum verum*	Aqueous	Refluxing	12 h	100	1 mg/ml	Han et al. (2016)
7	*Cinnamomum zeylanicum*	Aqueous	Refluxing	45 min	RT	1.6 mg/ml	Suresh et al. (2015b)
8	*Colocasia esculenta*	Aqueous	Ultrasonic, stirring, and refluxing	–	RT	0.5 mg/ml	Thakur and Karak (2012)
9	*Citrus sinensis*	Aqueous	Ultrasonic, Stirring, and refluxing	–	RT	0.5 mg/ml	Thakur and Karak (2012)
10	*Eucalyptus*	Aqueous	Refluxing	8 h	80	0.5 mg/ml	Jin et al. (2018)
11	*Ginkgo biloba*	Aqueous	Refluxing	12 h	95	0.5 mg/ml	Lee and Kim (2014)
12	*Hibiscus sabdariffa*	Aqueous	Stirring	1 h	RT	0.4 mg/ml	Chu et al. (2014)
13	*Melissa officinalis*	Aqueous	Stirring	12 h	RT	0.5 mg/ml	Elif et al. (2017)
14	*Mentha piperita*	Ethanolic	Stirring, refluxing	3 h	100	2.4mg/ml	Khojasteh et al. (2019)
15	*Mesua ferrea* Linn.	Aqueous	Ultrasonic, stirring, and refluxing	–	RT	0.5 mg/ml	Thakur and Karak (2012)
16	*Ocimum sanctum* L.	Aqueous	Stirring	4 h	70	1 mg/ml	Mahata et al. (2018)
17	*Origanum vulgare*	Ethanolic	Refluxing	48 h	70	–	Seyedi et al. (2018)
18	*Picrasma quassioides*	Aqueous	Stirring	12 h	50	–	Sreekanth et al. (2016)
19	*Rosa damascena*	Rose water	Autoclave	5 h	95	7 mg/ml	Haghighi and Tabrizi (2013)
20	*Salvadora persica*	Aqueous	Refluxing	24 h	98	5 mg/ml	Khan et al. (2015)
21	*Syzygium aromaticum*	Aqueous	Refluxing	30 min	100	1.6 mg/ml	Suresh et al. (2015a)
22	*Tribulus terrestris*	Ethanolic	Stirring, refluxing	3 h	100	2.4 mg/ml	Khojasteh et al. (2019)

Note: RT, room temperature.

8.5 APPLICATIONS OF GRAPHENE OXIDE

GO is a good tool for graphene-based usages in electronics, optics, energy storage, and biology (Dideikin and Vul, 2019). Before GO can be used for fundamental applications, it is critical to fully comprehend the structure of GO, which is made up of C, O, and H atoms, particularly when considering some of the biosensing characteristics. Since GO is a graphene derivative, it has the same layer structure as graphene and holds an sp3 hybridized carbon atom. GO has a variety of structures and characteristics depending on the synthesis procedure and various forms of oxygen species in groups, including carbonyl, carboxyl, epoxy, hydroxy, and oxidation (Inagaki and Kang, 2014). Figure 8.3 depicts the most significant GO functions.

8.5.1 MEMBRANES

Membrane technology is a rapidly developing field with a variety of applications, including desalination and water treatment. Based on the characteristics of the membrane and the species to be filtered, a membrane may allow the passage of some species while blocking the passage of others. Researchers have been developing this technology to create a more cost-effective and accurate membrane (Joshi et al., 2015). Graphene, generally, does not let anything pass through. GO is continuing to show its great membrane features and offers great potential for various usages (Joshi et al., 2015). GO has emerged as a great membrane material. Nair et al. (2012) demonstrated that the GO membrane allows unrestricted water permeation while blocking all else in the vapor form (Nair et al., 2012). The ease in forming atomically thin GO layers can be created in the form of a membrane, which gives it an advantage over other membranes in terms of practical applications.

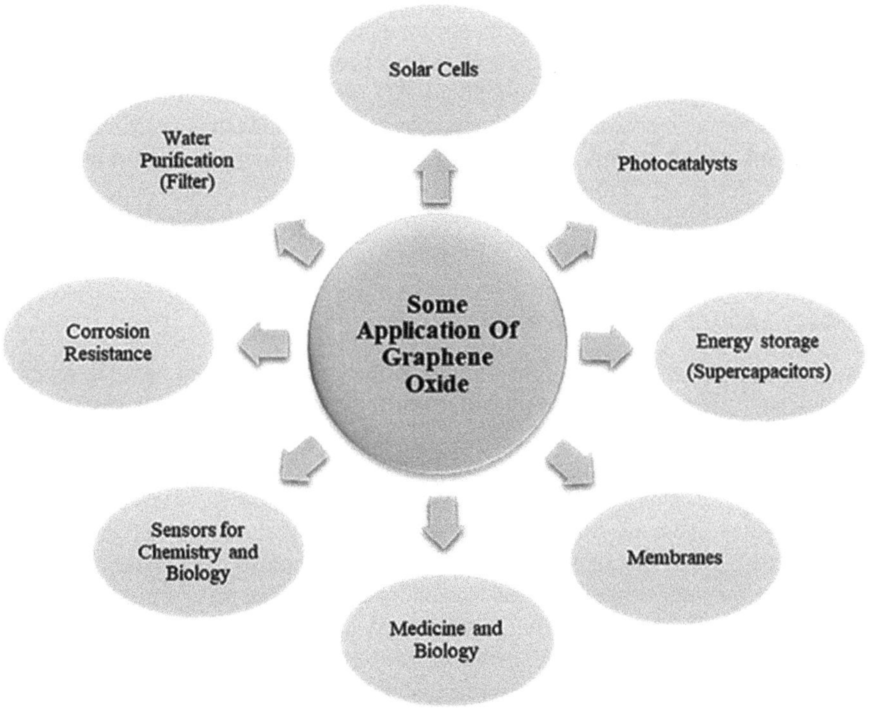

FIGURE 8.3 Contemporary and prospective applications of graphene oxide.

8.5.2 Sensors for Biology

Electronic biosensors made of graphene have been designed to monitor saccharides (Huang et al., 2010), proteins (Mao et al., 2010; Ohno et al., 2009, 2010; Yang and Gong, 2010), and DNAs in living organisms (Choi et al., 2010; Dong et al., 2010). Chen and colleagues created a chemical vapor deposition (CVD)-grown graphene sensor that can discover glucose and glutamate (Huang et al., 2010). The detection is carried out by functionalization by various enzymes. The process handled by enzymes makes free radicals, which are electron-withdrawing molecules. Therefore, the conductance of graphene layers will be enhanced (Liu et al., 2012).

8.5.3 Solar Cells

Widespread curiosity about energy supplies has sparked a boom in research into solar cells. The majority of today's solar cells are made of silicon (Si), which has the benefits of a wide spectrum of solar radiation absorption, plentiful resources on Earth, and developed manufacturing methods. Different methods have been investigated to decrease the cost and increase the performance of Si solar panels. To improve solar cell efficiency, a new graphene (Gr) structure on Si solar cells was created by embedding a GO interlayer between the monolayer Gr and Si. The GO interlayer will significantly enhance the voltage of solar panels while also suppressing interface recombination. The Gr/GO/Si solar panels have efficiency levels ranging from 5.10 to 6.18%.

8.5.4 Photocatalysts

Graphene-based semiconductor photocatalysts have inspired interest in recent years due to their enhanced photocatalytic efficiency (Li et al., 2011; Xiang et al., 2012a). Graphene is an important electron acceptor in photocatalysts, enhancing charge transfer and greatly reducing assimilation of electron-hole pairs in the composite to increase photocatalytic operation. The use of graphene-based semiconductor metal composites to improve photocatalytic operation was developed by Kamat (2010). Zhang demonstrated that a TiO_2-graphene nanocomposite improves organic pollutant photocatalytic degradation (Zhang et al., 2010a). Researchers examined various photocatalysts including graphene-C_3N_4 composite, graphene-modified TiO_2 nanosheet, and MoS_2-graphene-modified TiO_2 nanoparticles for the increased photocatalytic hydrogen manufacturing (Xiang et al., 2011a,b; Xiang et al., 2012b). Another study has exhibited that GO can function as an active photocatalyst for water splitting (Yeh et al., 2010). The inherent big bandgap of GO creates opportunities for photocatalytic applications, including CO_2 to methanol conversion, which harvests the sun's radiation while also reducing CO_2 (Hsu et al., 2013).

8.5.5 Energy Storage (Supercapacitors)

Due to the growing need for more environmentally friendly energy storage systems, supercapacitor and lithium-ion battery research has accelerated in recent years. Portable energy systems and hybrid vehicles have received the most attention. Despite all of the publicity, the creation of practical materials that improve the overall efficiency of the electrodes has been a challenge for these instruments. Although a wide range of energy storage materials utilized in Li-ion batteries and supercapacitors have been investigated, there are some limitations (Chidembo et al., 2012). Yoo et al. (2011) define a supercapacitor that could reserve energy by using electrical double-layer capacitors (EDLC) and pseudocapacitors. Compared with traditional capacitors, the supercapacitor has a higher power density and long-term cyclic constancy. The EDLC stores charge in a non-radical manner. On the opposite, pseudocapacitance results from the faradic reactions caused by the residual oxygen functionalities of reduced GO. To use GO in supercapacitor systems, the GO must be reduced to ensure electronic conjugation (Kuila et al., 2013). According to reports, after reduction, GO's specific capacitance rises to 100–300 F/g.

8.5.6 Corrosion Resistance

The corrosion rate of metals was greatly reduced by graphene films. The number of carboxyl groups changed on the GO surface affected how well it dispersed in acrylic resin. In a study, thermal reduction of GO yielded it with various carboxyl group contents (Chang et al., 2014a). The effect of carboxyl on the corrosion resistance of the photopolymerized polymethylmethacrylate-graphene composite coatings was investigated after that. Their findings revealed that the carboxyl group inhibited graphene agglomeration, supported graphene dispersion in acrylic resin, and provided an effective barrier to corrosive media. The corrosion resistance of the pure polymethylmethacrylate coatings was 27 times higher. Recent investigations have shown that incorporating functionalized GO nanosheets into polymeric coatings significantly improves their corrosion resistance efficiency. Because of their large surface area, GO sheets are impermeable to water, oxygen, and ions diffusion and provide a strong barrier (Ahmadi et al., 2016; Chang et al., 2014b). Furthermore, conducting polymers (CPs), which have attracted a lot of attention in recent years due to their unusual electrochemical features, can be used to covalently functionalize GO. Via a passivation process, CPs have been shown to properly protect metals from corrosion (Martins et al., 2009; Ogurtsov et al., 2004).

8.5.7 Water Purification (Filter)

In the past, activated carbon has been used in water purification technologies for decolorization (Santos et al., 2007) and heavy metal ion retention (Mohan et al., 2001). The major derivatives of graphene are GO and rGO. Various active functional groups are found on the surface of GO (Lujanienė et al., 2017; Wang and Chen, 2015). Because of its highly functionalized operative surface, GO may be used to remove toxins from water (Farooq and Jalees, 2020). Nair and colleagues were the first to discover that stacking GO films allows for a special water permeation pathway (Nair et al., 2012). GO is a new strategy for water treatment at a minimal cost. Numerous approaches for chemical derivatization of GNs have been published, including the flexible diazonium grafting chemistry; however, these methods focus on the conductivity and hydrophobic end product (Lomeda et al., 2008). The inability of GO to sequester water-soluble contaminants including heavy metals would be hampered by the poor accessibility of water molecules in these hydrophilic polymers. The nanostructured GO coating may notably develop water treatment industries (Gao et al., 2011).

8.5.8 Medicine

In recent years, since the negative impacts of chemical materials are recognized, there is a concern in utilizing natural materials, with a focus on the manufacture of nanomaterials containing natural extracts. Because of its great physical and chemical characteristics, GO is thought to be a great material for medical usages (Chung et al., 2013; Zhou and Liang, 2014). The usage of graphene-based substances in the biomedical industry, such as medication delivery, has developed rapidly in recent years. Graphene and GO have been thoroughly examined as new drugs nanocarrier for carrying drugs, such as antitumor medicines, non-soluble medicines, antibiotics, proteins, nucleic acids, and genes. Because of graphene and GO special characteristics, they have been widely promoted as great substances for biomedical usages. Wound healing and tissue engineering have also been investigated using graphene-based materials (Palmieri et al., 2020). Graphene with great mechanical characteristics can perform various functions on flat surfaces. As a result, it may be used in hydrogels, biodegradable films, and electrospun fibers, as a reinforcement material. Hydrogel scaffolds made of GO-chitosan have excellent mechanical characteristics and low degradability. Moreover, unlike neat chitosan, these scaffolds maintain their size and shape under various physiological and severe pH conditions (Depan et al., 2011).

8.6 GRAPHENE OXIDE USAGE IN AGRICULTURE

Nanomaterials were first used in agriculture in 2007, and the area has received a lot of attention since then (Lin and Xing, 2007; Liu et al., 2015; Mushtaq, 2011). Nanotubes loading into tomato seeds enhanced their germination (Novoselov et al., 2007). Lin and Xing found that nanoparticles imposed a notable inhibition impact on sprouting and root growth (Lin and Xing, 2007). Also, GO decreased dry and fresh weights and radix numbers and increased plant stress. The main applications of GO in agriculture are shown in Figure 8.4.

8.6.1 Assisted Promotion of Plant Growth

Controlling and promoting plant growth is a significant global problem. With the world's exponential population growth and climate change, effective food cultivation and food conservation have never been more important. As a result, different methods have been suggested to raise crop yields and protect crop and vegetation development in a variety of environmental environments, such as extreme temperature changes and natural disasters (Kim and Chung, 2018; Lesk et al., 2016; Ramankutty et al., 2018). Chemical-based methods have been widely used in seed production and plant biology among the different approaches. Park et al. (2020) devised a nanomaterial-assisted bionic technique for plant growth acceleration. In addition to the fact that nanomaterials are poisonous to plants, they reveal that GO can be used as a regulator to improve plant development. GO was inserted into the stem of the watermelon plant and applied to the growth medium of *Arabidopsis thaliana* L. Their findings revealed that GO had a positive impact on plant growth on growing root length, leaf area, number of leaves, and flower bud development when given in an acceptable quantity. Furthermore, GO influenced watermelon ripeness by growing the fruit's diameter and sugar content. GO could be utilized as a mechanism to speed up both plant growth and the ripening period of fruits.

8.6.2 Slow-Release Fertilizers

A significant quantity of fertilizer utilized in producing plants is a drift to the ecosystem and is unable to be used by herbs, resulting in contamination and also economic failure (Li et al., 2019;

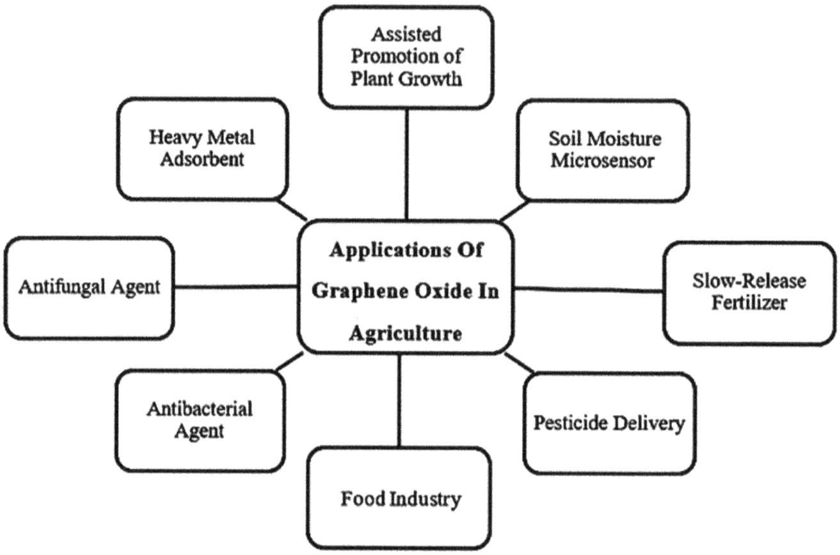

FIGURE 8.4 The main applications of graphene oxide in agriculture.

Tang et al., 2017a). Controlled-release fertilizers (CRFs), which support the critical minerals in herbs at a gentler rate than conventional fertilizers, have been proposed as a possible solution to these issues. CRFs can play a key role in raising crop production by correcting mineral deficiency over time and reducing fertilization frequency, lowering costs, and reducing pollution (Tang et al., 2017b; Zhang et al., 2017). CRFs are almost created by covering traditional fertilizers with safety frameworks, including different macromolecules, which causes the nourishment delivery speeds to be delayed. Zhang et al. used GO films for manufacturing controlled-release fertilizer (KNO_3) (Zhang et al., 2014). They showed that GO-coated KNO_3 showed a slow-release manner. According to the findings of that report, the reason was the low durability and penetrability of coatings. Also, various investigations have shown the importance of the enzymatic method that uses the peroxidase family to degrade graphene-based products (Allen et al., 2008; Kotchey et al., 2011; Lalwani et al., 2014). As a result, GO may be effectively degraded on the soil due to having the amount of lignin peroxidase in nature (Kabiri et al., 2017).

8.6.3 Pesticide Delivery

Pesticides are widely used to improve agricultural production because they are an efficient method to avoid and control pests. Industrial pesticide formulations have low target performance and non-target accumulation, resulting in environmental problems (Gerstl et al., 1998). Just 0.1% of pesticides applied affect the pest; the rest of the losses are caused by pesticide drift (Aktar et al., 2009). By covering more crop surface areas with nanoformulations, the target efficiency can be increased. With polymeric nanoparticles, a few nanoformulations for slow pesticide release have been developed (Loha et al., 2011; Sarkar et al., 2012). Pesticides can be adsorbed on the surface of GO, according to scanning electron microscope (SEM) findings. The synergistic mechanism may be that GO can act as a carrier for pesticides, allowing them to be adsorbed on mites' surfaces and thus increase pesticide efficacy and utilization performance. For the efficient control of plant pests, GO-pesticides (integrating pesticides and GO) may be a potential alternative to the shortcomings of traditional pesticides. However, there have been few examinations focusing on the usage of GO as a synergist to increase the toxicity of pesticides in pest control (Wang et al., 2019b).

8.6.4 Food Industry (Packaging Films)

Active food packaging substances are prepared to preserve the food from the surroundings and also to perform an important role by inhibiting various happenings that change the food quality (de Azeredo, 2013). Carbon nanoparticles stand out among the various nanoparticle families because of their exceptional characteristics, which allow the production of multifunctional polymer composites based upon carbon nanotubes, graphite nanoplatelets, and graphene (Gu et al., 2017; Huangfu et al., 2019; Wang et al., 2019a). Graphene and its derivatives, such as GO, have evolved into a new category of carbon-based nanoparticles with excellent characteristics that could be customized by their exfoliation rate and oxidation level. For example, graphene has recently enabled the production of various new nanocomposites and polymer composites with uses in supercapacitor electrode materials (Deng et al., 2018; Le et al., 2019; Zhang et al., 2019). Graphene plays an important role in packaging technology, based on how it is used (Ghanem et al., 2020). It could be used as a reinforcing material in a packaging process, as Loryuenyong et al. did when they enhanced the mechanical properties of poly(vinyl alcohol) packaging films. When small quantities of GO are added to a regulated polymer, the findings show a 49% enhancement in elastic strength (Loryuenyong et al., 2015). Since graphene has many oxygenated functional groups on its surface, which aids its dispersion in hydrophilic matrices (Ahmad et al., 2018; Gudarzi et al., 2016), it is commonly used in packaging applications as oxides. Nonetheless, due to its hydrophilic composition, GO has proven to be a significant obstacle when incorporated into hydrophobic polymers. To disperse GO in packaging material, especially hydrophobic film, an effective surface modification is needed.

8.6.5 Antifungal Agent

As their size-dependent characteristics, large surface-to-volume proportion, and unusual physicochemical characteristics, nanoparticle-based bactericidal potentials have developed as a multidisciplinary research area involving medicine, material science, biology, and chemistry. They exhibited good potential in herb health. For the first time, fungicidal potential against the *Fusarium graminearum* was evaluated using a GO-AgNPs nanocomposite. The examinations showed that the GO-AgNPs nanocomposite exhibited a 3- and 7-time enhancement of efficacy than pure AgNPs and GO suspension, respectively. A significantly lower concentration, 4.68 g/ml, induced spore germination inhibitory effect. The spores and hyphae were destroyed, which may be due to an antibacterial mechanism resulting from GO-AgNPs' impressive synergistic impact, that causes physical harm as well as chemically oxygen radicals production. More specifically, the strong bioactivity of GO-AgNPs may be due to the chemical reduction of GO mediated by fungi. In an experiment, the GO-AgNPs nanocomposite also demonstrated an important impact in regulating the leaf spot disease caused by *F. graminearum*. The GO-AgNPs nanocomposite produced in that study has the potential to be a promising material for the production of novel antimicrobial agents against pathogenic fungi or bacteria (Chen et al., 2016). An updated Hummers method was used to make reduced GO nanosheets. Three fungi, *Aspergillus niger* (*A. niger*), *Aspergillus oryzae* (*A. oryzae*), and *Fusarium oxysporum* (*F. oxysporum*), were used to monitor the antifungal activity of rGO nanosheets. The sharp edges of the rGO are thought to be responsible for inhibiting fungi mycelial development. The half-maximal inhibitory concentration (IC50), a measure of the efficacy of the rGO in inhibiting the fungi, was evaluated. The rGO has IC50 values of 50, 100, and 100 μgml^{-1} against *F. oxysporum*, *A. niger*, and *A. oryzae*, respectively (Sawangphruk et al., 2012).

8.6.6 Antibacterial Agent

Nanotechnology has been used in plant disease control in recent years, and it has shown to be an important method to reduce herb disorders. Minerals such as copper, zinc, and sulfur are effective biocides against plant pathogens (Graham et al., 2016; Kah and Hofmann, 2014; Rao and Paria, 2013). Many researchers have discovered that GO is a great antibacterial agent, with new possibilities for its use in crop disease control. Wang et al. (2013b) reported that by destroying the cell membrane, GO has great antimicrobial potential toward copper-resistant *Ralstonia solanaicearum*. Similarly, Chen et al. stated that intertwining and membrane perturbation could be responsible for GO's antibacterial mechanism against *Pseudomonas syringae* and *Xanthomonas campestris* pv. *undulosa* (Chen et al., 2014). Most notably, previous studies have shown that GO has little to no harmful effects on higher plants (Ocsoy et al., 2013; Zhao et al., 2015). Leaf blight, which is causes by *Xanthomonas oryzae* pv. *oryzae* (*Xoo*), is a disease that harms the growth of rice (*Oryza sativa* L.). Different chemicals had utilized to suppress this disorder, resulting in rice contamination, resistance among the *Xoo*, and serious threats to public health. The bactericidal characteristics of a GO/silver nanoparticles (GO-Ag) composite against *Xoo* were also evaluated. Comparing to pure AgNPs, the GO-Ag composite showed a nearly 4-times improvement. At a low concentration of 2.5 μg/ml, the bacteria were inactivated. The cell's integrity was affected, and intercellular contents were revealed out due to the slow release of silver from the GO-Ag composite. This composite caused the formation of reactive oxygen species and stopped DNA duplication at the molecular level. More significantly, the toxicity test revealed that the as-prepared GO-Ag composite was less harmful to herbs than AgNPs, indicating that it could be used to suppress crop disease (Liang et al., 2017). The physicochemical characteristics of GO have a key role in bactericidal properties by inducing various lethal interactions. The signaling mechanisms of GO in bacteria have been studied extensively (Zheng et al., 2018).

8.6.7 Soil Moisture Microsensor

Soil moisture monitoring is an important factor in agriculture for improving yield because plants steadily absorb the soil moisture. As a result, regular measuring and maintaining soil humidity is needed for optimal soil situations (Palaparthy et al., 2018). Sensors installed in the field for large-area farms must be low cost, and reliable during measurements, for routine monitoring of soil moisture content. To meet the farmer's needs, economical and stable sensors with good sensitivity are required (Palaparthy et al., 2013; Robinson et al., 2008; Lekshmi et al., 2014).

Graphene and its derivatives (graphene quantum dots, GO, rGO, graphene nanoribbons, and so on) have great carrier mobility. GO is one of the commonly used graphene derivatives for sensing functions, especially water molecule detection (Palaparthy et al., 2018). Furthermore, since GO contains oxygen functional units, it is electrically insulating (Park and Ruoff, 2009), making it suitable for use as a capacitive sensor (Bi et al., 2013). To test relative humidity (Palaparthy et al., 2018) and soil moisture (Kalita et al., 2016), graphene-based micro-sensors have resembled good sensitivity and stability (Bi et al., 2013). Finally, the use of GO and its products to measure moisture needs to be investigated more.

8.6.8 Heavy Metal Adsorbent

These days, metal contamination may be a public health concern. Cadmium (Cd), lead (Pb), and copper (Cu), as main heavy metals, are among the most hazardous contaminants as stated by Environmental Protection Agency (Demirbas, 2008). These metals are extremely toxic, even in trace amounts (Awual et al., 2014). Many attempts are being made to get rid of heavy metals due to their carcinogenicity and toxicity (Sadegh et al., 2017). In the context of water purification methods, adsorption is one of the hopeful techniques because of greater removal efficacy and environmentally friendly and relatively efficient and economical (Dubey et al., 2009; Farooq and Jalees, 2020). Because of its wide surface area, graphene could function as a suitable adsorbent for a variety of contaminants (Wang et al., 2013a). Unlike graphene, GO with various oxygen-containing functional groups is capable to be a binding site for heavy metals. Moreover, GO is hydrophilic and effectively dispersed in water and has fast kinetics and a higher surface area than graphene. As a result, GO with a good adsorption potential for heavy metals removal and can be a suitable candidate to improve water quality and removing salt from water (Elgengehi et al., 2020).

8.7 CONCLUSION

Toxic chemicals and environmentally safe reductant factors are the two major forms of reductant factors used in the chemical reduction of GO. Because of their negative environmental effects, compounds that are potent factors, like hydrazine, must be avoided. Environmentally safe reductant factors mostly come from fruits and vegetables. Sometimes, eco-friendly reducing agents are more beneficial than powerful reductant factors. In general, the GO reduction has been successfully demonstrated to mitigate environmental concerns using eco-acceptable reducing agents. However, there are still challenges with the green reduction procedures. As a result, extensive efforts have been focused on the creation and improvement of environmentally suitable reductant factors for reducing GO. Graphene-based composites are used in many technology fields, including graphene-based drug carriers to enhance drug safety, and metabolic kinetics, and many details remain unknown, requiring further study and discovery by scientists. As a result, graphene and GO will be thoroughly studied and widely used in the future.

REFERENCES

Ahmad, Hassan, Mizi Fan, and David Hui. "Graphene oxide incorporated functional materials: a review." *Composites Part B: Engineering* 145 (2018): 270–280.

Ahmadi, Anahita, Bahram Ramezanzadeh, and Mohammad Mahdavian. "Hybrid silane coating reinforced with silanized graphene oxide nanosheets with improved corrosion protective performance." *RSC Advances* 6, no. 59 (2016): 54102–54112.

Aktar, Wasim, Dwaipayan Sengupta, and Ashim Chowdhury. "Impact of pesticides use in agriculture: their benefits and hazards." *Interdisciplinary Toxicology* 2, no. 1 (2009): 1–12.

Allen, Brett L., Padmakar D. Kichambare, Pingping Gou, Irina I. Vlasova, Alexander A. Kapralov, Nagarjun Konduru, Valerian E. Kagan, and Alexander Star. "Biodegradation of single-walled carbon nanotubes through enzymatic catalysis." *Nano Letters* 8, no. 11 (2008): 3899–3903.

Al-Marri, Abdulhadi H., Mujeeb Khan, Merajuddin Khan, Syed F. Adil, Abdulrahman Al-Warthan, Hamad Z. Alkhathlan, Wolfgang Tremel, Joselito P. Labis, Mohammed Rafiq H. Siddiqui, and Muhammad N. Tahir. "*Pulicaria glutinosa* extract: a toolbox to synthesize highly reduced graphene oxide-silver nanocomposites." *International Journal of Molecular Sciences* 16, no. 1 (2015): 1131–1142.

Ankamwar, Balaprasad, Minakshi Chaudhary, and Murali Sastry. "Gold nanotriangles biologically synthesized using tamarind leaf extract and potential application in vapor sensing." *Synthesis and Reactivity in Inorganic, Metal-Organic, and Nano-Metal Chemistry* 35, no. 1 (2005): 19–26.

Awual, Md Rabiul, Mohamed Ismael, Md Abdul Khaleque, and Tsuyoshi Yaita. "Ultra-trace copper (II) detection and removal from wastewater using novel meso-adsorbent." *Journal of Industrial and Engineering Chemistry* 20, no. 4 (2014): 2332–2340.

Bhattacharya, Gourav, Shrawni Sas, Shikha Wadhwa, Ashish Mathur, James McLaughlin, and Susanta Sinha Roy. "Aloe vera assisted facile green synthesis of reduced graphene oxide for electrochemical and dye removal applications." *RSC Advances* 7, no. 43 (2017): 26680–26688.

Bi, Hengchang, Kuibo Yin, Xiao Xie, Jing Ji, Shu Wan, Litao Sun, Mauricio Terrones, and Mildred S. Dresselhaus. "Ultrahigh humidity sensitivity of graphene oxide." *Scientific Reports* 3, no. 1 (2013): 1–7.

Brodie, Benjamin C. "XIII. On the atomic weight of graphite." *Philosophical Transactions of the Royal Society of London* 149 (1859): 249–259.

Cao, Jun, Guo-Qiang Qi, Kai Ke, Yong Luo, Wei Yang, Bang-Hu Xie, and Ming-Bo Yang. "Effect of temperature and time on the exfoliation and de-oxygenation of graphite oxide by thermal reduction." *Journal of Materials Science* 47, no. 13 (2012): 5097–5105.

Cardinali, Marta, Luca Valentini, Paola Fabbri, and Josè M. Kenny. "Radiofrequency plasma-assisted exfoliation and reduction of large-area graphene oxide platelets produced by a mechanical transfer process." *Chemical Physics Letters* 508, no. 4–6 (2011): 285–288.

Chang, Kung-Chin, Wei-Fu Ji, Chih-Wei Li, Chi-Hao Chang, Yu-Yuan Peng, Jui-Ming Yeh, and Wei-Ren Liu. "The effect of varying carboxylic-group content in reduced graphene oxides on the anticorrosive properties of PMMA/reduced graphene oxide composites." *Express Polymer Letters* 8, no. 12 (2014a): 908–919.

Chang, Kung-Chin, Min-Hsiang Hsu, Hsin-I. Lu, Mei-Chun Lai, Pei-Ju Liu, Chien-Hua Hsu, and Wei-Fu Ji. "Room-temperature cured hydrophobic epoxy/graphene composites as corrosion inhibitor for cold-rolled steel." *Carbon* 66 (2014b): 144–153.

Chen, Juanni, Hui Peng, Xiuping Wang, Feng Shao, Zhaodong Yuan, and Heyou Han. "Graphene oxide exhibits broad-spectrum antimicrobial activity against bacterial phytopathogens and fungal conidia by intertwining and membrane perturbation." *Nanoscale* 6, no. 3 (2014): 1879–1889.

Chen, Juanni, Long Sun, Yuan Cheng, Zhicheng Lu, Kang Shao, Tingting Li, Chao Hu, and Heyou Han. "Graphene oxide-silver nanocomposite: novel agricultural antifungal agent against *Fusarium graminearum* for crop disease prevention." *ACS Applied Materials & Interfaces* 8, no. 36 (2016): 24057–24070.

Chettri, Prajwal, V. S. Vendamani, Ajay Tripathi, Anand P. Pathak, and Archana Tiwari. "Self-assembly of functionalized graphene nanostructures by one-step reduction of graphene oxide using aqueous extract of Artemisia vulgaris." *Applied Surface Science* 362 (2016): 221–229.

Chidembo, Alfred, Seyed H. Aboutalebi, Konstantin Konstantinov, Maryam Salari, Brad Winton, Sima A. Yamini, Ivan P. Nevirkovets, and Hua K. Liu. "Globular reduced graphene oxide-metal oxide structures for energy storage applications." *Energy & Environmental Science* 5, no. 1 (2012): 5236–5240.

Choi, Bong G., HoSeok Park, Min H. Yang, Young M. Jung, Sang Y. Lee, Won H. Hong, and Tae J. Park. "Microwave-assisted synthesis of highly water-soluble graphene towards electrical DNA sensor." *Nanoscale* 2, no. 12 (2010): 2692–2697.

Chu, Hwei-Jay, Chi-Young Lee, and Nyan-Hwa Tai. "Green reduction of graphene oxide by *Hibiscus sabdariffa* L. to fabricate flexible graphene electrode." *Carbon* 80 (2014): 725–733.

Chua, Chun K., and Martin Pumera. "Chemical reduction of graphene oxide: a synthetic chemistry viewpoint." *Chemical Society Reviews* 43, no. 1 (2014): 291–312.

Chung, Chul, Young-Kwan Kim, Dolly Shin, Soo-Ryoon Ryoo, Byung Hee Hong, and Dal-Hee Min. (2013). Biomedical applications of graphene and graphene oxide. *Accounts of Chemical Research*, 46(10), 2211–2224.

Cui, Xu, Chenzhen Zhang, Rui Hao, and Yanglong Hou. "Liquid-phase exfoliation, functionalization and applications of graphene." *Nanoscale* 3, no. 5 (2011): 2118–2126.

de Azeredo, Henriette M. C. "Antimicrobial nanostructures in food packaging." *Trends in Food Science & Technology* 30, no. 1 (2013): 56–69.

Demirbas, Ayhan. "Heavy metal adsorption onto agro-based waste materials: a review." *Journal of Hazardous Materials* 157, no. 2-3 (2008): 220–229.

Deng, Weiyuan, Tianhe Kang, Hu Liu, Jiaoxia Zhang, Ning Wang, Na Lu, Yong Ma, Ahmad Umar, and Zhanhu Guo. "Potassium hydroxide activated and nitrogen-doped graphene with enhanced supercapacitive behavior." *Science of Advanced Materials* 10, no. 7 (2018): 937–949.

Dideikin, Artur T., and Alexander Y. Vul. "Graphene oxide and derivatives: the place in graphene family." *Frontiers in Physics* 6 (2019): 149.

Dong, Xiaochen, Yumeng Shi, Wei Huang, Peng Chen, and Lain-Jong Li. "Electrical detection of DNA hybridization with single-base specificity using transistors based on CVD-grown graphene sheets." *Advanced Materials* 22, no. 14 (2010): 1649–1653.

Dong, Yongqiang, Jingwei Shao, Congqiang Chen, Hao Li, Ruixue Wang, Yuwu Chi, Xiaomei Lin, and Guonan Chen. "Blue luminescent graphene quantum dots and graphene oxide prepared by tuning the carbonization degree of citric acid." *Carbon* 50, no. 12 (2012): 4738–4743.

Depan, Dilip, Bhupendra Girase, Jinesh Shah, and Raja Devesh Kumar Misra. "Structure–process–property relationship of the polar graphene oxide-mediated cellular response and stimulated growth of osteoblasts on hybrid chitosan network structure nanocomposite scaffolds." *Acta Biomaterialia* 7, no. 9 (2011): 3432–3445.

Dubey, Shashi P., Krishna Gopal, and Jean-Luc Bersillon. "Utility of adsorbents in the purification of drinking water: a review of characterization, efficiency and safety evaluation of various adsorbents." *Journal of Environmental Biology* 30, no. 3 (2009): 327–332.

Elgengehi, Sara M., Sabry El-Taher, Mahmoud A. A. Ibrahim, Jacques K. Desmarais, and Khaled E. El-Kelany. "Graphene and graphene oxide as adsorbents for cadmium and lead-heavy metals: a theoretical investigation." *Applied Surface Science* 507 (2020): 145038.

Elif, Öztürk, Özbek Belma, and Şenel İlkay. "Production of biologically safe and mechanically improved reduced graphene oxide/hydroxyapatite composites." *Materials Research Express* 4, no. 1 (2017): 015601.

Emmanuel, Rohinton, Chelladurai Karuppiah, Shen-Ming Chen, Selvakumar Palanisamy, S. Padmavathy, and P. Prakash. "Green synthesis of gold nanoparticles for trace level detection of a hazardous pollutant (nitrobenzene) causing Methemoglobinaemia." *Journal of Hazardous Materials* 279 (2014): 117–124.

Farooq, Muhammad U., and Muhammad I. Jalees. "Application of magnetic graphene oxide for water purification: heavy metals removal and disinfection." *Journal of Water Process Engineering* 33 (2020): 101044.

Freudenberg, Karl, and Arthur C. Neish. "Constitution and biosynthesis of lignin." *Constitution and Biosynthesis of Ligni* 1 (1968): 142.

Galande, Charudatta, Wei Gao, Akshay Mathkar, Andrew M. Dattelbaum, Tharangattu N. Narayanan, Aditya D. Mohite, and Pulikel M. Ajayan. "Science and engineering of graphene oxide." *Particle & Particle Systems Characterization* 31, no. 6 (2014): 619–638.

Gao, Wei, Mainak Majumder, Lawrence B. Alemany, Tharangattu N. Narayanan, Miguel A. Ibarra, Bhabendra K. Pradhan, and Pulickel M. Ajayan. "Engineered graphite oxide materials for application in water purification." *ACS Applied Materials & Interfaces* 3, no. 6 (2011): 1821–1826.

Geim, Andre K. "Graphene prehistory." *Physica Scripta* 2012, no. T146 (2012): 014003.

Geim, Andre K., and Konstantin S. Novoselov "The rise of graphene." *Nanoscience and Technology* 30, no. 2 (2010): 11–19.

Geim, Andre K. "Graphene: status and prospects." *Science* 324, no. 5934 (2009): 1530–1534.

Gerstl, Zev, Amri Nasser, and Uri Mingelgrin. "Controlled release of pesticides into soils from clay–polymer formulations." *Journal of Agricultural and Food Chemistry* 46, no. 9 (1998): 3797–3802.

Ghanem, Ahmed F., Ahmed M. Youssef, and Mona H. Abdel Rehim. "Hydrophobically modified graphene oxide as a barrier and antibacterial agent for polystyrene packaging." *Journal of Materials Science* 55, no. 11 (2020): 4685–4700.

Graham, James H., Evan Johnson, Monty E. Myers, Mikaeel Young, Parthiban Rajasekaran, Smruti Das, and Swadeshmukul Santra. "Potential of nano-formulated zinc oxide for control of citrus canker on grapefruit trees." *Plant Disease* 100, no. 12 (2016): 2442–2447.

Gu, Junwei, Chaobo Liang, Xiaomin Zhao, Bin Gan, Hua Qiu, Yonqiang Guo, Xutong Yang, Qiuyu Zhang, and De-Yi Wang. "Highly thermally conductive flame-retardant epoxy nanocomposites with reduced ignitability and excellent electrical conductivities." *Composites Science and Technology* 139 (2017): 83–89.

Gudarzi, Mohsen M., Seyed H. Aboutalebi, and Farhad Sharif. "Graphene oxide-based composite materials." *Graphene Oxide* no. 1 (2016): 314–363.

Haghighi, Behzad, and Mahmoud A. Tabrizi. "Green synthesis of reduced graphene oxide nanosheets using rose water and a survey on their characteristics and applications." *RSC Advances* 3, no. 32 (2013): 13365–13371.

Han, Wei, Wen-Yi Niu, Bing Sun, Guang-Can Shi, and Xiu-Qin Cui. "Biofabrication of polyphenols stabilized reduced graphene oxide and its anti-tuberculosis activity." *Journal of Photochemistry and Photobiology B* 165 (2016): 305–309.

He, Heyong, Thomas Riedl, Anton Lerf, and Jacek Klinowski. "Solid-state NMR studies of the structure of graphite oxide." *The Journal of Physical Chemistry* 100, no. 51 (1996): 19954–19958.

Hernandez, Yenny, Valeria Nicolosi, Mustafa Lotya, Fiona M. Blighe, Zhenyu Sun, Sukanta De, Ignatius T. McGovern, et al. "High-yield production of graphene by liquid-phase exfoliation of graphite." *Nature Nanotechnology* 3, no. 9 (2008): 563–568.

Hong, Yanzhong, Zhiyong Wang, and Xianbo Jin. "Sulfuric acid intercalated graphite oxide for graphene preparation." *Scientific Reports* 3, no. 1 (2013): 1–6.

Hou, Dandan, Qinfu Liu, Xianshuai Wang, Ying Quan, Zhichuan Qiao, Li Yu, and Shuli Ding. "Facile synthesis of graphene via reduction of graphene oxide by artemisinin in ethanol." *Journal of Materiomics* 4, no. 3 (2018): 256–265.

Hsu, Hsin-Cheng, Indrajit Shown, Hsieh-Yu Wei, Yu-Chung Chang, He-Yun Du, Yan-Gu Lin, Chi-Ang Tseng, et al. "Graphene oxide as a promising photocatalyst for CO_2 to methanol conversion." *Nanoscale* 5, no. 1 (2013): 262–268.

Hu, Chuangang, Xiangquan Zhai, Lili Liu, Yang Zhao, Lan Jiang, and Liangti Qu. "Spontaneous reduction and assembly of graphene oxide into three-dimensional graphene network on arbitrary conductive substrates." *Scientific Reports* 3, no. 1 (2013): 1–10.

Huang, Yinxi, Xiaochen Dong, Yumeng Shi, Chang M. Li, Lain-Jong Li, and Peng Chen. "Nanoelectronic biosensors based on CVD grown graphene." *Nanoscale* 2, no. 8 (2010): 1485–1488.

Huangfu, Yiming, Kunpeng Ruan, Hua Qiu, Yuanjin Lu, Chaobo Liang, Jie Kong, and Junwei Gu. "Fabrication and investigation on the PANI/MWCNT/thermally annealed graphene aerogel/epoxy electromagnetic interference shielding nanocomposites." *Composites Part A* 121 (2019): 265–272.

Hummers, Jr, William S., and Richard E. Offeman. "Preparation of graphitic oxide." *Journal of the American Chemical Society* 80, no. 6 (1958): 1339–1339.

Inagaki, Michio, and Feiyu Kang. "Graphene derivatives: graphane, fluorographene, graphene oxide, graphene, and graphdiyne." *Journal of Materials Chemistry A* 2, no. 33 (2014): 13193–13206.

Jin, Xiaoying, Na Li, Xiulan Weng, Chengyang Li, and Zuliang Chen. "Green reduction of graphene oxide using eucalyptus leaf extract and its application to remove the dye." *Chemosphere* 208 (2018): 417–424.

Joshi, Rakesh K., Subbiah Alwarappan, Mashamichi Yoshimura, Veena Sahajwalla, and Yuta Nishina. "Graphene oxide: the new membrane material." *Applied Materials Today* 1, no. 1 (2015): 1–12.

Kabiri, Shervin, Fien Degryse, Diana N. H. Tran, Rodrigo C. da Silva, Mike J. McLaughlin, and Dusan Losic. "Graphene oxide: a new carrier for slow release of plant micronutrients." *ACS Applied Materials & Interfaces* 9, no. 49 (2017): 43325–43335.

Kah, Melanie, and Thilo Hofmann. "Nanopesticide research: current trends and future priorities." *Environment International* 63 (2014): 224–235.

Kalita, Hemen, Vinay S. Palaparthy, Maryam S. Baghini, and Mohammed Aslam. "Graphene quantum dot soil moisture sensor." *Sensors and Actuators B* 233 (2016): 582–590.

Kamat, P. V. Graphene-based nanoarchitectures. Anchoring semiconductor and metal nanoparticles on a two-dimensional carbon support. *The Journal of Physical Chemistry Letters* 1, no. 2 (2010): 520–527.

Karuppiah, Chelladurai, Kasithevar Muthupandi, Shen-Ming Chen, Mohammad Ajmal Ali, Selvakumar Palanisamy, Abraham Rajan, Periakaruppan Prakash, Fahad M. A. Al-Hemaid, and Bih-Show Lou. "Green synthesized silver nanoparticles decorated on reduced graphene oxide for enhanced electrochemical sensing of nitrobenzene in waste water samples." *RSC Advances* 5, no. 39 (2015): 31139–31146.

Khan, Mujeeb, Abdulhadi H. Al-Marri, Merajuddin Khan, Mohammed R. Shaik, Nils Mohri, Syed F. Adil, Mufsir Kuniyil, et al. "Green approach for the effective reduction of graphene oxide using *Salvadora persica* L. root (Miswak) extract." *Nanoscale Research Letters* 10, no. 1 (2015): 1–9.

Khanam, P. Noorunnisa, and Anwarul Hasan. "Biosynthesis and characterization of graphene by using a non-toxic reducing agent from *Allium Cepa* extract: Anti-bacterial properties." *International Journal of Biological Macromolecules* 126 (2019): 151–158.

Khojasteh, Hossein, Hamed Safajou, Sobhan Mortazavi-Derazkola, Masoud Salavati-Niasari, Kamran Heydaryan, and Morteza Yazdani. "Economic procedure for facile and eco-friendly reduction of graphene oxide by plant extracts; a comparison and property investigation." *Journal of Cleaner Production* 229 (2019): 1139–1147.

Kim, Kyung H., and Sun-Ok Chung. "Comparison of plant growth and glucosinolates of Chinese cabbage and kale crops under three cultivation conditions." *Journal of Biosystems Engineering* 43, no. 1 (2018): 30–36.

Kotchey, Gregg P., Brett L. Allen, Harindra Vedala, Naveena Yanamala, Alexander A. Kapralov, Yulia Y. Tyurina, Judith Klein-Seetharaman, Valerian E. Kagan, and Alexander Star. "The enzymatic oxidation of graphene oxide." *ACS Nano* 5, no. 3 (2011): 2098–2108.

Kuila, Tapas, Ananta K. Mishra, Partha Khanra, Nam H. Kim, and Joong H. Lee. "Recent advances in the efficient reduction of graphene oxide and its application as energy storage electrode materials." *Nanoscale* 5, no. 1 (2013): 52–71.

Kumawat, Mukesh K., Mukeshchand Thakur, Raju B. Gurung, and Rohit Srivastava. "Graphene quantum dots from *Mangifera indica*: application in near-infrared bioimaging and intracellular nanothermometry." *ACS Sustainable Chemistry & Engineering* 5, no. 2 (2017): 1382–1391.

Lalwani, Gaurav, Weiliang Xing, and Balaji Sitharaman. "Enzymatic degradation of oxidized and reduced graphene nanoribbons by lignin peroxidase." *Journal of Materials Chemistry B* 2, no. 37 (2014): 6354–6362.

Le, Kai, Zhou Wang, Fenglong Wang, Qi Wang, Qian Shao, Vignesh Murugadoss, Shide Wu, et al. "Sandwich-like NiCo layered double hydroxide/reduced graphene oxide nanocomposite cathodes for high energy density asymmetric supercapacitors." *Dalton Transactions* 48, no. 16 (2019): 5193–5202.

Lee, Geummi, and Beom S. Kim. "Biological reduction of graphene oxide using plant leaf extracts." *Biotechnology Progress* 30, no. 2 (2014): 463–469.

Lekshmi, S. U. Susha, Devendra Narain Singh, and Maryam S. Baghini. "A critical review of soil moisture measurement." *Measurement* 54 (2014): 92–105.

Lesk, Corey, Pedram Rowhani, and Navin Ramankutty. "Influence of extreme weather disasters on global crop production." *Nature* 529, no. 7584 (2016): 84–87.

Li, Qin, Beidou Guo, Jiaguo Yu, Jingrun Ran, Baohong Zhang, Huijuan Yan, and Jian R. Gong. "Highly efficient visible-light-driven photocatalytic hydrogen production of CdS-cluster-decorated graphene nanosheets." *Journal of the American Chemical Society* 133, no. 28 (2011): 10878–10884.

Li, Tiantian, Bin Gao, Zhaohui Tong, Yuechao Yang, and Yuncong Li. "Chitosan and graphene oxide nanocomposites as coatings for controlled-release fertilizer." *Water, Air, & Soil Pollution* 230, no. 7 (2019), 1–9.

Li, Ying, Tzu-Ying Wu, Shen-Ming Chen, M. Ajmal Ali, and Fahad MA AlHemaid. "Green synthesis and electrochemical characterizations of gold nanoparticles using leaf extract of *Magnolia kobus*." *International Journal of Electrochemical Science* 7, no. 12 (2012): 12742–12751.

Liang, You, Desong Yang, and Jianghu Cui. "A graphene oxide/silver nanoparticle composite as a novel agricultural antibacterial agent against *Xanthomonas oryzae* pv. *oryzae* for crop disease management." *New Journal of Chemistry* 41, no. 22 (2017): 13692–13699.

Lin, Daohui, and Baoshan Xing. "Phytotoxicity of nanoparticles: inhibition of seed germination and root growth." *Environmental Pollution* 150, no. 2 (2007): 243–250.

Liu, Shangjie, Hongmin Wei, Zhiyang Li, Shun Li, Han Yan, Yong He, and Zhihong Tian. "Effects of graphene on germination and seedling morphology in rice." *Journal of Nanoscience and Nanotechnology* 15, no. 4 (2015): 2695–2701.

Liu, Yuxin, Xiaochen Dong, and Peng Chen. "Biological and chemical sensors based on graphene materials." *Chemical Society Reviews* 41, no. 6 (2012): 2283–2307.

Loha, Kumelachew M., Najam A. Shakil, Jitendra Kumar, Manish K. Singh, Totan Adak, and Suresh Jain. "Release kinetics of β-cyfluthrin from its encapsulated formulations in water." *Journal of Environmental Science and Health Part B* 46, no. 3 (2011): 201–206.

Lomeda, Jay R., Condell D. Doyle, Dmitry V. Kosynkin, Wen-Fang Hwang, and James M. Tour. "Diazonium functionalization of surfactant-wrapped chemically converted graphene sheets." *Journal of the American Chemical Society* 130, no. 48 (2008): 16201–16206.

Loryuenyong, Vorrada, Charinrat Saewong, Chaiyud Aranchaiya, and Achanai Buasri. "The improvement in mechanical and barrier properties of poly (vinyl alcohol)/graphene oxide packaging films." *Packaging Technology and Science* 28, no. 11 (2015): 939–947.

Lujanienė, Galina, Sergej Šemčuk, A. Lečinskytė, Ieva Kulakauskaitė, Kęstutis Mažeika, Darius Valiulis, Vidas Pakštas, Martynas Skapas, and Stasys Tumėnas. "Magnetic graphene oxide-based nanocomposites for removal of radionuclides and metals from contaminated solutions." *Journal of Environmental Radioactivity* 166 (2017): 166–174.

Mahata, Suhasini, Anjumala Sahu, Prashant Shukla, Ankita Rai, Manorama Singh, and Vijai K. Rai. "The novel and efficient reduction of graphene oxide using *Ocimum sanctum* L. leaf extract as an alternative renewable bio-resource." *New Journal of Chemistry* 42, no. 24 (2018): 19945–19952.

Mahesh, Sankarapillai, Lalitha Lekshmi, Kizhisseri D. Renuka, and Kuruvilla Joseph. "Simple and cost-effective synthesis of fluorescent graphene quantum dots from honey: application as stable security ink and white-light emission." *Particle & Particle Systems Characterization* 33, no. 2 (2016): 70–74.

Malik, Sharali, Aravind Vijayaraghavan, Rolf Erni, Katsuhiko Ariga, Ivan Khalakhan, and Jonathan P. Hill. "High purity graphenes prepared by a chemical intercalation method." *Nanoscale* 2, no. 10 (2010): 2139–2143.

Mao, Shun, Ganhua Lu, Kehan Yu, Zheng Bo, and Junhong Chen. "Specific protein detection using thermally reduced graphene oxide sheet decorated with gold nanoparticle-antibody conjugates." *Advanced Materials* 22, no. 32 (2010): 3521–3526.

Mao, Shun, Haihui Pu, and Junhong Chen. "Graphene oxide and its reduction: modeling and experimental progress." *Royal Society of Chemistry Advances* 2, no. 7 (2012): 2643–2662.

Martins, José Inácio Ferrão De Paiva, Mohammed Bazzaoui, T. C. Reis, S. C. Costa, Marta Nunes, Luís Martins, and El Arbi Bazzaoui. "The effect of pH on the pyrrole electropolymerization on iron in malate aqueous solutions." *Progress in Organic Coatings* 65, no. 1 (2009): 62–70.

Mhamane, Dattakumar, Wegdan Ramadan, Manal Fawzy, Abhimanyu Rana, Megha Dubey, Chandrashekhar Rode, Benoit Lefez, Beatrice Hannoyer, and Satishchandra Ogale. "From graphite oxide to highly water dispersible functionalized graphene by single-step plant extract-induced deoxygenation." *Green Chemistry* 13, no. 8 (2011): 1990–1996.

Mohan, Dinesh, Vinod Kumar Gupta, Suresh Srivastava, and Subhash Chander. "Kinetics of mercury adsorption from wastewater using activated carbon derived from fertilizer waste." *Colloids and Surfaces A* 177, no. 2–3 (2001): 169–181.

Mushtaq, Yasmeen K. "Effect of nanoscale Fe_3O_4, TiO_2 and carbon particles on cucumber seed germination." *Journal of Environmental Science and Health* 46, no. 14 (2011): 1732–1735.

Nair, Rahul Raveendran Nair, Hengan Wu, Parthipan N. Jayaram, Irina V. Grigorieva, and A. K. Geim. "Unimpeded permeation of water through helium-leak-tight graphene-based membranes." *Science* 335, no. 6067 (2012): 442–444.

Namasivayam, Chinnaiya, Neerathilingam Muniasamy, K. Gayatri, Monika V. Usha Rani, and Kuppusamy Ranganathan. "Removal of dyes from aqueous solutions by cellulosic waste orange peel." *Bioresource Technology* 57, no. 1 (1996): 37–43.

Nasrollahzadeh, Mahmoud, Mehdi Maham, Akbar Rostami-Vartooni, Mojtaba Bagherzadeh, and Seyed M. Sajadi. "Barberry fruit extract assisted in situ green synthesis of Cu nanoparticles supported on a reduced graphene oxide–Fe_3O_4 nanocomposite as a magnetically separable and reusable catalyst for the *O*-arylation of phenols with aryl halides under ligand-free conditions." *Royal Society of Chemistry Advances* 5, no. 79 (2015): 64769–64780.

Novoselov, K. S., S. V. Morozov, T. M. G. Mohinddin, L. A. Ponomarenko, D. C. Elias, R. Yang, I. I. Barbolina, et al. "Electronic properties of graphene." *Physica Status Solidi (B)* 244, no. 11 (2007): 4106–4111.

Novoselov, Kostya S., Andre K. Geim, Sergei V. Morozov, Dingde Jiang, Yanshui Zhang, Sergey V. Dubonos, Irina V. Grigorieva, and Alexandr A. Firsov. "Electric field effect in atomically thin carbon films." *Science* 306, no. 5696 (2004): 666–669.

Ocsoy, Ismail, Mathews L. Paret, M. A. Ocsoy, Muserref Arslan, Sanju Kunwar, Tao Chen, Mingxu You, and Weihong Tan. Nanotechnology in plant disease management: DNA-directed silver nanoparticles on graphene oxide as an antibacterial against *Xanthomonas perforans*. *ACS Nano* 7, no. 10 (2013): 8972–8980.

Ogurtsov, Nikolay A., Alexander A. Pud, Peter Kamarchik, and Galina S. Shapoval. "Corrosion inhibition of aluminum alloy in chloride mediums by undoped and doped forms of polyaniline." *Synthetic Metals* 143, no. 1 (2004): 43–47.

Ohno, Yasuhide, Kenzo Maehashi, and Kazuhiko Matsumoto. "Label-free biosensors based on aptameri-modified graphene field-effect transistors." *Journal of the American Chemical Society* 132, no. 51 (2010): 18012–18013.

Ohno, Yasuhide, Kenzo Maehashi, Yusuke Yamashiro, and Kazuhiko Matsumoto. "Electrolyte-gated graphene field-effect transistors for detecting pH and protein adsorption." *Nano Letters* 9, no. 9 (2009): 3318–3322.

Palaparthy, Vinay S., Hemen Kalita, Sandeep G. Surya, Maryam S. Baghini, and Mohammed Aslam. "Graphene oxide-based soil moisture microsensor for in situ agriculture applications." *Sensors and Actuators B* 273 (2018): 1660–1669.

Palaparthy, Vinay S., Maryam S. Baghini, and Devendra N. Singh. "Review of polymer-based sensors for agriculture-related applications." *Emerging Materials Research* 2, no. 4 (2013): 166–180.

Palmieri, Valentina, Marco De Spirito, and Massimiliano Papi. "Graphene-based scaffolds for tissue engineering and photothermal therapy." *Nanomedicine* 15, no. 14 (2020): 1411–1417.

Paredes, J. I., Silvia Villar-Rodil, Amelia Martínez-Alonso, and Juan M. D. Tascon. "Graphene oxide dispersions in organic solvents." *Langmuir* 24, no. 19 (2008): 10560–10564.

Park, Sungjin, and Rodney S. Ruoff. "Chemical methods for the production of graphenes." *Nature Nanotechnology* 4, no. 4 (2009): 217–224.

Park, Sunho, Kyoung S. Choi, Sujin Kim, Yonghyun Gwon, and Jangho Kim. "Graphene oxide-assisted promotion of plant growth and stability." *Nanomaterials* 10, no. 4 (2020): 758.

Pendolino, Flavio, and Nerina Armata. *Graphene Oxide in Environmental Remediation Process.* Switzerland: Springer, 2017.

Poniatowska, Aleksandra, Maciej Trzaskowski, and Tomasz Ciach. "Production and properties of top-down and bottom-up graphene oxide." *Colloids and Surfaces A* 561 (2019): 315–324.

Ramanathan, Subramanian, Elaiyappillai Elanthamilan, Asir Obadiah, Arulappan Durairaj, Johnson P. Merlin, Subramanian Ramasundaram, and Samuel Vasanthkumar. "*Aloe vera* (L.) Burm. f. extract reduced graphene oxide for supercapacitor application." *Journal of Materials Science* 28, no. 22 (2017): 16648–16657.

Ramankutty, Navin, Zia Mehrabi, Katharina Waha, Larissa Jarvis, Claire Kremen, Mario Herrero, and Loren H. Rieseberg. "Trends in global agricultural land use: implications for environmental health and food security." *Annual Review of Plant Biology* 69 (2018): 789–815.

Rao, Korukonda Jagajjanani, and Santanu Paria. "Use of sulfur nanoparticles as a green pesticide on *Fusarium solani* and *Venturia inaequalis* phytopathogens." *RSC Adv* 3 (2013): 10471–10478.

Ray, Sekhar C. Application and uses of graphene oxide and reduced graphene oxide. *Applications of Graphene and Graphene Oxide-Based Nanomaterials* 1 (2015): 39–55.

Ren, Peng-Gang, Ding-Xiang Yan, Xu Ji, Tao Chen, and Zhong-Ming Li. "Temperature dependence of graphene oxide reduced by hydrazine hydrate." *Nanotechnology* 22, no. 5 (2010): 055705.

Robinson, David A., Colin S. Campbell, Jan W. Hopmans, Brian K. Hornbuckle, Scott B. Jones, Rosemary Knight, Fred Ogden, John Selker, and Ole Wendroth. "Soil moisture measurement for ecological and hydrological watershed-scale observatories: a review." *Vadose Zone Journal* 7, no. 1 (2008): 358–389.

Roy, Swarup, and Tapan K. Das. "Plant mediated green synthesis of silver nanoparticles-A." *International Journal of Plant Biology & Research* 3, no. 3 (2015): 1044.

Sadegh, Hamidreza, Gomaa A. M. Ali, Vinod K. Gupta, Abdel S. H. Makhlouf, Ramin Shahryari-ghoshekandi, Mallikarjuna N. Nadagouda, Mika Sillanpää, and Elżbieta Megiel. "The role of nanomaterials as effective adsorbents and their applications in wastewater treatment." *Journal of Nanostructure in Chemistry* 7, no. 1 (2017): 1–14.

Salas, Everett C., Zhengzong Sun, Andreas Lüttge, and James M. Tour. "Reduction of graphene oxide via bacterial respiration." *Royal Society of Chemistry Nano* 4, no. 8 (2010): 4852–4856.

Santos, Aurora, Pedro Yustos, Sergio Rodríguez, Félix Garcia-Ochoa, and Miguel de Gracia. "Decolorization of textile dyes by wet oxidation using activated carbon as catalyst." *Industrial & Engineering Chemistry Research* 46, no. 8 (2007): 2423–2427.

Sarkar, Dhruba J., Jitendra Kumar, Najam A. Shakil, and Suresh Walia. "Release kinetics of controlled release formulations of thiamethoxam employing nano-ranged amphiphilic PEG and diacid based block polymers in soil." *Journal of Environmental Science and Health* 47, no. 11 (2012): 1701–1712.

Sathishkumar, Muthuswamy, Krishnamurthy Sneha, Sung-wook Won, Chul-woong Cho, Sok Kim, and Yeoung-Sang Yun. "*Cinnamon zeylanicum* bark extract and powder mediated green synthesis of nano-crystalline silver particles and its bactericidal activity." *Colloids and Surfaces B* 73, no. 2 (2009): 332–338.

Sawangphruk, Montree, Pattarachai Srimuk, Poramane Chiochan, Tanas Sangsri, and Patcharaporn Siwayaprahm. "Synthesis and antifungal activity of reduced graphene oxide nanosheets." *Carbon* 50, no. 14 (2012): 5156–5161.

Seyedi, Neda, Kazem Saidi, and Hassan Sheibani. "Green synthesis of Pd nanoparticles supported on magnetic graphene oxide by *Origanum vulgare* leaf plant extract: catalytic activity in the reduction of organic dyes and Suzuki–Miyaura cross-coupling reaction." *Catalysis Letters* 148, no. 1 (2018): 277–288.

Shiv Shankar, Sangaru, Absar Ahmad, Renu Pasricha, and Murali Sastry. "Bioreduction of chloroaurate ions by geranium leaves and its endophytic fungus yields gold nanoparticles of different shapes." *Journal of Materials Chemistry* 13, no. 7 (2003): 1822–1826.

Shiv Shankar Sangaru, Akhilesh Rai, Absar Ahmad, and Murali Sastry. "Rapid synthesis of Au, Ag, and bimetallic Au core–Ag shell nanoparticles using Neem (*Azadirachta indica*) leaf broth." *Journal of Colloid and Interface Science* 275, no. 2 (2004): 496–502.

Shehab, Mona, Shaker Ebrahim, and Moataz Soliman. "Graphene quantum dots prepared from glucose as optical sensor for glucose." *Journal of Luminescence* 184 (2017): 110–116.

Shin, Hyeon-Jin, Ki K. Kim, Anass Benayad, Seon-Mi Yoon, Hyeon K. Park, In-Sun Jung, Mei H. Jin "Efficient reduction of graphite oxide by sodium borohydride and its effect on electrical conductance." *Advanced Functional Materials* 19, no. 12 (2009): 1987–1992.

Shubha, Priya Babu, Keerthiraj Namratha, Huligerepura Sosalegowda Aparna, N. R. Ashok, Mohammed Shafiul Mustak, Jit Chatterjee, and Kullaiah Byrappa. Facile green reduction of graphene oxide using Ocimum sanctum hydroalcoholic extract and evaluation of its cellular toxicity. *Materials Chemistry and Physics* 198 (2017): 66–72.

Singh, A., D. Jain, M. K. Upadhyay, N. Khandelwal, and H. N. Verma. "Green synthesis of silver nanoparticles using *Argemone mexicana* leaf extract and evaluation of their antimicrobial activities." *Digest Journal of Nanomaterials and Biostructures* 5, no. 2 (2010): 483–489.

Smith, Andrew T., Anna M. LaChance, Songshan Zeng, Bin Liu, and Luyi Sun. "Synthesis, properties, and applications of graphene oxide/reduced graphene oxide and their nanocomposites." *Nano Materials Science* 1, no. 1 (2019): 31–47.

Sreekanth, Thupakula Venkata Madhukar, Min-Ji Jung, and In-Yong Eom. "Green synthesis of silver nanoparticles, decorated on graphene oxide nanosheets and their catalytic activity." *Applied Surface Science* 361 (2016): 102–106.

Srivastava, Sarvesh K., Chiaki Ogino, and Akihiko Kondo. "Green synthesis of thiolated graphene nanosheets by alliin (garlic) and its effect on the deposition of gold nanoparticles." *RSC Advances* 4, no. 12 (2014): 5986–5989.

Stankovich, Sasha, Dmitriy A. Dikin, Richard D. Piner, Kevin A. Kohlhaas, Alfred Kleinhammes, Yuanyuan Jia, Yue Wu, SonBinh T. Nguyen, and Rodney S. Ruoff. "Synthesis of graphene-based nanosheets via chemical reduction of exfoliated graphite oxide." *Carbon* 45, no. 7 (2007): 1558–1565.

Staudenmaier, L. "Verfahren zur darstellung der graphitsäure." *Berichte der deutschen chemischen Gesellschaft* 31, no. 2 (1898): 1481–1487.

Stergiou, Anastasios, Georgia Pagona, and Nikos Tagmatarchis. "Donor–acceptor graphene-based hybrid materials facilitating photo-induced electron-transfer reactions". *Beilstein Journal of Nanotechnology* 5, no. 1 (2014): 1580–1589.

Sun, Ling. "Structure and synthesis of graphene oxide." *Chinese Journal of Chemical Engineering* 27, no. 10 (2019): 2251–2260.

Sun, Zhengzong, Zheng Yan, Jun Yao, Elvira Beitler, Yu Zhu, and James M. Tour. "Growth of graphene from solid carbon sources." *Nature* 468, no. 7323 (2010): 549–552.

Suresh, Doddavenkatanna, H. Nagabhushana, and S. C. Sharma. "Clove extract mediated facile green reduction of graphene oxide, its dye elimination, and antioxidant properties." *Materials Letters* 142 (2015a): 4–6.

Suresh, Doddavenkatanna, Mungara Anil Pavan Kumar, H. Nagabhushana, and S. C. Sharma. "Cinnamon supported a facile green reduction of graphene oxide, its dye elimination, and antioxidant activities." *Materials Letters* 151 (2015b): 93–95.

Sutter, Peter W., Jan-Ingo Flege, and Eli A. Sutter. "Epitaxial graphene on ruthenium." *Nature Materials* 7, no. 5 (2008): 406–411.

Suvarnaphaet, Phitsini, Chandra S. Tiwary, Jutaphet Wetcharungsri, Supanit Porntheeraphat, Rassmidara Hoonsawat, Pulickel M. Ajayan, I-Ming Tang, and Piyapong Asanithi. "Blue photoluminescent carbon nanodots from limeade." *Materials Science and Engineering: C* 69 (2016): 914–921.

Tang, Yafu, Xinying Wang, Yuechao Yang, Bin Gao, Yongshan Wan, Yuncong C. Li, and Dongdong Cheng. "Activated-lignite-based super large granular slow-release fertilizers improve apple tree growth: synthesis, characterizations, and laboratory and field evaluations." *Journal of Agricultural and Food Chemistry* 65, no. 29 (2017a): 5879–5889.

Tang, Yafu, Yuechao Yang, Dongdong Cheng, Bin Gao, Yongshan Wan, and Yuncong C. Li. "Value-added humic acid derived from lignite using novel solid-phase activation process with Pd/CeO_2 nanocatalyst: a physiochemical study." *ACS Sustainable Chemistry & Engineering* 5, no. 11 (2017b): 10099–10110.

Thakur, Suman, and Niranjan Karak. "Green reduction of graphene oxide by aqueous phytoextracts." *Carbon* 50, no. 14 (2012): 5331–5339.

Upadhyay, Ravi K., Navneet Soin, Gourav Bhattacharya, Susmita Saha, Anjan Barman, and Susanta S. Roy. "Grape extract assisted green synthesis of reduced graphene oxide for water treatment application." *Materials Letters* 160 (2015): 355–358.

Wang, Jun, and Baoliang Chen. "Adsorption and coadsorption of organic pollutants and a heavy metal by graphene oxide and reduced graphene materials." *Chemical Engineering Journal* 281 (2015): 379–388.

Wang, Lei, Lixin Chen, Ping Song, Chaobo Liang, Yuanjin Lu, Hua Qiu, Yali Zhang, Jie Kong, and Junwei Gu. "Fabrication on the annealed $Ti_3C_2T_x$ MXene/Epoxy nanocomposites for electromagnetic interference shielding application." *Composites Part B* 171 (2019a): 111–118.

Wang, Liang, Weitao Li, Bin Wu, Zhen Li, Shilong Wang, Yuan Liu, Dengyu Pan, and Minghong Wu. "Facile synthesis of fluorescent graphene quantum dots from coffee grounds for bioimaging and sensing." *Chemical Engineering Journal* 300 (2016): 75–82.

Wang, Shaobin, Hongqi Sun, Ha-Ming Ang, and Moses O. Tadé. "Adsorptive remediation of environmental pollutants using novel graphene-based nanomaterials." *Chemical Engineering Journal* 226 (2013a): 336–347.

Wang, Xiuping, Haicui Xie, Zhenying Wang, and Kanglai He. "Graphene oxide as a pesticide delivery vector for enhancing acaricidal activity against spider mites." *Colloids and Surfaces B* 173 (2019b): 632–638.

Wang, Xiuping, Xueqin Liu, and Heyou Han. "Evaluation of antibacterial effects of carbon nanomaterials against copper-resistant *Ralstonia solanacearum*." *Colloids and Surfaces B* 103 (2013b): 136–142.

Wang, Yi, Pu Zhang, Chun F. Liu, Lei Zhan, Yuan F. Li, and Cheng Z. Huang. "Green and easy synthesis of biocompatible graphene for use as an anticoagulant." *RSC Advances* 2, no. 6 (2012): 2322–2328.

Wang, Zhaofeng, Jingjing Liu, Weixing Wang, Haoran Chen, Zhihong Liu, Qingkai Yu, Huidan Zeng, and Luyi Sun. "Aqueous phase preparation of graphene with low defect density and adjustable layers." *Chemical Communications* 49, no. 92 (2013c): 10835–10837.

Wissler, Mathis. "Graphite and carbon powders for electrochemical applications." *Journal of Power Sources* 156, no. 2 (2006): 142–150.

Xiang, Quanjun, Jiaguo Yu, and Mietek Jaroniec. "Enhanced photocatalytic H_2-production activity of graphene-modified titania nanosheets." *Nanoscale* 3, no. 9 (2011a): 3670–3678

Xiang, Quanjun, Jiaguo Yu, and Mietek Jaroniec. "Graphene-based semiconductor photocatalysts." *Chemical Society Reviews* 41, no. 2 (2012a): 782–796.

Xiang, Quanjun, Jiaguo Yu, and Mietek Jaroniec. "Preparation and enhanced visible-light photocatalytic H_2-production activity of graphene/C_3N_4 composites." *The Journal of Physical Chemistry C* 115, no. 15 (2011b): 7355–7363

Xiang, Quanjun, Jiaguo Yu, and Mietek Jaroniec. "Synergetic effect of MoS_2 and graphene as cocatalysts for enhanced photocatalytic H_2 production activity of TiO_2 nanoparticles." *Journal of the American Chemical Society* 134, no. 15 (2012b): 6575–6578

Yang, Minghui, and Shaoqin Gong. "Immunosensor for the detection of cancer biomarker based on percolated graphene thin film." *Chemical Communications* 46, no. 31 (2010): 5796–5798.

Yeh, Te-Fu, Jhih-Ming Syu, Ching Cheng, Ting-Hsiang Chang, and Hsisheng Teng. "Graphite oxide as a photocatalyst for hydrogen production from water." *Advanced Functional Materials* 20, no. 14 (2010): 2255–2262.

Yoo, Jung J., Kaushik Balakrishnan, Jingsong Huang, Vincent Meunier, Bobby G. Sumpter, Anchal Srivastava, Michelle Conway "Ultrathin planar graphene supercapacitors." *Nano Letters* 11, no. 4 (2011): 1423–1427.

Yu, Wang, Li Sisi, Yang Haiyan, and Luo Jie. "Progress in the functional modification of graphene/graphene oxide: a review." *RSC Advances* 10, no. 26 (2020): 15328–15345.

Zhang, Hao, Xiaojun Lv, Yueming Li, Ying Wang, and Jinghong Li. "P25-graphene composite as a high-performance photocatalyst." *ACS Nano* 4, no. 1 (2010a): 380–386.

Zhang, Jiaoxia, Zhuangzhuang Zhang, Yueting Jiao, Hongxun Yang, Yuqing Li, Jing Zhang, and Peng Gao. "The graphene/lanthanum oxide nanocomposites as electrode materials of supercapacitors." *Journal of Power Sources* 419 (2019): 99–105.

Zhang, Long, Xuan Li, Yi Huang, Yanfeng Ma, Xiangjian Wan, and Yongsheng Chen. "Controlled synthesis of few-layered graphene sheets on a large scale using chemical exfoliation." *Carbon* 48, no. 8 (2010b): 2367–2371.

Zhang, Shugang, Yuechao Yang, Bin Gao, Yuncong C. Li, and Zhiguang Liu. "Superhydrophobic controlled-release fertilizers coated with bio-based polymers with organosilicon and nano-silica modifications." *Journal of Materials Chemistry A* 5, no. 37 (2017): 19943–19953.

Zhang, Yakun, Liangguo Yan, Weiying Xu, Xìaoyao Guo, Limei Cui, Liang Gao, Qin Wei, and Bin Du. "Adsorption of Pb(II) and Hg(II) from aqueous solution using magnetic $CoFe_2O_4$-reduced graphene oxide." *Journal of Molecular Liquids* 191 (2014): 177–182.

Zhao, Shengqing, Qianqian Wang, Yunli Zhao, Qi Rui, and Dayong Wang. "Toxicity and translocation of graphene oxide in *Arabidopsis thaliana*." *Environmental toxicology and pharmacology* 39, no. 1 (2015): 145–156.

Zheng, Huizhen, Ronglin Ma, Meng Gao, Xin Tian, Yong-Qiang Li, Lingwen Zeng, and Ruibin Li. "Antibacterial applications of graphene oxides: structure-activity relationships, molecular initiating events, and biosafety." *Science Bulletin* 63, no. 2 (2018): 133–142.

Zhou, Xianfeng, and Feng Liang. "Application of graphene/graphene oxide in biomedicine and biotechnology." *Current Medicinal Chemistry* 21, no. 7 (2014): 855–869.

Zhu, Yanwu, Shanthi Murali, Meryl D. Stoller, Aruna Velamakanni, Richard D. Piner, and Rodney S. Ruoff. "Microwave-assisted exfoliation and reduction of graphite oxide for ultracapacitors." *Carbon* 48, no. 7 (2010): 2118–2122.

9 Graphene from Sugarcane Bagasse

Synthesis, Characterization, and Applications

Gunjan Varshney, Eksha Guliani, and Christine Jeyaseelan

CONTENTS

DOI: 10.1201/9781003169741-9

9.1 INTRODUCTION

Graphene is essentially a two-dimensional (2D) layer of carbon atoms bonded collectively in a hexagonal lattice structure (Nebol'sin, Galstyan, and Silina 2020). Despite the fact that it was isolated for the first time in 2004, in recent years, it has conventionally achieved a significant deal of attention. In 2010, Andre Geim and Konstantin Novoselov were honored with a noble prize in Physics for their innovative work on graphene. The prompt increase of attention in graphene is due to mainly enormous exceptional properties that it possesses (Cooper et al. 2012). Due to the exclusive properties of graphene, primarily the chemical and physical properties that include mechanical strength, stiffness, and elasticity, also intensely great thermal and electrical conductivity, it is defined as a consequential alternative to several conventional resources in a variety of applications, with the potential to assist many unmanageable innovations and assuredly present markets (Ghany, Elsherif, and Handal 2017).

9.1.1 Graphene: Structure and Shape

It is a carbon allotrope with a 2D structure. It is composed entirely of carbon atoms arranged in a hexagonal pattern resembling chicken wire (Lu and Li 2013) see Figure 9.1. It is basically a carbon allotrope with a 0.142-nm molecular bond length and sp^2-bonded atoms plane.

A single graphene sheet is made by a single layer of carbon atoms organized in a honeycomb structure (Allen, Tung, and Kaner 2009). Multilayer graphene is made up of numerous sheets loaded one on top of the other, till the point at which the substance converts to graphite (this generally happens over about 30 layers, with an interplanar spacing of 0.335 nm).

In graphene, each carbon atoms are bonded to the other three carbon atoms by covalent bond. Graphene is extremely stable and has a very high tensile strength because of the strength of the covalent bonds that exist between carbon atoms (Zhu et al. 2020). As graphene is flat, so each and every atom is on the surface. It is accessible from both sides and allows for more interaction

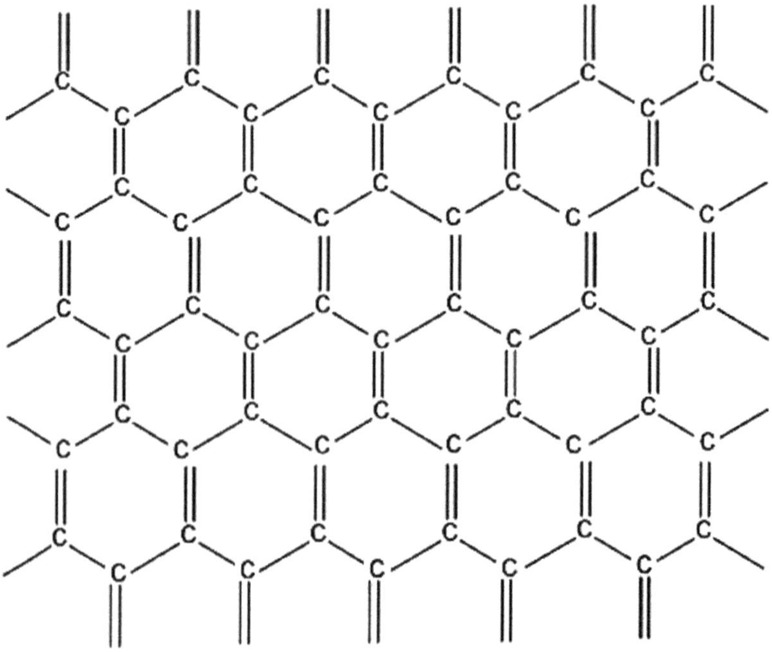

FIGURE 9.1 Structure of graphene [2D].

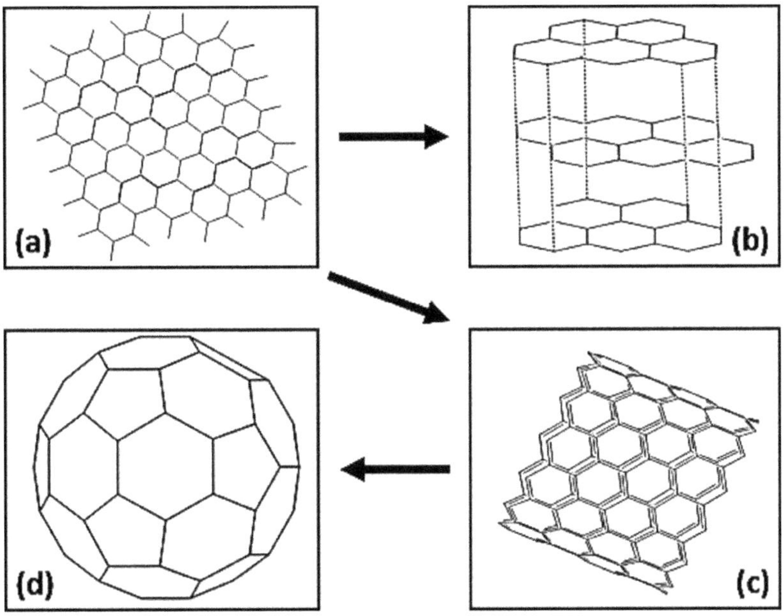

FIGURE 9.2 Graphene carbon structures.

with surrounding molecules. Furthermore, the carbon atoms are only bonded to three other atoms, despite the fact that they have the ability to show bonding with other atoms. This property is then combined with graphene's high surface area to volume ratio and tensile strength (Papageorgiou, Kinloch, and Young 2017). It allows it to be engaging for use in composite materials.

9.1.2 Graphene on the Basis of Other Carbon Structures

Graphene can also serve as a parent form for a number of carbon structures (Zhen and Zhu 2018), for example, the graphite, carbon nanotubes (which can be thought of as rolled-up sheets of graphene made into tubes), and buckyballs (these are spherical structures with a cage-like structure that are formed from graphene only with few of the hexagonal rings that are substituted by the pentagonal rings) see Figure 9.2.

9.1.3 Characteristics of Graphene

Electronic Properties

- At room temperature, electron mobility is extremely high, with reported values exceeding 15,000 cm^2/Vs.
- Intrinsic graphene is a semiconductor or semimetal with a zero gap.
- Low resistivity, improved current capacity, and temperature conditions.
- Graphene is thought to operate at tetrahertz frequencies, or trillions of operations per second.

Optical Properties

- It has an exceptionally high capacity for an atomic monolayer, absorbing $\pi\alpha = 2.3\%$ of white light, where α is the fine structure.
- Because of its universal optical absorption, graphene can be easily saturated in the visible to near-infrared range under strong excitation.

Mechanical Properties

- Great Young's modulus (1.100 GPa), highest breaking strength (42 N/m).
- 0.77-mg/m^2 density.
- The thinnest material conceivable (0.345 nm).
- More brittle than diamond.
- Capable of retaining its original size after strain elasticity.
- It is flexible and can be extended to 20% of its original size.

9.1.4 Advantages of Graphene

- It is ultralight and extremely tough.
- It is about 200 times stronger than the steel and is even incredibly flexible.
- It is one of the thinnest materials possible and is even completely transparent being able to transmit more than 90% of the light.
- It is the ultimate conductor and can function as a perfect barrier. Helium, too, cannot pass through graphene.
- When compared to silicon, it can transfer electrons at a much faster rate. It can travel at a speed of 1000 km/s, which is approximately 30 times faster than silicon.
- It can be used in clothing that incorporates graphene-based photovoltaic cells as well as super conductors. As a result, tablets and smartphones can be charged in minutes while still in their pockets.
- It has numerous applications mainly in the flexible displays (OLEDs, i.e., organic light-emitting diode and LCDs, i.e., liquid crystal display), energy efficient transistors, RAM, spintronics, thermal management, and so on.

9.1.5 General Applications of Graphene

Graphene is an extraordinary material that is 2D and has piqued the interest of branches of both the academia and industry (Yang et al. 2015). It has incredible electrical and thermal conductivity, as well as good mechanical behavior, making it a suitable material for electronic equipment, super-conductors, flexible transparent displays, composites, photovoltaic cells, capacitors, and detectors. Several techniques have been used to produce pristine or doped graphene. Mechanical exfoliation, chemical vapor deposition (CVD), SiC decomposition, and pulsed laser deposition (PLD), liquid-phase exfoliation are a few examples. PLD, which is frequently used for developing complex oxide thin films, has verified to be an alternative to the more broadly reported CVD method for manufacturing graphene thin films due to its advantages (Bleu et al. 2018).

9.2 SUGARCANE PRODUCTION

Sugarcane is India's most important cash crop, covering 5.06 Mha and producing 341.20 mt. Brazil, India, China, Thailand, Pakistan, and Mexico are the major sugarcane-producing countries (Nebol'sin, Galstyan, and Silina 2020). Only Brazil and India produce about half of the world's sugarcane. Production & Consumption rate of sugarcane in India (2004-2020) is mentioned in Figure 9.3. Sugarcane is grown by more than 45 million people, and sugarcane industries employ approximately 7.5% of the rural population. Sugarcane contributes 80% of the sugar, while sugar beet contributes the remaining 20%. For the production of sugar, India is primarily reliant on sugarcane. However, sugar beet cultivation for sugar is gaining popularity as a result of the commercialization of tropical sugar beet (Panella 2010).

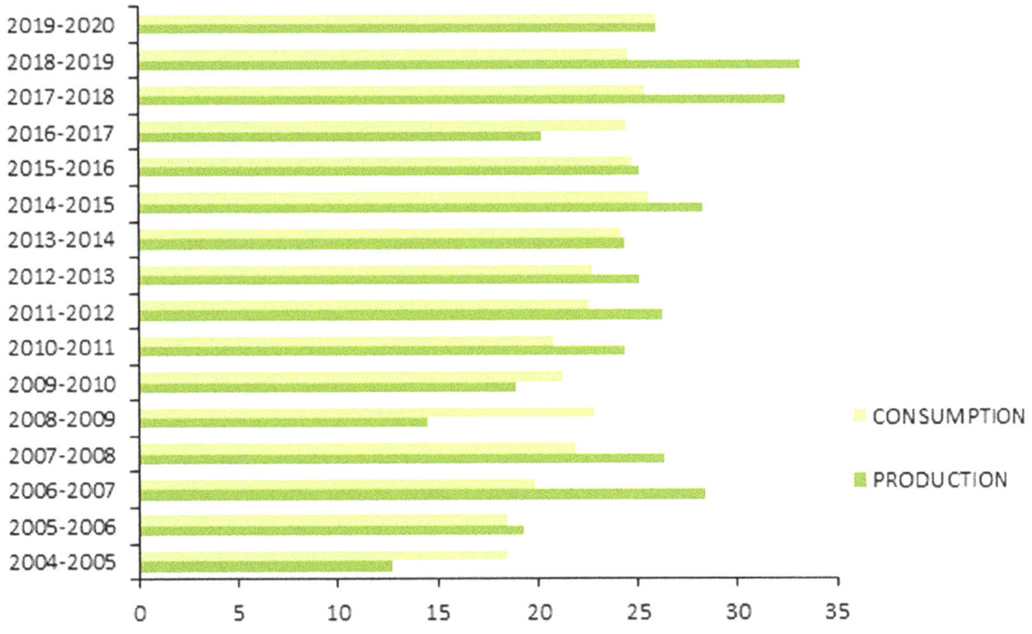

FIGURE 9.3 Production and consumption data of sugarcane in India (2004–2020) in million tons.

9.2.1 Sugarcane Bagasse (SBE)

Bagasse is the residual fiber left over after extracting sugarcane juice. Bagasse is classified into two types:

1. **Factory bagasse** is produced through a series of extraction steps in an industrial setting. Bagasse is a fibrous byproduct produced during the milling of sugarcane stalks for juice extraction. The fiber is strained to eliminate fine particles that can later be used as a filter aid or as a feedstuff ("pith bagasse") (Hussein, Amer, and Sawsan 2008). Much of the bagasse is used to generate the energy needed to run the factory.
2. **Pressed cane stalks**, also well-known as "farm bagasse," are obtained by fractionating cane on-farm or in a small-scale factory using only two or three crushers. Because of the lower extraction efficiency, it contains more sugar-rich juice and is more valuable to ruminants (50% vs. 70% extraction rate).

Dry bagasse, like other fibrous materials, is frequently used as pig and poultry litter. After that, the bagasse is recycled into organic fertilizer.

9.2.2 Sugarcane Bagasse Composition

SBE is mainly composed of hemicellulose, cellulose, and lignin, just like any other lignocellulosic material (Pereira and Arantes 2018). These three constituents account for more than 90% of the dry weight of the fiber. In general, the ash content is low (Melati et al. 2017). SBE contains a variety of minerals, including Ca, P, K, Na, Mg, Zn, Cu, and Fe (Tyagi et al. 2019). Nonfiber compounds, also known as extractives, may account for a sizable portion of the dry weight in the raw material. Raw bagasse samples were isolated in the current study using water, ethanol, and water followed by

TABLE 9.1
The Main Composition of Sugarcane Bagasse (Rezende et al. 2011)

	Water (H_2O)	Ethanol (C_2H_5OH)	Water + Ethanol ($H_2O + C_2H_5OH$)
Cellulose (%)	42.51	45.26	46.83
Hemicellulose (%)	24.89	27.30	26.85
Lignin (%)	20.91	18.94	19.77
Ash (%)	1.65	1.65	1.65
Extractives (%)	5.84	5.65	9.39
Sum (%)	95.76	98.76	104.45

ethanol (Xu et al. 2020) see Table 9.1. The chemical compositions of the extracted bagasse were then determined using a series of analyses based on Browning's (1967) methods with modifications. Extraneous materials were removed from bagasse fibers by water extraction in 5.8% (Hamzeh et al. 2013). The ethanol extraction resulted in a similar extractive content (5.6%). The successive extraction with both solvents reduced the raw material weight by 9.4%, signifying that these two solvents could dissolve structurally in different compounds. The chemical composition of sugarcane bagasse, amount of minerals present and its nutritive values are mentioned in Table 9.2, 9.3 and 9.4 respectively.

TABLE 9.2
Chemical Composition of Sugarcane Bagasse, Fresh and Nutritional Value

Main Analysis	Unit	Average Value	Standard Deviation	Minimum Value	Maximum Value	No. of Values (Samples) Used
Dry matter	% as fed	46.1	10.6	31.4	63.0	18
Crude protein	% DM	1.9	0.4	1.5	2.5	19
Crude fiber	% DM	46.0	3.8	35.9	50.4	20
Neutral detergent fiber	% DM	87.0	6.2	72.7	92.0	9
Acid detergent fiber	% DM	58.5	2.5	55.2	62.3	9
Lignin	% DM	12.6	1.2	11.1	13.7	7
Ether extracted	% DM	0.7	0.3	0.5	0.8	4
Ash	% DM	6.0	2.1	2.8	10.7	19

TABLE 9.3
Amount of Minerals Present in the Sugarcane Bagasse

Minerals	Unit	Average Value	Standard Deviation	Minimum Value	Maximum Value	No. of Values (Samples) Used
Ca	g/kg DM	1.5	1.9	0.4	7.2	15
P	g/kg DM	0.7	0.9	0.3	3.4	15
K	g/kg DM	1.4	1.1	0.7	3.4	11
Na	g/kg DM	0.2	–	0.2	0.2	3
Mg	g/kg DM	0.9	0.4	0.5	1.5	14
Zn	mg/kg DM	104	–	–	–	1.1
Cu	mg/kg DM	13	–		–	1.1
Fe	mg/kg DM	328	–	–	–	1.1

TABLE 9.4
Nutritive Values of Sugarcane Waste

Nutritive Values	Unit	Average Value
OM digestibility, ruminant	%	49.8
Energy digestibility, ruminant	%	46.6
DE ruminant	MJ/kg DM	8.7
ME ruminant	MJ/kg DM	7.1
Nitrogen degradability	%	31
Energy digestibility, growing pig	%	18.1
DE growing pig	MJ/kg DM	3.4

9.2.3 Methods of Sugarcane Bagasse Disposal

1. ***Sugarcane Waste Composting:*** The use of sugarcane waste byproducts for composting is becoming more common around the world. It aids the sugar industry in reducing the storage problem associated with sugarcane industry waste (Meghana and Shastri 2020). Figure 9.4 shows the processing of sugarcane waste composting. The following are some of the reasons why sugarcane waste by-products can be widely used for composting:
   - Composting is used to handle a large amount of sugarcane waste by-product, which greatly reduces the sugar factory's processing burden. Composting technology significantly reduces pollution and maximizes resource utilization.
   - A large amount of sugarcane industry waste has the potential to be used in agriculture to reduce fertilizer requirements and save money on chemical fertilizers. It can also be combined with inorganic chemical fertilizers and can be packed and marketed along with commercial fertilizer for a specific cropping system.
   - Sugarcane waste contains a variety of chemical elements, such as nitrogen (N), phosphorus (P), potassium (K), iron (Fe), zinc (Zn), manganese (Mn), and copper (Cu), which can improve soil structure, increase organic matter, promote soil permeability, and improve soil quality and yield (Christofoletti et al. 2013).

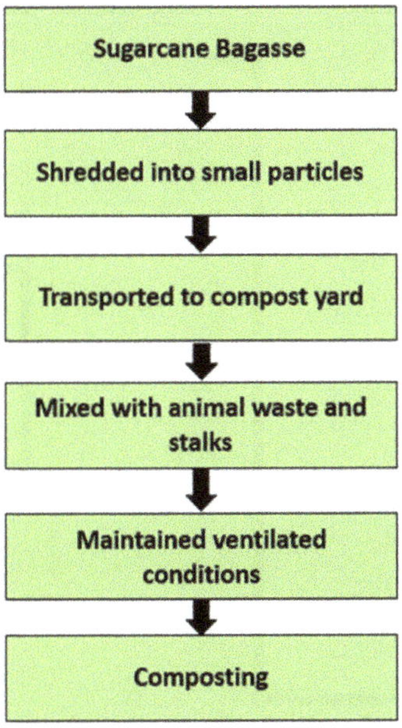

FIGURE 9.4 Processing of sugarcane waste composting.

In terms of composting technology, a variety of methods, such as windrow composting, in-vessel composting, and other indoor composting methods, have been widely adopted.

2. ***SBE Charcoal Machine:*** SBE is a waste product that can only be disposed of by feeding it to animals. The traditional method of removing SBE does not maximize the value of sugarcane waste. Beston bagasse charcoal machine has entirely exploited the surplus value of SBE in order to achieve resource recycling. Beston Machinery Company's sugarcane charcoal machine can convert SBE waste to artificial charcoal via smoke volatilization, high-temperature heating, charcoal enrichment, and sulfur release as shown in Figure 9.5.

   Features:

   - Raw materials should have a moisture content of less than 20% and a particle size of less than 30 mm.
   - The ideal temperature range is 400°C–600°C. The raw material then undergoes sharp thermal decomposition and destructive distillation to produce charcoal.
   - The charcoal product is always maintained at a ratio of 3:1 or 4:1. The yield varies depending on the raw materials and batches used.

The following are the benefits of a SBE charcoal manufacturing machine:

- It is an environmentally friendly process that can ensure long-lasting enlargement.
- The complete sugarcane charcoal carbonization process is sealed, so no pollution is released. Then the surface of furnace is therefore covered by the help of casing, which protects the workers from scalding. It also reduces heat loss, resulting in lower fuel consumption and cost.
- The production of high charcoal yield can assist us in generating additional economic benefits, resulting in an excellent market evaluation.
- The Beston biomass sugarcane charcoal making plant can not only process sugarcane waste but also municipal solid wastes (MSW), sewage, and other biomass materials such as palm shell, rice hull, wood, and sawdust.

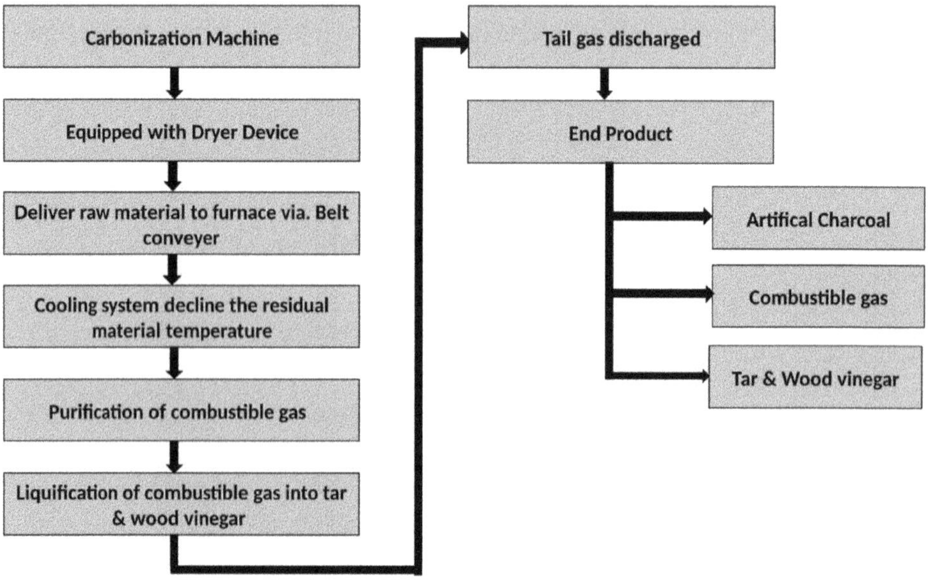

FIGURE 9.5 Process of formation of charcoal through sugarcane bagasse.

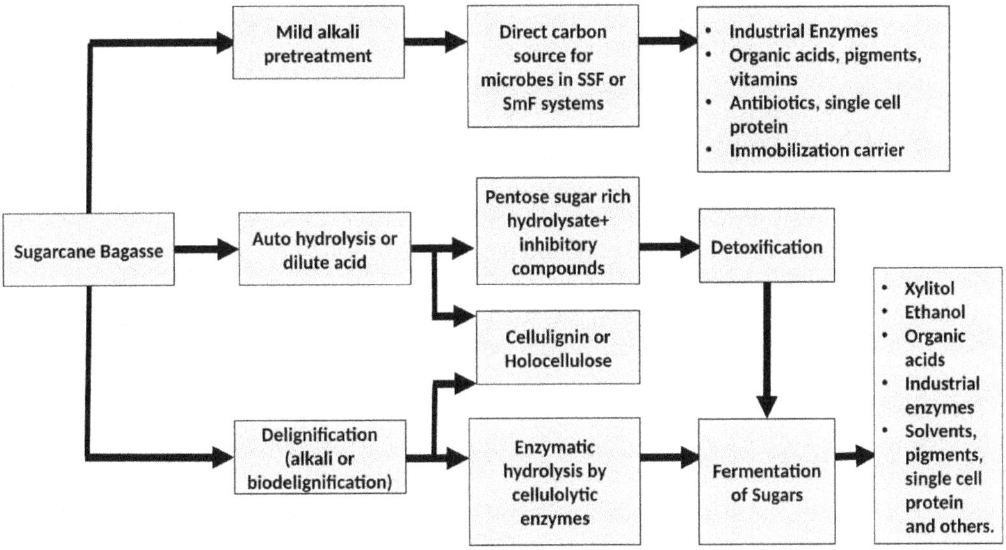

FIGURE 9.6 Other disposal ways of sugarcane bagasse.

- ***Other Methods:*** Bagasse can also be utilized to manufacture a variety of industrially important products such as industrial enzymes, antibiotics, xylitol, ethanol, organic acids, pigments, and vitamins (Sindhu et al. 2016) as shown in Figure 9.6. But, on the other hand, it will be subjected to a variety of conditions in order to be converted into these industrial products, including mild alkali pretreatment, auto hydrolysis or dilute acid, and delignification.

9.2.4 Processing and Extraction of Sugarcane Bagasse

The process of extraction of sugarcane bagasse from sugarcane plants is mentioned in Figure 9.7.

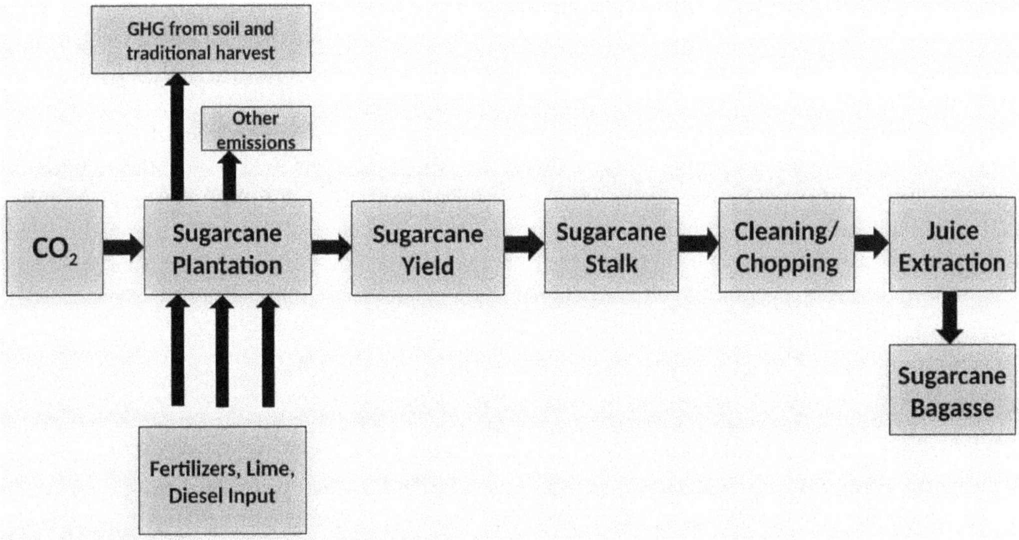

FIGURE 9.7 Extraction of sugarcane bagasse.

9.3 SYNTHESIS OF GRAPHENE FROM SUGARCANE BAGASSE

There are several ways to extract graphene from the sugarcane waste.

9.3.1 Acidic Synthesis of Graphene

In this process of synthesis, deionized water is used to wash the SBE carefully, and the washings were performed three times. Then it was put into an oven with a temperature of 60°C for about 24 hours. Later, it was ground to obtain the powered form of the sample. After this step, sulfuric acid (about 300 mL) was added to the SBE-powdered solution (about 400 mL). Then it was heated at 60°C and after 5 hours, the mixture obtained from the above process was then filtered with the help of 0.45-mm filter so that extract could be obtained. The filtrate obtained was stored at 4°C.

The classical procedure for carrying out the process of chemical conversion of graphene oxide to reduced graphene oxide can be explained as follows. It could be carried out by taking 300 mg of dried graphene oxide and dispersing it into 600 mL of deionized water. The prepared solution is then exfoliated with the help of an ultrasonicator for about 30 minutes which consists of certain hydrophilic groups. These are situated at the corners and the basal planes to ensure that graphene oxide is easily disposed of throughout the water layer after the process is completed.

The graphene oxide nanosheets were prepared using very fine dispersion of graphene oxides, and it was supported by the addition of 6000 mL of ammonia and stored at 95°C for about 12 hours. Then the extremely reduced graphene oxide (rGO) was composed by the filtration process, and with the help of deionized water, it is washed three times, so that any excess extract residue is removed if present, and the redistribution takes place into the water for carrying out the process of sonication. The suspension formed is then centrifuged for about 30 minutes at the rate of 4000 rpm. After the process completion, the precipitate achieved is collected and desiccated in a vacuum chamber and frozen for about 48 hours (Li et al. 2018).

9.3.2 Synthesis from Organic Molecules

In this process, the SBE (freshly prepared) is taken as raw material. The main components of SBE basically consist of cellulose, hemicellulose, lignin, ash, and waxes. The first step is to grind the SBE into a mesh of 0.45-mm powder. It is then followed by a process of drying in the loft drier. This is done for about 2 hours at 120°C. The SBE powder that is prepared is then soaked in the ethanol-water solution of 55 wt.% at pH 3–4. It is then followed by a process of heating at 200°C for about 100 minutes in a boiler that is designed especially for this purpose.

The initial step of this type of synthesis involves the breaking of the chemical bonds that are present among the lignin and hemicellulose. After that, the lignin gets dissolved in the solution of ethanol-water. The hemicellulose and cellulose that have been left undissolved are removed from the solution and are then transferred to the reactor in which the hemicellulose is further hydrolyzed to xylose. It is carried out at 140°C for about 20 minutes in acidic conditions, and this separates with the cellulose. The lignin, which is dissolved in the solution of ethanol-water, is recovered by the process of solvent vaporization (Tang et al. 2018).

9.3.3 Synthesis from Agro-Based Sugarcane Bagasse

In this process, graphene oxide is obtained from the agro kind of the SBE. It is basically that type of an agricultural waste that is also known as agro waste. First, the juice is extracted from it and then the fiber remaining as the residue is taken. The fiber obtained is finally crushed and grounded, so that a fine powder is produced from it. The process of crushing and separation is repeated till fine powder is obtained.

Approximately, 0.5 g of the SBE powder is first mixed with a small amount, i.e., about 0.1 g of ferrocene and is placed in a crucible. It is then put directly in a muffle furnace. It is kept at a temperature of 300°C for about 10 minutes under the atmospheric conditions. The next step is the production of the black solid obtained from the previous process. The black solid obtained as a product is collected at room temperature. This is further subjected to the different kinds of analysis, mainly spectroscopic analysis (Krishna et al. 2015).

9.3.4 Hummer's Method

This type of synthesis method is very well adapted from the modified hummer's method. Initially, 0.5 g (approx.) of charred SBE was first dispersed in nearly about 125 mL of the 98% sulfuric acid contained in an ice bath, and then slow addition of $KMnO_4$ is carried out into the mixture with the help of a glass rod with continuous stirring and the temperature conditions are controlled below 20°C. The mixture obtained is then transferred into a water bath having a temperature of 40°C with constant stirring and is carried on for 3 hours that produces a brownish gray color type of paste. Then a process of mixing is done by adding 100 mL of deionized water and constant stirring is carried out for 15 minutes at 95°C. The addition of 100 mL of deionized water addition is done one more time, then 25 mL of 30% hydrogen peroxide is slowly added, so that any excess of $KMnO_4$ gets removed. It is observed that the sample color is gets changed from dark brown to yellowish shade, and after that, the process of filtration is carried out under warm conditions. Several washings further are given with the help of distilled water to ensure that the pH becomes neutral. The sample in the dissolved form is obtained and is then dried in an oven having a temperature of 80°C for 6 hours. The solid sample that is dried is now ultrasonically removed with the help of an ultrasonic bath for several minutes and then the analysis is carried out (Chen et al. 2013).

9.3.5 Graphene Synthesis with Urea

Initially, 5 g of SBE powder was treated with 6% NaOH (100 mL) solution, and the temperature conditions were kept at 60°C for the time duration of 30 minutes. Later, it was settled down and cooled at room temperature. Then, the vacuum filtration was carried out and washings were done many times with distilled water till the solution that was obtained, and it became neutral. The solid was then dried in an oven set at a temperature of 120°C for about 2 hours, and it was observed that a light-yellow-colored SBE fiber was obtained as the product. Then a mixture was prepared of distilled water, NaOH, and urea in the ratio of 85:10:5, and this solution was kept at −10°C for about 8 hours. The SBE that was prepared above was mixed in 200 mL solution 0°C, and it was kept for 6 hours with a little bit of stirring. The temperature was raised to room temperature, and 0.5% basic SBE fiber solution was obtained. Then, about 5-mg/mL graphene oxide suspension was then added to the solution as prepared above, and this suspension was proceeded for filtration and washings were given with the help of distilled water. The wet solid obtained was therefore dried in an oven at a 60°C, and finally, the graphene-oxide-modified SBE was obtained as the product (Aruna et al. 2021).

9.4 CHARACTERIZATION TECHNIQUES

9.4.1 UV-Visible (UV-Vis) Spectra

The graphene oxide prepared was first analyzed by the UV-Vis spectra of reduced graphene oxide. The data collected after the analysis explained that the absorption peak of the SBE extract was at 196 nm, and for graphene oxide, it was at 232.5 nm see Figure 9.8. This kind of absorption revealed that there is

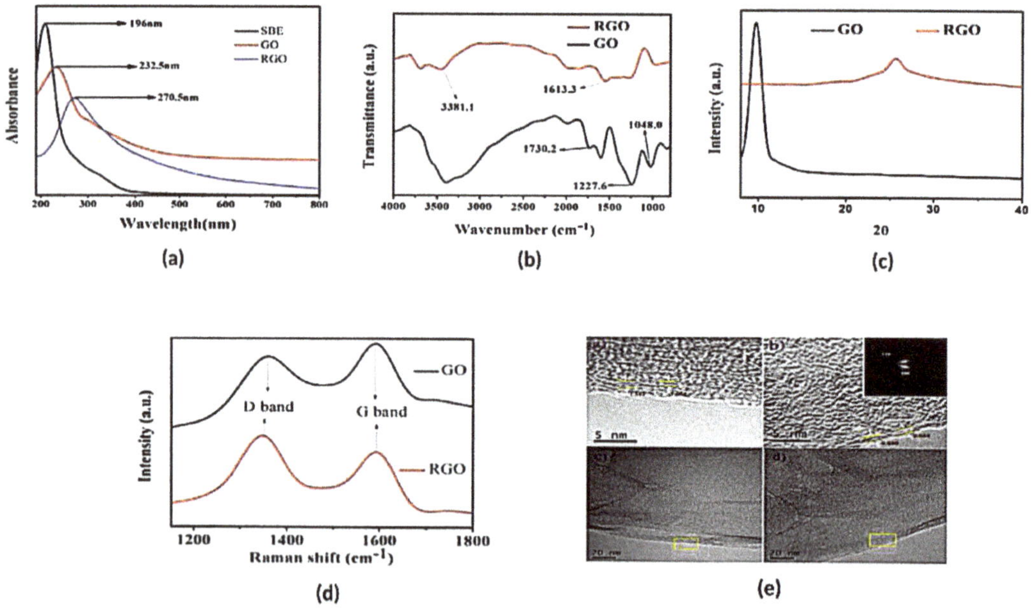

FIGURE 9.8 Characterization of the synthesized graphene: (a) UV, (b) IR, (c) XRD, (d) Raman spectra, and (e) HRTEM.

a π→π* transition that is occurring because of the presence of the aromatic C-C bonds. Also, the weak absorption band that was spotted at 300 nm deduced that there is a type of n→π* transition of the C=O bonds that are found in the graphene oxide (Li et al. 2018).

9.4.2 Fourier Transform Infrared (FTIR)

It was observed from the FTIR data of graphene oxide that there exists O-H stretching vibration, and in one of the strong bands, which was a wide band, it was recorded at 3381.1 cm^{-1} as shown in Figure 9.8. Also, a narrow peak was observed at the absorption value 1730.2 cm^{-1}. This is due to the stretching of C=O from the COOH group. The bands were observed at 1613.3 cm^{-1} and 1048 cm^{-1}. These corresponded to the presence of C=C and C=O, respectively. The peak that was marked at the absorption value 1594 cm^{-1} was allotted to the skeletal vibrations of the graphene oxide. Also, the band marked at 1060 cm^{-1} revealed the presence of the epoxy groups (C-O) in the compound.

9.4.3 High Resolution Transmission Electron Microscopy (HRTEM)

HRTEM was also brought into use for the determination of reduced graphene-oxide-layered structure. It was observed that the graphene oxide layer was folded in some parts as mentioned in Figure 9.8. When the observation was done at a higher magnification, the average thickness of the reduced graphene oxide layer was recorded at 0.34 nm.

9.4.4 X-Ray Diffraction (XRD) Analysis

XRD analysis showed that the diffraction peak of graphene oxide was observed near $2\Theta = 10.80$. It basically represents the d-spacing of about 0.83 nm. The value of this d-spacing is much larger than the value of the d-spacing of the original graphite as its value is 0.34 nm with $2\Theta = 26.50$. This happened due to the presence of the oxidized functional groups onto the sheets of carbon. After the reduction was performed, the peak observed for reduced graphene oxide was at $2\Theta = 26.50$ see Figure 9.8. The peak for graphene oxide at $2\Theta = 10.80$ disappeared, which reduced the interlayer distance (Li et al. 2018).

9.4.5 Scanning Electron Microscopy (SEM)

The samples of the graphene oxide synthesized were rigid and hard, so they were first grounded into fine powder. The SEM images of the sample were recorded, and it revealed that the presence of the sharp edges and curved surface were observed. These were very similar to those of the ceramics. It also tells that graphene oxide had a layered and a sheet-like structure. As there were epoxy and the hydroxy groups present, the layer appears to be set apart from one another. The particles when observed closely had a well-defined and isotropic shape. The hydrogen bonds along with covalent bonds contributed to the surface properties that gradually increased the ability of the absorption.

9.4.6 Raman Spectra

Raman analysis was also carried out efficiently for the identification of the structure of graphene. Two types of peaks were observed nearly at 1350 cm^{-1} (high D band) and 1580 cm^{-1} (low G band) as shown in Figure 9.8. The intensity ratio of the reduced graphene oxide corresponded to that of the graphene sheet obtained was higher and kind of disordered. Also, the size of the sp^2 in the plane domain got declined after the process of reduction took place. It led to the confirmation that the reduction of the graphene oxide has led to the fragmentation and produced many reduced graphene oxide domains.

9.4.7 Fluorescence

The excitation of the graphene particles was done at 350 nm. Under the UV filter, a blue-colored fluorescence was observed. On carrying out the reduction of the graphene oxide substance, π domains became isolated because they formed π-electrons that were sp^2 hybridized. The quantum dots that were carbon-based had C=C as their core along with as the functional groups (O and H) were present on the surface. These have the energy levels that are the reason for series emission. The functional groups that are oxygen related and are present at the boundaries are mainly responsible for emitting the blue emission of carbon-based quantum dots after the excitation process was carried out.

9.5 APPLICATIONS OF GRAPHENE SYNTHESIZED FROM SUGARCANE BAGASSE

9.5.1 Membranes

Due to the impermeability of graphene, a single layer of atoms can act as a perfect barrier when commerce with liquids and gases (Berry 2013). However, graphene pores (defects) can selectively permit the passage of water and gas molecules, letting graphene to be tuned to allow selective gas permeability.

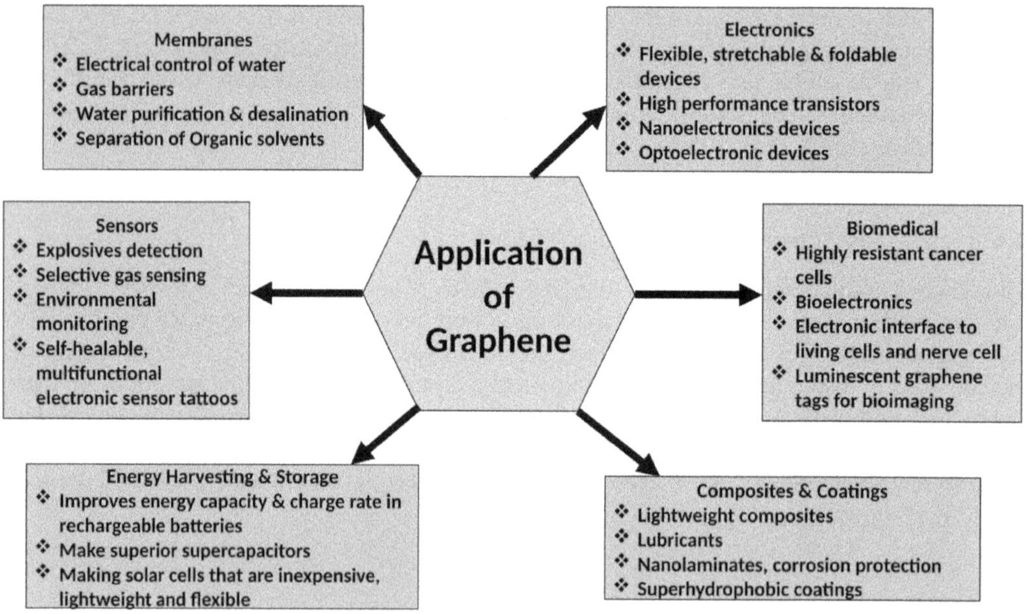

FIGURE 9.9 Various application of graphene obtained from sugarcane bagasse.

For example, the GO-FeNPs coated RO membranes (Armendariz-Ontiveros et al. 2018), as mentioned in Figure 9.10.

Applications

- Using electricity to control the flow of water through graphene membranes.
- Gas barriers, such as those found in food packaging.
- Demineralization and purification of water.
- Organic solvent separation from water.

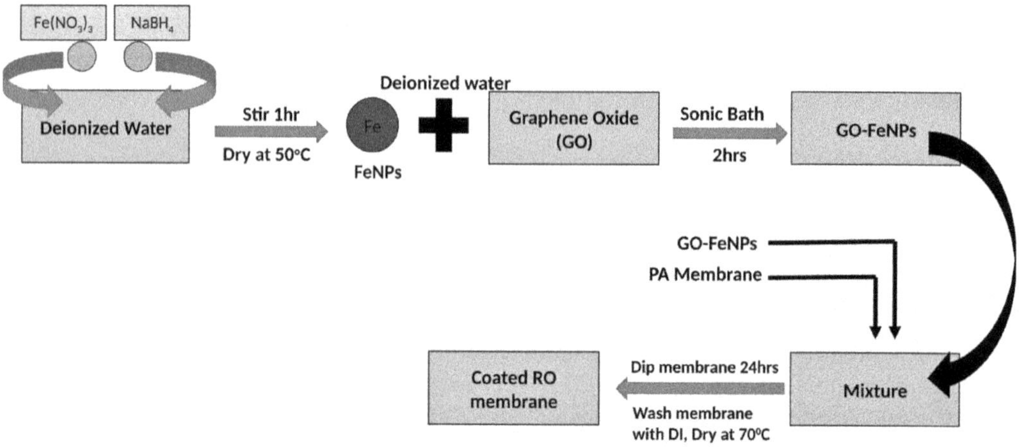

FIGURE 9.10 The GO-FeNPs-coated RO membranes are depicted schematically.

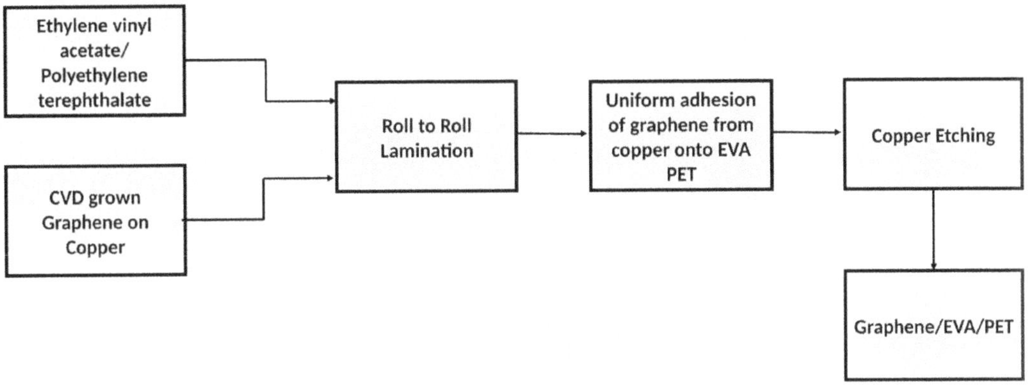

FIGURE 9.11 Fabrication flow chart of top layer of the graphene-based triboelectric nanogenerator.

9.5.2 Electronics

The atomic organization of the carbon atoms in graphene is very exclusive as it allows its electrons to easily move even at enormously high speed without affecting scattering, saving the lost energy observed in other conductors (Trivedi, Lobo, and Ramakrishna Matte 2019). The graphene system electronic properties are determined by the number of graphene layers present, as well as the coupling effects from the primary substrate.

For example, fabrication of flow chart of top layer of the graphene-based triboelectric nanogenerator (Shankaregowda et al. 2016), as mentioned in Figure 9.11.

Applications

- Devices that are stretchable, flexible, and foldable devices.
- Printable electronics at a low cost.
- Electronic equipment with high frequencies.
- Transistors with superior efficiency.
- Nano-electronics thermal management and heat dissipation.
- Because graphene's optical properties can be governed through doping, it's quite well for optoelectronic devices (electrical devices that source, detect, and control light).

9.5.3 Biomedical Technologies

Due to its electron (e^-) mobility, functionalization capability, and high surface area, graphene can improve/boost up the innovative biomedical technologies. Graphene bioelectronics (transistors and electrode arrays) provide exciting potential for the formation of new types of biosensors capable of sustaining excellent interfaces with soft tissue (Kireev and Offenhäusser 2018).

For example, tumor killing therapy using graphene-based nanocomposites activated by neutrons (Ghorai 2019), as mentioned in Figure 9.12.

Applications

- Thermal ablation of cancer cells with high resistance.
- Detection and neutralization of cancer stem cells.
- Bioelectronics (bionics).
- Electronics can contact with living cells and nerve tissue.
- Bioimaging with luminescent graphene tags.

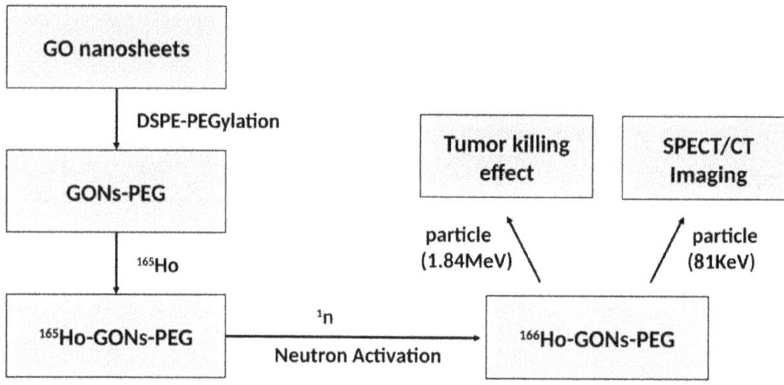

FIGURE 9.12 Tumor killing therapy using graphene-based nanocomposites activated by neutrons.

9.5.4 Sensors

In graphene, every atom is liable to its surroundings, and it is a novel material for chemical, gas, and biological sensors (Nag, Mitra, and Mukhopadhyay 2018).

For example, key steps for fabricating the ZnO nanorod-graphene hybrid sensor (Yi, Lee, and Park 2011) as mentioned in Figure 9.13.

Applications

- Detection of explosives.
- Biosensor for detecting Parkinson's disease biomarkers and bacteria.
- Specific gas detection.
- Tattoos with self-healing, multifunctional electronic sensors.
- Observing the environment (Aïssa et al. 2015).

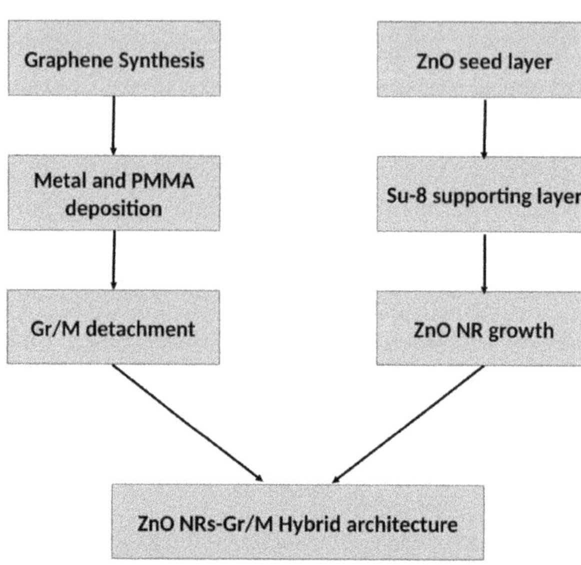

FIGURE 9.13 The key steps for fabricating the ZnO nanorod-graphene hybrid sensor architectures are depicted schematically.

9.5.5 Energy Harvesting and Storage

Graphene and graphene-based nanomaterials have an influence on four energy-related areas. Photovoltaics, graphene batteries, superconductors, and catalysis, mostly for fuel cells, are examples of these (Mamvura and Simate 2019).

For example, flow chart to fabricate Ag/graphene oxide/p-silicone/Ti-Au heterojunction photovoltaic cells (Mahala, Gupta, and Singh 2020), as mentioned in Figure 9.14.

Applications

- Graphene increases the energy capacity as well as the charge rate of rechargeable batteries.
- Activated graphene essentially produces superior supercapacitors, which are required for energy storage.
- Graphene electrodes could pave the way for a promising approach to producing solar cells that are low cost, flexible, and lightweight.
- Experimentation designs for graphene-based solar cells in which graphene serves as various components of the cell.
- Multifunctional graphene mats are promising catalytic system substrates.
- Proton transport in graphene holds promise for mimicking photosynthesis artificially.

9.5.6 Composites and Coatings

Graphene's mechanical properties, such as its low loading and low mass requirements, distinguish it as a reinforcing agent in composites. Graphene-coated objects and graphene-reinforced composites can also serve a variety of functions (Das and Prusty 2013).

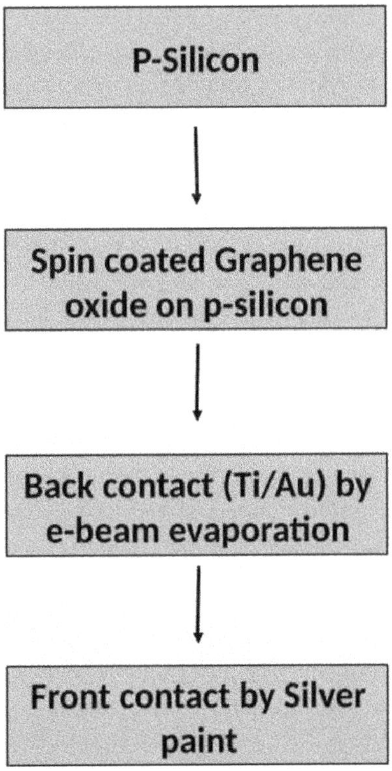

FIGURE 9.14 Flow chart to fabricate Ag/graphene oxide/*p*-silicon/Ti-Au heterojunction photovoltaic cells.

Applications

- Body structures made of lightweight composites.
- Lubricants with improved anti-wear characteristics.
- Nano-laminates as high-permeation barrier films.
- Corrosion resistance.
- Translucent conductive coating for solar devices.
- Protection from lightning strikes and radiation shielding.
- Coatings that are superhydrophobic.
- Thin films that are translucent, conductive, and flexible.
- Highly effective electromagnetic interference (EMI) shielding sheets.

9.6 CONCLUSION AND FUTURE PROSPECTS

This chapter demonstrates a fundamentally simple and quick method for effectively transforming solid sugarcane waste into valuable graphene oxide. The resulting graphene has a well-graphitized structure. SBE is recyclable and a great source of carbon-based nanomaterials. Graphene was synthesized using several different methods. In comparison to commonly used methods, the methods discussed in this chapter are more eco-friendly because they avoid the emission of toxic gases during the synthesis. This enormously simple and inexpensive synthesis method could be exposed up with new possibilities for the low-cost manufacture of graphene-based materials for gas sensors and other functional devices along with energy storage (Mohan et al. 2018).

Graphene-based hybrid nanostructures are presently being considered as ideal materials for fields such as biomedical engineering, physics, green chemistry, nanotechnology, and material science, due to their tunable physical properties, elevated electronic, high surface area, and thermal properties. As a consequence, graphene and its derivatives have revealed excellent commercial applications in nanomedicines, composites, bioimaging, and nanoelectronics, in a comparatively short period of time. Functionalized graphene nanosheets, for example, have demonstrated an upgraded interfacial interface along with the adhesive properties of protein, mammalian cells, and microbial, which makes graphene a value-added nanosystem for the next-generation multi-functional in bioengineering applications. The manufacturing, on the one hand, is inexpensive, and research on ultrapure pristine graphene layers, on the other hand, is still in progress. As a result, finding a simple and delectable way to yield graphene is a vast challenge and problem for materials scientists. Likewise, time-dependent compatibility and interfaces of graphene and its derivatives in vitro and in vivo conditions are one of the most challenging tasks for graphene researchers employed on several features of graphene. Consequently, it can be determined that graphene has exhibited a great competency in practically every division of science and technology. An abundant support for supplementary research is vital from the governments and industries to tackle the complete prospective of graphene and its derivatives for catalyst, bioimaging, waterproof coating, UV lens, charge conductor, sound transducers, optoelectronics, radio wave absorption, coolant additive, frequency multiplier, Hall effect sensors, conductive ink, spintronics, and piezoelectric applications (Wang 2020).

Since graphene assists as a model for new kinds of 2D materials like germanene, phosphorene, and silicene, and various transition metals such as chalcogenides, hexagonal boron nitride, and boron nitride nanosheets are appearing as new and innovative nanostructures for next-generation nanotechnology and nanoscience research (Pan et al. 2014). It has been expressed that in some cases, these 2D nanomaterials outperform graphene. Deep investigation on these new nanosystems is currently underway, with increasing success for prototype applications.

REFERENCES

Aïssa, Brahim, Nasir K Memon, Adnan Ali, and Marwan K Khraisheh. 2015. "Recent Progress in the Growth and Applications of Graphene as a Smart Material: A Review." *Frontiers in Materials* 2: 58. https://doi.org/10.3389/fmats.2015.00058

Allen, Matthew, Vincent Tung, and Richard Kaner. 2009. "Honeycomb Carbon: A Review of Graphene." *Chemical Reviews* 110 (August): 132–45. https://doi.org/10.1021/cr900070d

Armendariz-Ontiveros, Maria, Alejandra Garcia, Sergio de los Santos-Villalobos, and Gustavo Fimbres Weihs. 2018. "Biofouling Performance of RO Membranes Coated with Iron NPs on Graphene Oxide." *Desalination* 451 (July). https://doi.org/10.1016/j.desal.2018.07.005

Aruna, Nisha B, Ashok K Sharma, and Surender Kumar. 2021. "A Review on Modified Sugarcane Bagasse Biosorbent for Removal of Dyes." *Chemosphere* 268: 129309. https://doi.org/10.1016/j.chemosphere.2020.129309

Baweja, Himani, and Kiran Jeet. 2019. "Economical and Green Synthesis of Graphene and Carbon Quantum Dots from Agricultural Waste." *Materials Research Express* 6 (August): 0850g8. https://doi.org/10.1088/2053-1591/ab28e5

Berry, Vikas. 2013. "Impermeability of Graphene and Its Applications." *Carbon* 62 (June). https://doi.org/10.1016/j.carbon.2013.05.052

Bleu, Yannick, Florent Bourquard, Teddy Tite, A.-S Loir, Chirandjeevi Maddi, Christophe Donnet, and F Garrelie. 2018. "Review of Graphene Growth from a Solid Carbon Source by Pulsed Laser Deposition (PLD)." *Frontiers in Chemistry* 6 (November). https://doi.org/10.3389/fchem.2018.00572

Chen, Ji, Bowen Yao, Chun Li, and Gaoquan Shi. 2013. "An Improved Hummers Method for Eco-Friendly Synthesis of Graphene Oxide." *Carbon* 64: 225–29. https://doi.org/10.1016/j.carbon.2013.07.055

Christofoletti, Cintya, Janaína Escher, Jorge Correia, Julia Marinho, and Carmem Fontanetti. 2013. "Sugarcane Vinasse: Environmental Implications of Its Use." *Waste Management (New York, N.Y.)* 33 (September). https://doi.org/10.1016/j.wasman.2013.09.005

Cooper, Daniel R, Benjamin D'Anjou, Nageswara Ghattamaneni, Benjamin Harack, Michael Hilke, Alexandre Horth, Norberto Majlis, et al. 2012. "Experimental Review of Graphene." In, edited by Y Kopelevich and S Bud'ko. *ISRN Condensed Matter Physics*, 501686. https://doi.org/10.5402/2012/501686

Das, Tapan, and Smita Prusty. 2013. "Graphene-Based Polymer Composites and Their Applications." *Polymer-Plastics Technology and Engineering* 52 (March): 319–33. https://doi.org/10.1080/03602559.2012.751410

Ghany, Nabil A A, Safaa A Elsherif, and Hala T Handal. 2017. "Revolution of Graphene for Different Applications: State-of-the-Art." *Surfaces and Interfaces* 9: 93–106. https://doi.org/10.1016/j.surfin.2017.08.004

Ghorai, Tanmay K. 2019. "9 – Graphene Oxide-Based Nanocomposites and Biomedical Applications." In, edited by Sabyasachi Maiti and Sougata B T – *Functional Polysaccharides for Biomedical Applications Jana*, 305–28. Woodhead Publishing. https://doi.org/10.1016/B978-0-08-102555-0.00009-1

Hamzeh, Yahya, Alireza Ashori, Zeinab Khorasani, Ali Abdulkhani, and Ali Abyaz. 2013. "Pre-Extraction of Hemicelluloses from Bagasse Fibers: Effects of Dry-Strength Additives on Paper Properties." *Industrial Crops and Products* 43: 365–71. https://doi.org/10.1016/j.indcrop.2012.07.047

Hussein, M, A A Amer, and I I Sawsan. 2008. "Oil Spill Sorption Using Carbonized Pith Bagasse: 1. Preparation and Characterization of Carbonized Pith Bagasse." *Journal of Analytical and Applied Pyrolysis* 82 (2): 205–11. https://doi.org/10.1016/j.jaap.2008.03.010

Kireev, Dmitry, and Andreas Offenhäusser. 2018. "Graphene & Two-Dimensional Devices for Bioelectronics and Neuroprosthetics." *2D Materials* 5 (4): 42004. https://doi.org/10.1088/2053-1583/aad988

Krishna, Mohana V, Thirunavukkarasu Somanathan, Karthika Prasad, Kostya Ostrikov, and A Saravanan. 2015. "Graphene Oxide Synthesis from Agro Waste." *Nanomaterials* 5 (May): 826–34.

Li, Beibei, Xiaoying Jin, Jiajiang Lin, and Zuliang Chen. 2018. "Green Reduction of Graphene Oxide by Sugarcane Bagasse Extract and Its Application for the Removal of Cadmium in Aqueous Solution." *Journal of Cleaner Production* 189: 128–34. https://doi.org/10.1016/j.jclepro.2018.04.018

Lu, Haigang, and Si-Dian Li. 2013. "Two-Dimensional Carbon Allotropes from Graphene to Graphyne." *Journal of Materials Chemistry C* 1 (May): 3677–80. https://doi.org/10.1039/C3TC30302K

Mahala, Pramila, Navneet Gupta, and Sumitra Singh. 2020. "Silicon Photovoltaic Cell Based on Graphene Oxide as an Active Layer." *Microsystem Technologies*, March. https://doi.org/10.1007/s00542-020-04763-3

Mamvura, Tirivaviri A, and Geoffrey S Simate. 2019. "Chapter 4 – The Potential Application of Graphene Nanotechnology for Renewable Energy Systems." In, edited by Mohammad Jawaid, Akil Ahmad, and David B T – *Graphene-Based Nanotechnologies for Energy and Environmental Applications Lokhat (Micro and Nano Technologies)*, 59–80. Elsevier. https://doi.org/10.1016/B978-0-12-815811-1.00004-1

Meghana, Munagala, and Yogendra Shastri. 2020. "Sustainable Valorization of Sugar Industry Waste: Status, Opportunities, and Challenges." *Bioresource Technology* 303: 122929. https://doi.org/10.1016/j.biortech.2020.122929

Melati, Ranieri, Alison Schmatz, Fernando Pagnocca, Jonas Contiero, and Michel Brienzo. 2017. "Sugarcane Bagasse: Production, Composition, Properties, and Feedstock Potential." In *Sugarcane: Production Systems, Uses and Economic Importance*, 1–38, Nova Science Publisher.

Mohan, Velram B, Kin-tak Lau, David Hui, and Debes Bhattacharyya. 2018. "Graphene-Based Materials and Their Composites: A Review on Production, Applications and Product Limitations." *Composites Part B: Engineering* 142: 200–220. https://doi.org/10.1016/j.compositesb.2018.01.013

Nag, Anindya, Arkadeep Mitra, and Subhas C Mukhopadhyay. 2018. "Graphene and Its Sensor-Based Applications: A Review." *Sensors and Actuators A: Physical* 270: 177–94. https://doi.org/10.1016/j.sna.2017.12.028

Nebol'sin, V A, V Galstyan, and Y E Silina. 2020. "Graphene Oxide and Its Chemical Nature: Multi-Stage Interactions between the Oxygen and Graphene." *Surfaces and Interfaces* 21: 100763. https://doi.org/10.1016/j.surfin.2020.100763

Pan, Yi, Lizhi Zhang, Li Huang, Linfei Li, Lei Meng, Min Gao, Qing Huan, et al. 2014. "Construction of 2D Atomic Crystals on Transition Metal Surfaces: Graphene, Silicene, and Hafnene." *Small* 10 (June). https://doi.org/10.1002/smll.201303698

Panella, Lee. 2010. "Sugar Beet as an Energy Crop." *Sugar Tech* 12 (December). https://doi.org/10.1007/s12355-010-0041-5

Papageorgiou, Dimitrios G, Ian A Kinloch, and Robert J Young. 2017. "Mechanical Properties of Graphene and Graphene-Based Nanocomposites." *Progress in Materials Science* 90: 75–127. https://doi.org/10.1016/j.pmatsci.2017.07.004

Pereira, Bárbara, and Valdeir Arantes. 2018. "Chapter 9 – Nanocelluloses From Sugarcane Biomass." In, edited by Anuj Kumar Chandel and Marcos Henrique B T – *Advances in Sugarcane Biorefinery Luciano Silveira*, 179–96. Elsevier. https://doi.org/10.1016/B978-0-12-804534-3.00009-4

Rezende, Camila A, Marisa A de Lima, Priscila Maziero, Eduardo R deAzevedo, Wanius Garcia, and Igor Polikarpov. 2011. "Chemical and Morphological Characterization of Sugarcane Bagasse Submitted to a Delignification Process for Enhanced Enzymatic Digestibility." *Biotechnology for Biofuels* 4 (1): 54. https://doi.org/10.1186/1754-6834-4-54

Shankaregowda, Smitha A, B N Chandrashekar, Xiao-Liang Cheng, Mayue Shi, Zhong-fan Liu, and Hai-Xia Zhang. 2016. "A Flexible and Transparent Graphene Based Triboelectric Nanogenerator." *IEEE Transactions on Nanotechnology* 15 (May): 1. https://doi.org/10.1109/TNANO.2016.2540958

Sindhu, Raveendran, Edgard Gnansounou, Parameswaran Binod, and Ashok Pandey. 2016. "Bioconversion of Sugarcane Crop Residue for Value Added Products – An Overview." *Renewable Energy* 98: 203–15. https://doi.org/10.1016/j.renene.2016.02.057

Tang, Pei-Duo, Qi-Shi Du, Da-Peng Li, Jun Dai, Yan-Ming Li, Fang-Li Du, Si-Yu Long, Neng-Zhong Xie, Qing-Yan Wang, and Ri-Bo Huang. 2018. "Fabrication and Characterization of Graphene Microcrystal Prepared from Lignin Refined from Sugarcane Bagasse." *Nanomaterials*. https://doi.org/10.3390/nano8080565

Trivedi, Shivam, Kenneth Lobo, and H S S Ramakrishna Matte. 2019. "Chapter 3 – Synthesis, Properties, and Applications of Graphene." In, edited by Morgan Hywel, Chandra Sekhar Rout, and Dattatray J B T – *Fundamentals and Sensing Applications of 2D Materials Late (Woodhead Publishing Series in Electronic and Optical Materials)*, 25–90. Woodhead Publishing. https://doi.org/10.1016/B978-0-08-102577-2.00003-8

Tyagi, Swati, Kui-Jae Lee, Sikandar I Mulla, Neelam Garg, and Jong-Chan Chae. 2019. "Chapter 2 - Production of Bioethanol From Sugarcane Bagasse: Current Approaches and Perspectives." In, edited by Pratyoosh B T – *Applied Microbiology and Bioengineering Shukla*, 21–42. Academic Press. https://doi.org/10.1016/B978-0-12-815407-6.00002-2

Vitti, Godofredo C, Pedro H de Cerqueira Luz, and Wellington S Altran. 2015. "Chapter 4 – Nutrition and Fertilization." In, edited by Fernando Santos, Aluízio Borém, and Celso B T – *Sugarcane Caldas*, 53–88. San Diego: Academic Press. https://doi.org/10.1016/B978-0-12-802239-9.00004-9

Wang, Nannan. 2020. "Graphene Research and Their Outputs: Status and Prospect," October. https://doi.org/10.1016/j.jsamd.2020.01.006

Xu, Chao, Fen Liu, Md Asraful Alam, Huanjun Chen, Yu Zhang, Cuiyi Liang, Huijuan Xu, Shushi Huang, Jingliang Xu, and Zhongming Wang. 2020. "Comparative Study on the Properties of Lignin Isolated from Different Pretreated Sugarcane Bagasse and Its Inhibitory Effects on Enzymatic Hydrolysis." *International Journal of Biological Macromolecules* 146 (March): 132–40. https://doi.org/10.1016/j.ijbiomac.2019.12.270

Yang, Guohai, Chengzhou Zhu, Dan Du, Jun-Jie Zhu, and Yuehe Lin. 2015. "Graphene-like Two-Dimensional Layered Nanomaterials: Applications in Biosensors and Nanomedicine." *Nanoscale* 7 (July). https://doi.org/10.1039/C5NR03398E

Yi, Jaeseok, Jung M Lee, and Won Il Park. 2011. "Vertically Aligned ZnO Nanorods and Graphene Hybrid Architectures for High-Sensitive Flexible Gas Sensors." *Sensors and Actuators B: Chemical* 155 (1): 264–69. https://doi.org/10.1016/j.snb.2010.12.033

Zhen, Zhen, and Hongwei Zhu. 2018. "1 – Structure and Properties of Graphene." In, edited by Hongwei Zhu, Zhiping Xu, Dan Xie, and Ying B T - *Graphene Fang*, 1–12. Academic Press. https://doi.org/10.1016/B978-0-12-812651-6.00001-X

Zhu, Xin, Jianglei Luo, Li Fan, Feng Chen, Chunmei Li, Guannan Li, and Zhiqian Chen. 2020. "Regulation of Graphane by Strain: First-Principles Study." *Journal of Solid State Chemistry* 291: 121626. https://doi.org/10.1016/j.jssc.2020.121626

10 Graphene Synthesis, Characterization and Applications

Hamidreza Bagheri, Marzieh Fatehi, and Ali Mohebbi

CONTENTS

10.1 INTRODUCTION

Carbon derives its name of the word carbo, and its meaning is charcoal (Bonaccorso et al., 2012). Carbon is significant element due to its significant electronic structure that allows for hybridization to shape up sp, sp^2 and sp^3 networks and, therefore, to form stable allotropes compared to other elements (Huang et al., 2011). Graphite is the most popular carbon allotropic, and it is a natural mineral. Carbon element was found out in the all known life forms (Wanno and Tabtimsai, 2014). Carbon individuality is established in several allotropes in that it happens. This element shows smoothness such as graphite in pencil, to hardest known element in diamond (Trivedi et al., 2019). The other allotropes of carbon are nanotubes, fullerenes, glassy carbon and amorphous carbon (Figure 10.1). The main parameters of carbon-based materials of various dimensionalities are given in Table 10.1. Forth suggested naming conventions are also given in Table 10.2. Indeed, the phrase *graphene* must be used for isolated monolayer hexagonally form that organized bonded of carbon in configuration of sp^2 (Feriancikova and Xu, 2012). The term *graphite oxide* is the material in solid state, provided with oxidation of graphite to functionalize fundamental planes and enhance inter-layer spacing. The term *graphene oxide* is graphite oxide exfoliated form and it is usually provided with dispersing this very solvable material in an aqueous solvent. To end, *reduced graphene oxide* consequences from reduction of graphene oxide (GO) (Lee et al., 2008).

The carbon catenation property is important and it allows atoms of carbon to bond with the other atoms to procedure structures or long chains (Zhu et al., 2011). Graphite contains of sp^2 hybridized

DOI: 10.1201/9781003169741-10

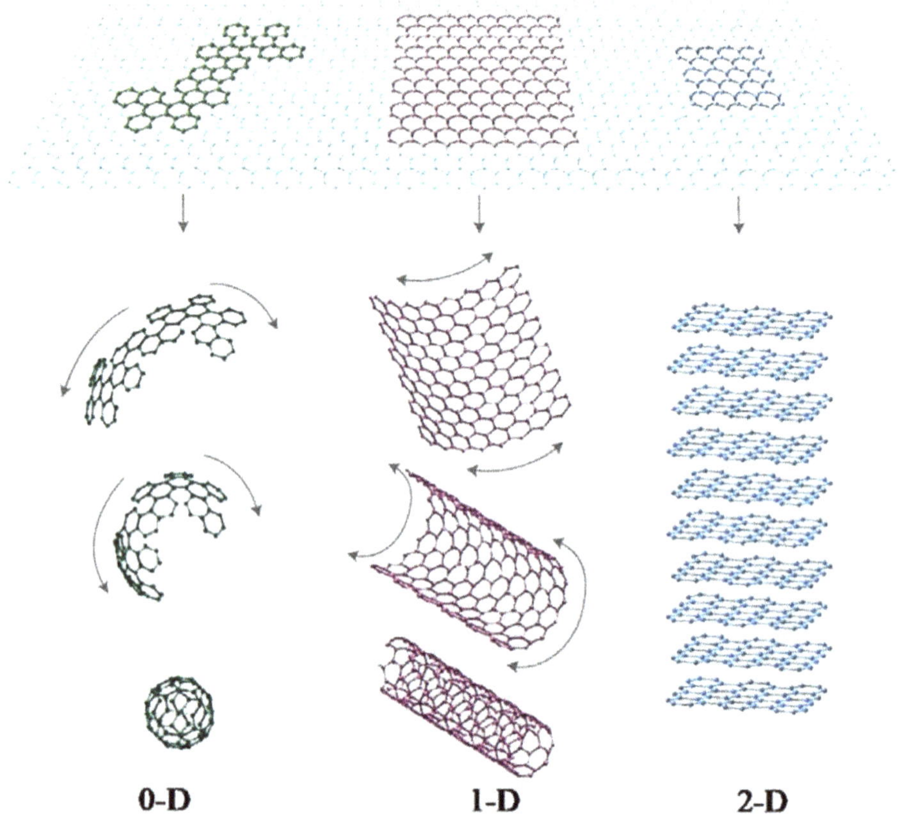

FIGURE 10.1 Different forms of carbon in various dimensions. (Reprinted with permission of Trivedi et al. (2019).)

carbon element layers that are stacked with van der Waals forces. The carbon atoms single layers strongly packed in a two-dimensional honeycomb crystal framework is called graphene. This name was introduced in 1994 (Subrahmanyam et al., 2011). Graphene is a two-dimensional sheet with single carbon atoms layer and hexagonal packed structure. Graphite shows an amazing anisotropic treatment by regard to electrical and thermal conductivity properties (Wang et al., 2009). Graphite is very conductive into direction parallel to graphene layers due to in-plane character, but graphite indicates weak conductivity in direction vertical to layers due to the weak van der Waals forces interactions among them (Schniepp et al., 2006). Graphite amounts interplanar spacing to 3.3×10^{-10} m and is not enough huge to inorganic components, organic molecules and host ions. But, numerous intercalation plans

TABLE 10.1
Significant Carbon Materials Parameters of Various Dimensionalities

Dimension	3-Dimensional	2-Dimensional	1-Dimensional	0-Dimensional
Isomer	Diamond	Graphene	Nanotube	Fullerene
Hybridization	sp^3	sp^2	sp^2	sp^2
Density (kg/m^3)	3515	2260	1200–2000	1720
Bond length	1.54 (C-C)	1.42 (C=C)	1.44 (C=C)	1.40 (C=C), 1.46 (C-C)
Electronic properties	Insulator	Zero-gap semiconductor	Metal/Semiconductor	Semiconductor

Source: Modified after Gong et al. (2013).

TABLE 10.2
Nomenclature Applied for Graphene Family Substances

Substance	Abbreviation	Explanation
Graphene	–	2D sp^2 bonded carbon layer in a hexagonal arrangement by single atom
Graphene oxide	GO	GO exfoliated form
Graphite oxide	–	Graphite that has been oxidized to prepare O_2 functional groups on basal planes and enhance inter-layer space
Reduced graphene oxide	rGO	Product of reduction (solvothermal, thermal-chemical) of GO

Source: Modified after Zhu et al. (2011).

have been used to increase the graphite interlayer galleries from 3.3×10^{-10} m to up, that can attain more than 1-nm value in some situations, depending on guest components size (Evanoff et al., 2011).

Graphene has been considered by about 60 years; the first investigations on the graphite properties were presented between 1946 and 1959 and they are still studied as a critical reference in describing the features of different carbon-based components (Zhang et al., 2013). The graphene was supposed to be unstable with respect to the curved structures forms like nanotubes and fullerenes and also supposed not to be in a free state (Peres, 2010). In 1930s, two important papers were presented which, according to theoretical investigations, contended that severely two-dimensional crystals were thermodynamically unstable consequently, it could not exist. Mermin (1968) developed the instability argument of the two-dimensional crystal growth using many experimental investigations. However, there have been some intentions to extract single sheets of graphene. The graphene was first detected in 1960s as a "disordered structure" on surfaces of heated platinum in ultrahigh vacuum and recently recognized as the single graphitic sheet deposition (Morgado and Silveirinha, 2018). With the mentioned unique structural features, graphene has indicated special physical features that have attracted vast research interest in engineering and scientific communities. One of the most significant graphene features is graphene charge carriers treat as Dirac fermions or massless particles and under surrounding conditions graphene can move with little scattering (Ramanathan et al., 2019). This significant treatment has led to some unique phenomena for graphene (Eigler, 2016): 1 – graphene is a zero-bandgap two-dimensional semiconductor with a small overlap between conduction and valence bands. 2 – graphene shows a strong ambipolar electric field influence; therefore, the concentrations of charge carrier up to 1.01×10^7 m^{-2} and $T = 298$K mobilities of 6×10^9 m^2/min are calculated. 3 – a scarce half integer quantum Hall influence for both hole carriers and electron in the graphene has been seen using adjusting chemical potential by electric area influence. Furthermore, graphene is extremely transparent, by 2.4% absorption rate to observable light (Torres et al., 2020). Graphene thermal conductivity is calculated by an amount of 500 kW/m. K for a single-layer sheet at ambient condition. The basic mechanical features of freestanding monolayer graphene membranes were calculated using nano-indentation by an atomic force microscope (AFM). The strength of breaking is 4.2×10^{-4} kN/cm and Young's modulus is 1024 GPa, representing graphene is one of the strongest substances ever measured. In addition, the two-dimensional graphene-based sheets and other graphene-related substances like zero-dimensional graphene quantum dots (GQD), one-dimensional graphene nanoribbons and graphene nanomeshes are synthesized. These materials are expected to possess various optical and electrical features, because of the changes in geometry and size and also, the presence of a large value of edge deficiencies (Choi et al., 2010). For instance, the one-dimensional graphene nanoribbons and graphene nanomeshes have shown band opening by improved on/off ratios in area influence transistors in comparison with two-dimensional graphene sheets (Abd Ali, 2019). Introducing significant chemical and physical features and having dependable synthetic techniques for solution-phase and solid processes, graphene material and graphene material derivatives have been combined in a number of functional components to construct composites and have been applied as construction blocks for different classes of utilizations, counting field-effect transistors, photovoltaic devices photocatalysis, memories, sensors, intracellular imaging and cell cultures (Tewari et al., 2019). The main reason of unique features of graphene is its energy band spectrum. Graphene is known as a semi-metal material, with no bandgap just at the graphene Brillouin zone six K-points, where the bands of energy follow to linear dispersion connection. A nonsquare-like dispersion law consequences in zero actual masses for holes and electrons (Li et al., 2017).

Magnetic materials find out different performances in technology, for instance, power production and data storage (Chandu et al., 2018). Conventionally applied magnetics materials contain Heussler alloys, rare earth metal, nickel, cobalt and iron. The mentioned metals are able of showing magnetism owing to their *f* and *d* electrons that are capable to equal their spins to setup powerful magnetic materials (Basu and Bhattacharyya, 2012). Magnetic features of carbon-based substances have involved a great consideration. The presence of *p* and *s* electrons only makes graphene counterintuitive to expect magnetism (Wang et al., 2012). The magnetic substances according to carbon are an area of significant attention owing to less energy-intensive production, biocompatibility and abundance of carbon. Magnets flexible and lightweight can be fabricated by *C* element for storage apparatuses (Elemike et al., 2019). Magnetic carbon; moreover, finds out performances in bio-sensing and bio-imaging (Mohan et al., 2018). The magnetic vulnerability of the different carbon forms is dependent on carbon forms band structure that can be improved using presenting defects, impurities and using adsorbed entities interactions (Sun et al., 2020). The understanding in the graphene magnetic features is according to density functional theory. One of the most graphene properties is its high surface area. This parameter is determined using Brunauer-Emmett-Teller (BET) method and it depends on number of layers and the synthesis technique (Elemike et al., 2019). The more details about the surface area features graphene-based substances are given in Table 10.3.

Graphene-based substances have the surface area in the range of 3×10^6 to 3×10^7 cm^2/g. The single-layer graphene surface area is calculated theoretically as 2.63×10^7 cm^2/g. Also, graphene thermal exfoliation usually yields a very particular surface area because of single- and few-layer graphene (Kurmarayuni et al., 2020). In graphene, the atoms of carbon are bonded by other atoms in a hexagonal framework. Maximum of stress, which a substance can tolerate before breaking, is related to material defect free framework (Mindivan and Göktaş, 2020). Deficiencies in framework lead to decrease the strength of tensile. Graphene basic tensile strength is 0.127 TPa that is related to strength of C-C bond. Consequently, graphene is the strongest substance by a strength power around 199 than steel. Young's modulus of monolayer graphene was also reported to be 1 TPa (Saeed et al., 2016).

TABLE 10.3
The Surface Area Features of Graphene-Based Substance

Graphene-Based Materials	Surface Area (cm^2/kg)	Synthesis Method
Graphene oxide	3.2×10^8	Gas-based hydrazine reduction
Functionalized graphene	$(0.7–1.5) \times 10^9$	Thermal exfoliation of graphene oxide
G-sheets	6.7×10^8	Exfoliation of graphene oxide
	2.7×10^8	Arc graphite deposition under hydrogen
G-sheets	6.7×10^8	Exfoliation of graphene oxide
	2.1×10^8	Arc graphite deposition under hydrogen
G-sheets	9.4×10^8	Graphene oxide subjected to thermal shock to exfoliate and then reduced by H_2
Porous carbon	3.1×10^9	Exfoliated GO chemical activation
Porous three-dimensional graphene material	3.6×10^9	Hydrothermal followed by activation using potassium hydroxide

Source: Modified after Trivedi et al. (2019).

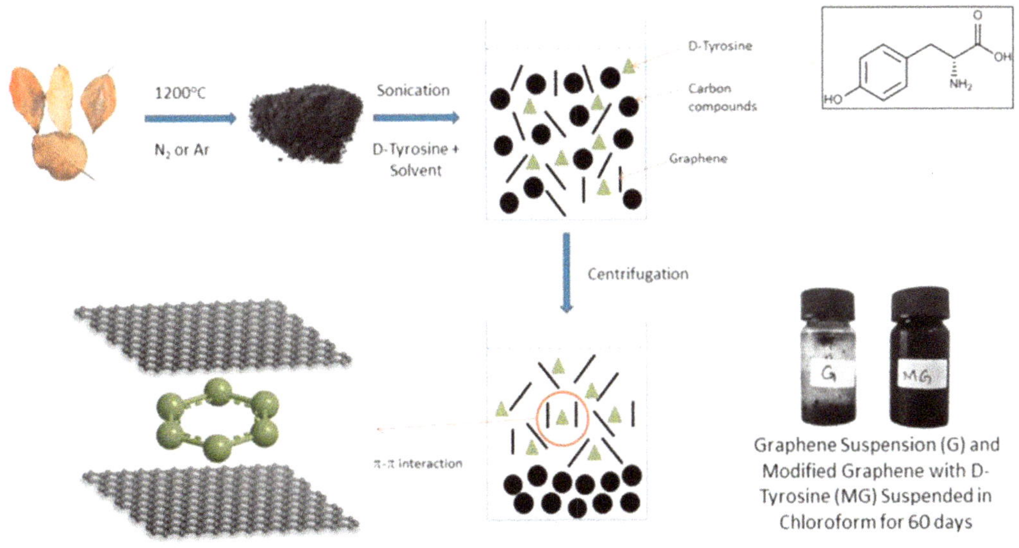

FIGURE 10.2 Scheme of graphene synthetic from leaf waste. (Reprinted with permission of Shams et al. (2015).)

Waste is a substance class that any purpose and no longer serves is usually thrown away (Elsen et al., 2019). Organic wastes act a main role in the producing contamination because of human activities, such as fermentation and landfill. In the organic waste, the main waste portion is fallen dry leaves from trees (Voros et al., 2019). The leaf wastes are not handled correctly and because of human ignorance, leaf wastes are not disposed correctly. The contaminants that are generated of the leaf wastes enter in atmosphere that consequences in greenhouse gas influence as a whole, in addition to particulate matter formation of (González et al., 2020). Besides the chemical contaminants released, mound spores are dispersed in the fire plume. These particulates can reach deep in lung tissue and reason chest pain, wheezing, coughing and sometimes long-term problems (Gokulkumar et al., 2020). One of methods to reduce the mentioned human and environmental hazards is to convert them to graphene (Figure 10.2).

The main aim of this chapter was a comprehensive graphene synthesis from leaf wastes. Consequently, at the first step, the synthesis methods for graphene production by leaves are discussed in detail. The characterization and morphology of produced graphene based on scanning electron microscope (SEM), transmission electron microscope (TEM), high-resolution transmission electron microscope (HRTEM), selected area electron diffraction (SAED), AFM, X-ray analysis (EDX), Fourier transform infrared (FTIR) spectroscopy, UV-Vis absorption spectra, or photoluminescence (PL) spectroscopy were discussed. Because of unique properties of graphene-based materials, the applications of them in various fields are reviewed.

10.2 GRAPHENE SYNTHESIS AND CHARACTERIZATION METHODS

The proteins, carbohydrates and fibers are the main components of dead leaves. Hemi-cellulose, lignin and cellulose of fibers and carbohydrates are mainly considered as significant carbon source (Biswal et al., 2013). Therefore, researchers are convinced to use these natural carbon sources for graphene production and have provided various methods for its synthesis. The leaves of green tea were chosen as the carbon resource to synthesis of graphene by Roy et al. (2021) due to the labile linkages of its structure and its polyphenols content. The dried leaves were pyrolyzed at 900°C for 3 h and cooled to ambient temperature under inert atmosphere (nitrogen gas) to prevent graphene

TABLE 10.4
Carbon Content of Produced Sample (G11) during Pyrolysis Process at 1100°C from X-Ray Photoelectron Spectroscopy Analysis

Group	Position of Peak	Content (%)	Group	Position of Peak	Content (%)
C=C	284.3	75	C=O carbonyl	530.5	23
C-C	284.9	12	C=O carboxyl	531.5	18
C-O	286.0	9	C-O hydroxyl	532.0	38
C=O	287.8	4	C-O carboxyl	533.3	20

Source: Modified after Roy et al. (2021).

oxidation and the obtained product was named G9. In second experimental set, pyrolysis of dried leaves was done at 1100°C for 3 h (obtained product: G11). The field emission scanning electron microscopy (FESEM) and Raman spectroscopy were applied to characterize the production of these experimental sets. The FESEM image of G9 indicated the lump-like and porous structure, but the flake-like structure of graphene was not observed. By increasing the temperature in the second pyrolysis, the sheet-like and rock-like structure was formed after 30 min of graphitization process. According to these results, the optimal temperature to synthesis of graphene from green tea leaf is 1100°C. The X-ray photoelectron spectroscopy (XPS) analysis confirmed the high carbon and low oxygen contents in G11 samples, i.e., 92.75% and 6.81%, respectively. The carbon content of G11 was determined in detail by XPS, as provided in Table 10.4.

According to these results, the highest content is for sp^2 carbon (75.2%) while sp^3 carbon had low content percentage (11.6%) in G11 samples. The intensity ratio of D band to G band (I_D/I_G ratio) and number of layers of G11 were predicted by Raman spectroscopy and were 0.88 and 3, respectively, The intensity ratio of D band to G band (*ID/IG* ratio) and number of layers of G11 were predicted by Raman spectroscopy and were 0.88 and 3, respectively. These values are similar to the reported values for graphene. HRTEM and SAED analyses established the crystallinity of the graphene sheets for G11, as shown in Figure 10.3. In SAED pattern of G11, sixfold symmetry in graphene's structure is distinctly visible (six clear dots).

Formed graphene fringes in samples identified crystalline domain of obtained graphene. The distance between graphene fingers was measured by using of the atomic resolution images, which was 0.37 nm for G11. They also used AFM images to predict the thicknesses, lateral sizes and number

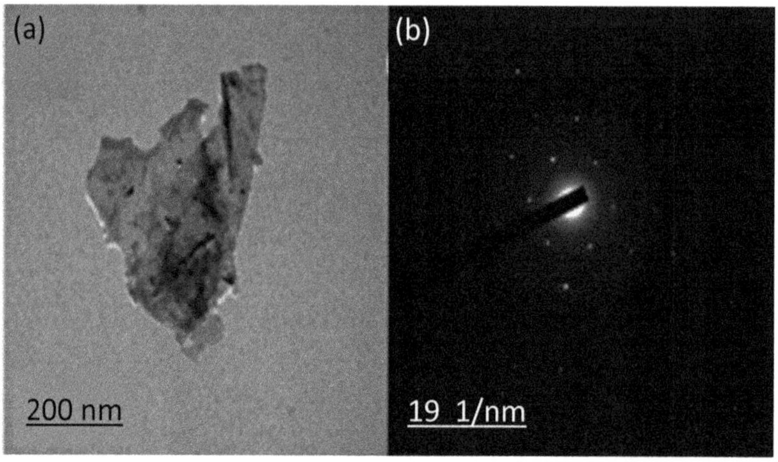

FIGURE 10.3 The high-resolution transmission electron microscope image of produced sample (G11) during pyrolysis process at 1100°C (a) and selected area electron diffraction pattern of G11 (b). (Reprinted with permission of Roy et al. (2021).)

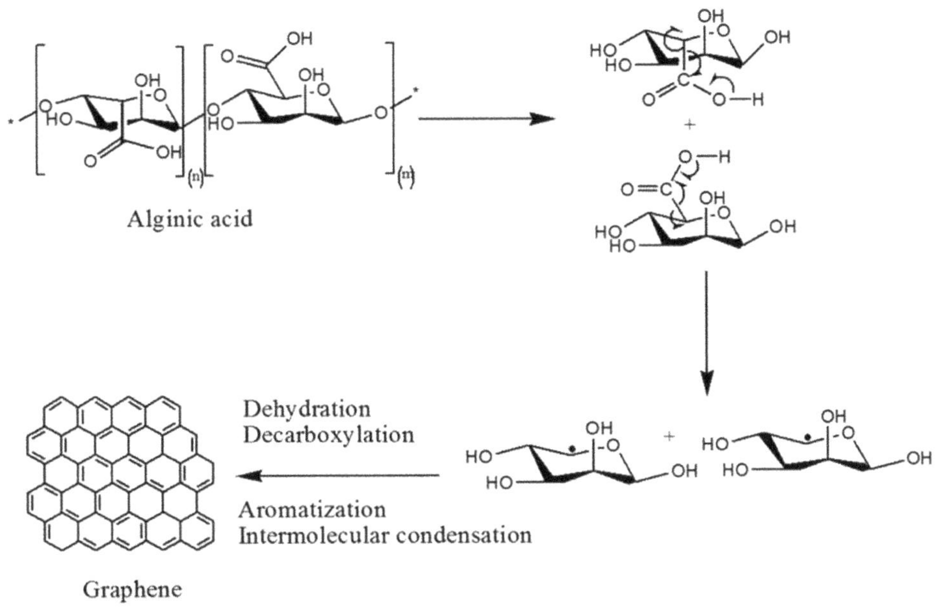

FIGURE 10.4 The plausible mechanism of alginic acid pyrolysis to produce graphene. (Reprinted with permission of Roy et al. (2021).)

of layers of the graphene sheets, which were 1.5, 403 and 3 nm, respectively. Roy et al. (2021) also used tannic and alginic acid powder as carbonaceous materials to produce graphene. The tannic and alginic acids were produced from leaves of oak tree and seaweeds, respectively. Two pyrolysis sets were performed for these acids in the same condition and manner as for dried green tea. The plausible mechanism of pyrolysis to produce graphene from alginic acid is illustrated in Figure 10.4.

The achieved results indicated that leaf derivatives are also an appropriate source for graphene production. The characterization analyses specified that a bilayer and few layers graphene are fabricated from alginic and tannic acids, respectively, after 3 h pyrolysis at 1100°C (the products of this experimental set were named A11 and T11). The produced graphene from these acids had high quality and sixfold crystalline symmetry structure, which characterized by FESEM, XPS, HRTEM, AFM images, SAED pattern and Raman spectroscopy. As a result, nonconventional carbonaceous sources like green tea leaf, tannic acid and alginic acid were qualified and were suitable resources to produce graphene due to their structures, contents and abundance. However, a lot of effort is required to provide a facile synthesis method with optimal conditions.

In another study, Chen et al. (2016) converted wheat straw to graphene sheets because of its multilayer structure and polymeric content using an eco-friendly and low-cost method. They combined graphitization and hydrothermal processes for this purpose. After hydrothermal treatment with hot KOH solution to convert wheat straw to cellulose fibers, the resulting product was carbonized at high temperature (800°C) and in an inert atmosphere and then the obtained carbon nanosheets thermally treated at 2600°C to produce graphene film (Figure 10.5).

The morphology, surface structure, the thickness between the nanosheets, porous structure, number of graphene layers, space between carbon layers and graphitization degree were analyzed. The obtained results indicated the successful production of porous layered graphene. According to the SEM images (Figure 10.6(a and b)), the thickness of graphene sheets is about 1–4 nm and the ripple structure of graphene is quite obvious. The interlocked porous structure of obtained graphene was confirmed by analyzing TEM images (Figure 10.6(c–e)); in addition, the red arrows in Figure 10.6(c) show some meso pores (2–20 nm) in the graphene structure. As indicated in Figure 10.6(d and e), the derived graphene from wheat straw can have 2–10 layers, which the distance between layers is about

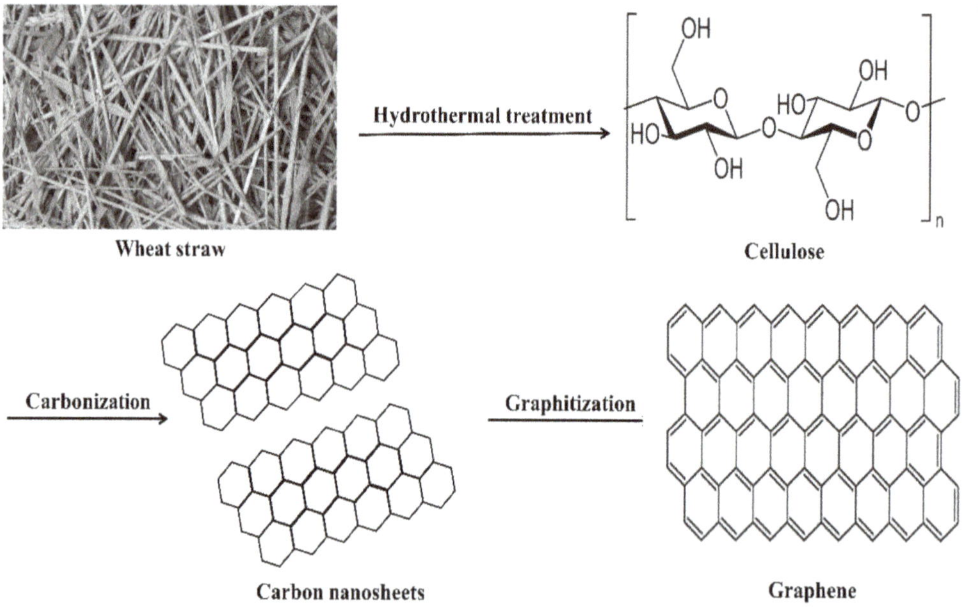

FIGURE 10.5 Schematic of the presented method to fabricate graphene from wheat straw. (Modified after Chen et al. (2016).)

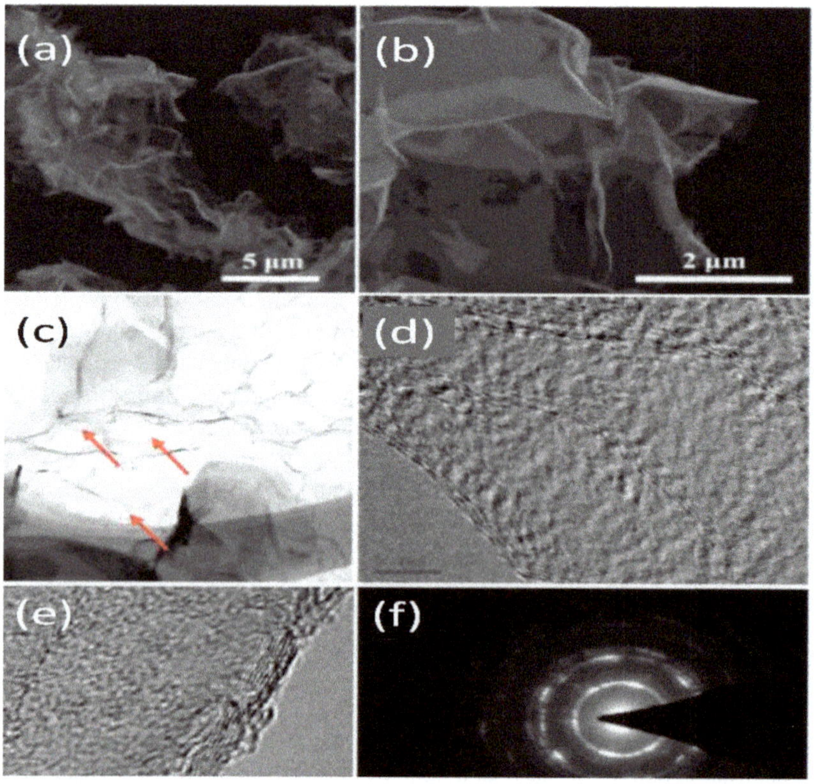

FIGURE 10.6 The scanning electron microscopy images (a) and (b), transmission electron microscopy images (c–e) and selected area electron diffraction pattern (f) of the derived graphene from wheat straw. (Reprinted with permission of Chen et al. (2016).)

0.34 nm. The crystalline structure of graphene was analyzed by SAED pattern (Figure 10.6(f)), which indicated the sixfold symmetry and its high graphitization degree.

The prepared graphene had high graphitization degree (90.7%), which indicates its excellent conductivity behavior. In addition, the specific surface area (SSA) and the volume of pores were 35.5 m^2/g and 0.10 cm^3/g, respectively. To investigate electrochemical potential, the prepared graphene was used as anode of lithium-ion battery. The result of charge-discharge profile showed higher reversible capacity (502 mAh/g) than graphite capacity but irreversible capacity was 295.8 mAh/g. This reduction in capacity is common for carbonaceous materials. Cycling performance of graphene during 40 cycles was significant and more than the control sample (graphite). The appropriate chemical and physical features of synthesized graphene using a novel, low-cost and green process, can show a new way to use large amount of wheat straw output in the world.

In the study of shams et al. (2015), dead leaf of camphor was used as carbon source to fabricate graphene by a facile, scalable and ecological method. They cleaned the leaves from dirty by acetone and dried at 60°C for 4 h. Then fragmented samples were pyrolyzed at 1200°C under nitrogen atmosphere and finally cooled to ambient temperature (Figure 10.7).

After this single-stage pyrolysis at high temperature, the sample was dissolved in the solution of D-Tyrosine and trichloromethane and treated during 15-min sonication. The graphene layers were suspended in solution after centrifugation and other carbon compounds deposited at the bottom. D-Tyrosine was washed from graphene layers and obtained graphene was characterized. The thermal gravimetric analysis, which was similar to conventional pyrolysis curve indicated that hemicellulose is the largest carbon source in camphor leaves. The analysis of TEM images revealed the formation of graphene layers from the leaves, which the thickness and number of them were estimated by AFM images and were 2.37 nm and about 7, respectively. On the one hand the I_D/I_G ratio (0.99) was calculated using Raman spectroscopy data, which was close to the presented values for pure graphene. On the other hand, the nano porous structure of obtained graphene was confirmed by BET test and its surface area was 296 m^2/g. Also, the production efficiency of graphene from camphor leaves was 0.8%. These results indicated that the leaf waste can be used as a potential bio resource to produce graphene without complex method, chemicals and catalysts, even on a large scale.

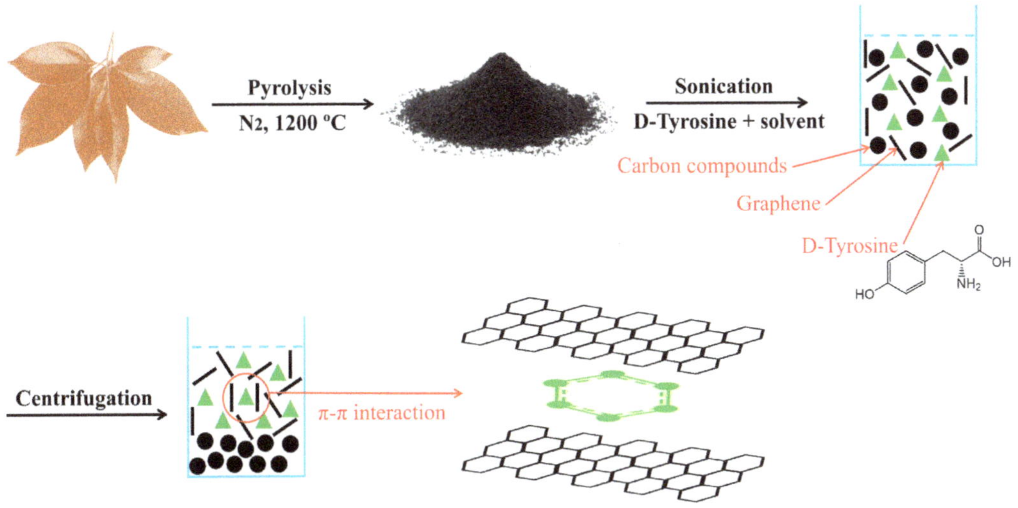

FIGURE 10.7 Synthetic scheme of graphene form dead leaf of camphor. (Modified after Shams et al. (2015).)

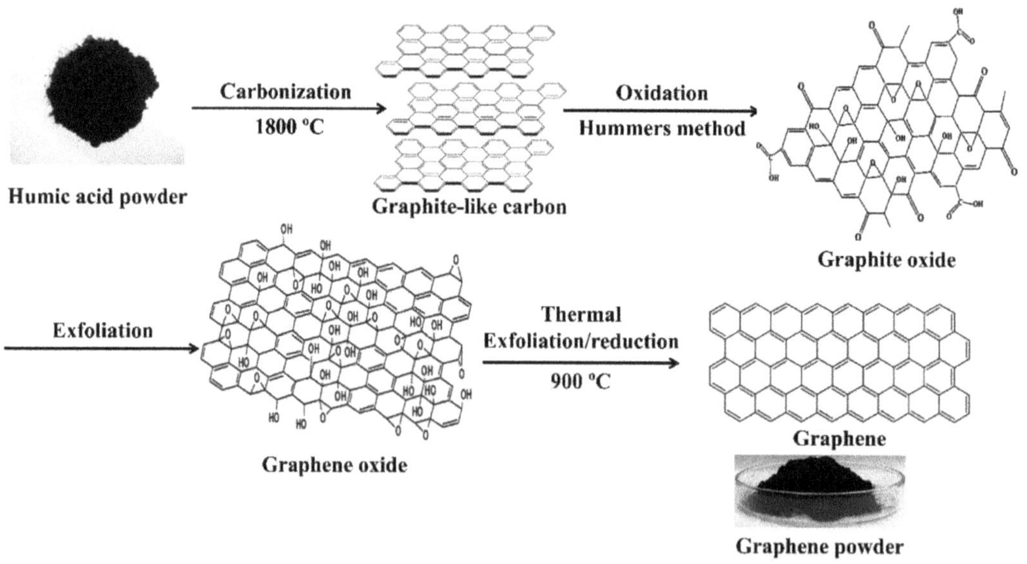

FIGURE 10.8 Synthesis method of graphene from humic acid. (Modified after Xing et al. (2017).)

The leaf waste derivatives are also a valuable source of carbon to fabricate graphene. Humic acids are important organic redox-active components, which are formed through the decomposition of plant and animal materials. These eco-friendly complex molecules exist abundantly in nature (Jayarathne et al., 2017). The chemical structure of humic acids and its similarity to graphene structure attract the attention to produce graphene from these organic acids. Graphene is traditionally synthesized using natural or synthetic graphite but these resources are of high-cost and nonrenewable. In 2017, a facile, eco-friendly and inexpensive synthesis method was reported by Xing et al. (2017) to produce graphene sheets using humic acid. They used Hummer's method to obtain layered structure of graphite at high temperature and then oxidation-exfoliation-thermal reduction strategy was applied to synthesize nanosheets of graphene. Figure 10.8 briefly illustrates the synthesis of graphene from humic acid. The SEM and TEM images are used for the characterization of surface structure of the obtained graphene, which showed curved, wrinkled and transparent structure with a large number of interconnected nanopores. These results indicated that the surface morphology of obtained graphene from carbonization of humic acid is very similar to the synthesized samples using natural graphite. Furthermore, the pore morphology and surface area of obtained graphene were analyzed by nitrogen adsorption/desorption technique, which confirmed the high volume of pores (2.987 cm^3/g) and considerable SSA (495 m^2/g). They also used density functional theory to analyze pore size distribution and concluded that the obtained graphene has excellent capacitance and high electrochemical performance. This work emphasized that green, abundant and renewable humic acid is a deserving alternative for low-cost production of graphene with unique properties such as excellent SSA, high volume of pores, interconnected nanopores morphology and as a result its favorable performance in supercapacitors.

Graphene derivatives also have a wide range of applications and have recently attracted much attention. Despite all the outstanding properties of graphene, it is a nonluminescent material but some of its derivatives do not have this shortcoming. GQD is one of the important alternatives to graphene, which possesses remarkable magnetic features because of quantum confinement and edge effects. Microwave-assisted hydrothermal, cyclic voltammetry and electron beam lithography methods are typical synthesis approach to fabricate GQD. The size

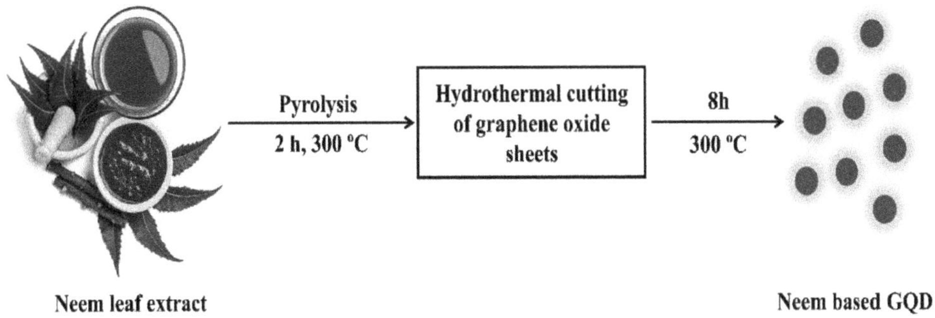

FIGURE 10.9 Schematic of graphene quantum dots production from Neem leaves. (Modified After Roy et al. (2014).)

of GQD influences its magnetic features. The synthesis of GQD with sizes under 20 nm is complex and costly and providing an effective method for resolving existing issues is one of the researcher's interests (Chuang et al., 2016). Roy et al. (2014) presented a novel and facile method to synthesize GQD from Neem and Fenugreek leaves without any organic solvent and chemical agents. Formation of Neem-derived GQD using hydrothermal method, is presented in Figure 10.9.

During pyrolysis process, the temperature has main effect on carbonization so that sufficient carbonization of leaves does not occur under 300°C and higher temperature destroys their structure and surface. The average sizes of GQD were estimated by HRTEM and were 5 and 7 nm for Neem and Fenugreek leaves, respectively. In optical analysis, Neem-derived GQD had higher quantum yields (41.2%) than Fenugreek-derived one (38.9%), which indicates that Neem and Fenugreek leaves are appropriate renewable resources to prepare GQD, which can be used in nanolasers and solar cells. In the continuation of this study, the properties and intrinsic magnetic feature of prepared GQD from Neem leaves were investigated (Chuang et al., 2016). According to the HRTEM and fast Fourier transform (FFT) images of Neem-derived GQD, the orientation of edge in GQD structure was zigzag, which indicates the mechanism of GQD formation from GO sheets. Superconducting quantum interference device was used to investigate the GQD magnetization at different temperatures (2–300K). The results showed that the temperature affects the GQD magnetization and the GQD spin effect is changed by thermal energy. For instance, paramagnetic behavior was changed to diamagnetic behavior at 100K. In fact, paramagnetic electron spin moments change and decrease to zero with an increase in temperature. This valuable information can give an accurate comprehension of GQD properties, functions and applications. During the recent years, significant efforts have been made by many researchers to produce carbon quantum dots from leaf wastes and green sources using facile methods. The synthesis of leaf-derived GQD with different methods is briefly provided in Table 10.5.

One of the most important graphene-related materials is GO, which is widely used in material, pharmaceuticals, biomedical and biological sciences. This single atomic layer material is produced by exfoliation and chemical vapor deposition methods (Georgakilas et al., 2012; Mao et al., 2013; Somanathan et al., 2015). In order to fabricate GO or reduced graphene oxide (rGO), graphite is generally oxidized by various methods like Hummer's method and modified one (Gopinath et al., 2018). Reduction of GO is one of the top-down methods to graphene production from graphite derivatives. Graphene with high conversion yield (about 99%) and carbon-to-oxygen ratio can be produced by reducing GO. Graphite derivatives are oxidized with oxidizing agents to fabricate GO. The electrical conductivity and thermal properties of graphene are affected by the presence

TABLE 10.5
Synthesis of Graphene Quantum Dots from Leaf Wastes by Different Methods

Key Material	Final Product	Methods	Results	Reference
Dead Neem leaves	Amine terminated GQD (Am-GQD)	Pyrolysis and chemical treatments with H_2SO_4 and HNO_3	UV-Vis absorption spectra and PL spectroscopy are used to investigate about optical properties of GQD and Am-GQD. Photoluminescence intensity and water dispersity of the modified GQD were higher than GQD. Am-GQD showed a strong fluorescence quenching for some metal ions such as Cu^{2+}, Ni^{2+}, CO^{2+}, Fe^{2+}, Fe^{3+}, Hg^{2+} and Pb^{2+}	Suryawanshi et al. (2014)
Mango leaves	GQD	Perpetration of leaf extract with ethanol and heating with domestic microwave oven	The prepared GQD with facile and green method showed bright red fluorescence. This photostable GQD had magnificent cellular sorption (100%) and exhibited selective fluorescence emission in the near-infrared region. Also, the result of temperature sensing analysis indicated its stability after several cycles of temperature switching	Kumawat et al. (2017)
Therapeutic guava leaf	Red-fluorescent GQD (G-GQD)	Solvothermal treatment	The excitation wavelength of prepared G-GQD was 300-420 nm, which depended on the fluorescence intensity and pH. Selective quantification of mercury (II) in an aqueous solution was investigated by the G-GQD emission property. The obtained results confirmed the ability of prepared GQD for sensitive and rapid response fluorescence for detection of Hg^{2+}	Khose et al. (2021)

GQD: Graphene quantum dots; PL: photoluminescence.

of oxygen after oxidation reaction. The deoxygenating of GO solves this problem by using some reduction methods such as thermal or chemical reduction and photocatalytic. In thermal reduction, hydroxyl, carboxyl and epoxy groups are removed from GO by heating at high temperature (900°C) under an inert atmosphere, as well as various light sources are used in photocatalytic reduction of GO. In deoxygenating process of GO with chemical reduction, chemical reducing agents are used and this method is facile and reliable for industrial scale production. But these reductants are toxic; therefore, rGO cannot be used in biomedical applications (Ismail, 2019). Recently, the development and promotion of green GO reduction processes have attracted the attention of many researchers. Scientific literatures indicated that leaf extracts can play reducing and capping roles instead of chemical agents in the chemical reduction of GO, but some parameters must be considered like reduction duration, temperature and concentration of GO. In the study of Chuah et al. (2020), phyto-extract of Eclipta prostrata was used as reducing agent to fabricate rGO. They used leaves of Eclipta prostrata to produce an aqueous phytoextract by drying, powdering, agitating at 50°C and filtering. Graphite was used for GO preparation and then the prepared phytoextract was added to the solution of GO and water. After dispersing rGO, phytoextract was separated using centrifugation and distilled water washing. The morphology and structural characteristics of prepared GO and rGO were

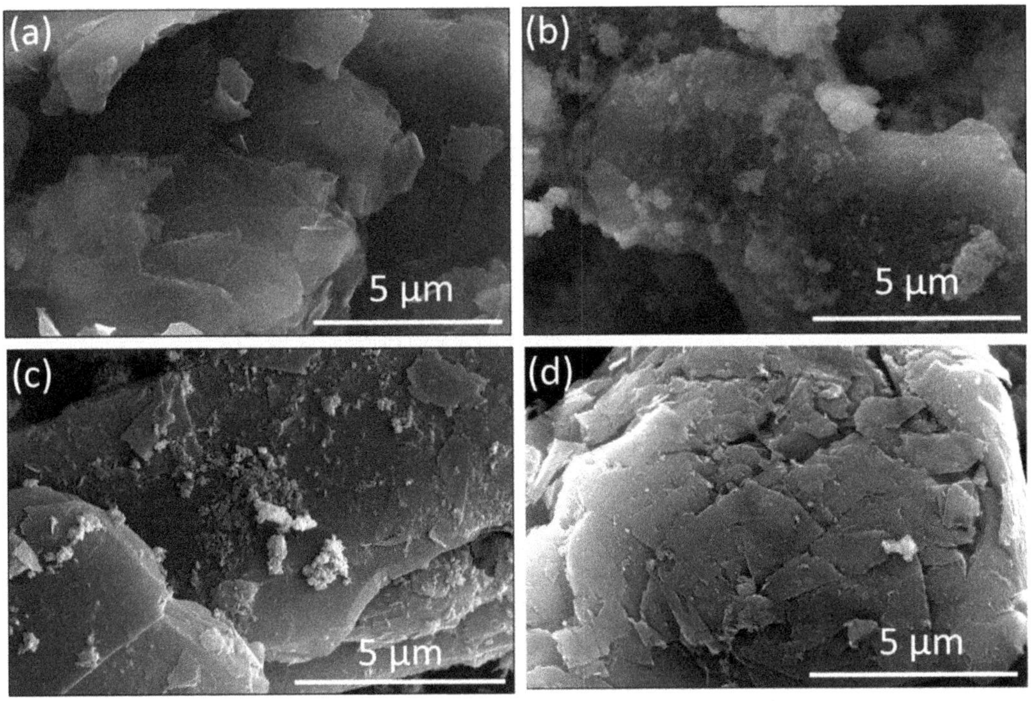

FIGURE 10.10 The field emission scanning electron microscope images of (a) graphite, (b) graphite oxide, (c) graphene oxide and (d) reduced graphene oxide at magnification-scale: 5 µm. (Reprinted with permission of Chuah et al. (2020).)

analyzed by FESEM images. The results indicated that the produced rGO has smooth and sinuate surface with lattice fringes (Figure 10.10(d)). As shown in Figure 10.10, the shape of edges and corners of graphite oxide changes from round to sharp after graphite exfoliation. Also, 3D, porous and interconnected structure of GO is observed in Figure 10.10(c). In additionally, the TEM images confirmed these results.

Furthermore, the results of energy dispersive EDX and FTIR spectroscopy proved that the oxygen content of rGO decreased after GO reducing with Eclipta prostrata phytoextract. This study indicated that phytoextract of leaves can be used as reducing agent for nontoxic rGO production. Elemike et al. (2019) reported an innovative approach to synthesize developed rGO using zinc oxide (ZnO) and silver nanoparticles. They fabricated GO from graphite flakes using Hummer's method and then the leaf extract of Stigmaphyllon ovatum was used as reductant agent to produce rGO. The obtained rGO was sonicated in water and then $AgNO_3$ was added to it. Following that, ZnO particles and leaves extract were added to this solution in order to produce developed rGO, which SEM images and EDX spectroscopy confirmed. Nanoparticles of metals enhanced photocatalytic and optical applications of rGO. The bandgap energy and photocatalytic efficiency were 2.75 eV and 68.1% (higher than rGO), respectively. This reported reaction for developed rGO production is eco-friendly, low-cost and without any toxicity. Some valuable researches are summarized in Table 10.6, which indicated that leaf extract has the ability and competence to be used in the bio-rGO production.

According to the reviewed literature, the leaf wastes and their derivatives can be used as eco-friendly sources and reduction agents to produce graphene and its derivatives. But the synthesis conditions must be considered and optimized to provide facile methods to produce high quality green products.

TABLE 10.6
Summary of Several Studies about the Ability of Leaf Extract as Reductant Agent for Reduced Graphene Oxide Production

Reducing Agent (leaf extract)	Methods	Results	Ref
Plectranthus amboinicus	Hydrothermal treatment of GO and $AgNO_3$ and reduction with leaf extract of Plectranthus amboinicus	The X-ray analysis, SEM and TEM image, FTIR and UV-Vis spectroscopy confirmed the structure of rGO/Ag nanocomposite. The obtained rGO/Ag nanocomposite had excellent electrochemical properties and good selectivity as electrode surface for electrochemical sensor	Zheng et al. (2016)
Lotus garcinii	Three steps method for rGO/Fe_2O_3 fabrication (Hummer's method to produce GO, reduction of GO in colloidal suspension, fabrication of rGO/Fe_2O_3 by coprecipitation method), finally adding leaf extract and Ag nanoparticles to biosynthesis of rGO/Ag/Fe_2O_3	The leaf extract of Lotus garcinii played important roles as reductant and stabilization agent to synthesis of rGO/Ag/Fe_2O_3. This modified rGO was used as catalyst to removal organic pollutant from waste water, which exhibited remarkable catalyst activity in comparison with other reported nanocatalysts	Maham et al. (2017)
Eucalyptus	Fabrication of GO by Hummer's method and reduction with leaf extract	Leaf extract of Eucalyptus was used as reductant and stabilization agent, which enhanced the electrochemical, stability and biocompatibility features of rGO. The morphology analysis results showed that the thickness of rGO was 0.807–1.129 nm, which is higher than GO. Eucalyptus biomolecules increased the thickness of rGO and improved its biocompatibility features	Li et al. (2017)
Amla	Hydrothermal treatment to produce Ag/graphene and reduction with green reductant	The spectroscopy analyses confirmed and characterized the structure of obtained Ag/rGO. Minimum inhibition concentrations of Ag/rGO were calculated to investigate about its antimicrobial efficiency. *Xanthomonas campestris*, *Candida albicans*, *E. coli* and *Bacillus megaterium* were used to analysis of antimicrobial activity. The results showed that this green nanocomposite killed bacterial colonies	Chandu et al. (2018)
Eichhornia crassipes	Treating obtained rGO (with leaf extract) with ZnO to fabricate rGO/ZnO nanocomposites	The rGO/ZnO concentration influenced on the dye degradation efficiency and enhanced dye adsorption capacity and removal efficiency. The obtained rGO/ZnO by leaf extract of Eichhornia crassipes had higher degradation efficiency than rGO/ZnO produced without this green reductant	Ramanathan et al. (2019)

(Continued)

TABLE 10.6 (Continued)
Summary of Several Studies about the Ability of Leaf Extract as Reductant Agent for Reduced Graphene Oxide Production

Reducing Agent (leaf extract)	Methods	Results	Ref
Palm Oil	Sonicating graphite oxide solution to produce GO, refluxing GO solution and leaf extract mixture, centrifuging and drying to fabricate rGO	The structural analysis results indicated that interspace distance of rGO planes depends on the oxidization progress. The distance increased during graphite oxidization and enhanced from 0.33 to 0.84 nm. Hydroxyl group of GO removed by using palm oil leaf extract as green reductant and also changed the amorphotization of structure of sp^2 carbon. C/O ratios increased from 1:1 to 3:1 during reducing process. Finally, voltammetry analysis of obtained rGO confirmed its electrical conductivity	Faiz et al. (2020)
Erythrina senegalensis	Reduction of produced GO by leaf extract using a facile one-pot synthesis	The reduction of GO was confirmed using X-ray and UV-Vis spectroscopy. Further, the TEM images showed sheet-like and transparent of produced rGO. A concentration dependent toxicity to SICH cells was observed for produced rGO, which shows its ability to use in medical applications	Qi et al. (2021)

FTIR: Fourier transform infrared; GO: graphene oxide; rGO: reduced graphene oxide; SEM: scanning electron microscope; SICH: cardiac cell lines of *Catla catla*; TEM: transmission electron microscope.

10.3 GRAPHENE APPLICATIONS

Recently, graphene-based substances have been established for many applications, counting plans for energy conversion and storage, absorption agent, electrochemical sensor and functional composite substances (Chabot et al., 2014). Graphene stimulated researchers' enthusiasm, which are exploring application of graphene in different areas. The overview of recent consequences in critical application of graphene is presented (Rao et al., 2014). The immediate performance of graphene is in the polymer nanocomposite component that indicates a strange improvement of some important features, like thermal stability, electrical conductivity, tensile strength and elastic modulus upon integration of GO or graphene platelets (Phukan et al., 2019).

10.3.1 Absorber

Because of the significant features of materials based on graphene, they can be applied as forming composite materials and an additive substance, by some preferred features and performance (Chabot et al., 2014). Some graphene sponges are powerful in compressive stress; however, fracture in higher loads. In certain concentration conditions or lower cross-linked, shaped graphene sponges can show powerful reversible compression loads, however, graphene sponges still continue delicate through handling (Phukan and Sahu, 2020). But high strength and big surface area of sponge structure in the optimal mass ratios have made the mentioned properties desirable in preparing polymer matrices reinforcement by increased tailored and strength features. Consequently, graphene sponge matrices can prepare great conductivity, high strength and superhydrophobicity, in addition to, reversible elongation or flexibility to some compositing usages (Siong et al., 2019). GO and three-dimensional graphene structures have vast surface area and can be made freestanding that must prepare a good surface area for absorption process (Huang et al., 2019).

10.3.1.1 Gas Absorption

Wide functionality on GO acts as a well-organized site for, resistive detection, leading to physical and reduction gases events. The instance of mentioned phenomena complicated GO application to catalyze sulfur dioxide oxidation to sulfur trioxide adsorbed on gas decreased GO at 298K (Li et al., 2019a). Later brown sponges, gas exposure twisted black and trapped sulfur trioxide could be changed to H_2SO_4 on exposure to H_2O, permitting for simple restore GO structure impurities filtration (Lazar et al., 2013). Otherwise, high quality sheets of graphene show some main features that make graphene sheets appropriate to ultrasensitive gas sensing, depressed to single molecule adsorption measures for nitrogen dioxide (Shakak et al., 2020). The ammonia trace sensitivity, hydrogen cyanide, carbon monoxide, chlorine and water have also been indicated or modeled using principal measurements. Further, electronic features indicate powerful dependence on absorbents surface counting molecules of gas and significant 2D crystal framework can result in high sensitivity and low electrical noise (Huang et al., 2017). Indeed, graphene sheets are potential for sensing is hindered using individual graphene sheet susceptibility to chemical agents and extraneous environment that affect the device reliability and repeatability (Wu et al., 2016). Yavari et al. (2011) presented applying of robust using chemical vapor deposition technique, to sense toxic nitrogen dioxide and ammonia gases. In another study, sponge sheets of graphene were cut in 5×10^{-5} m^2 strips and also, fixed to a five chips by interconnects of copper element for making low reducing electrical noise and contact resistance (Chen et al., 2019). The robust and simple sensor plan was capable to consistently sense resistance variation on exposure to 19 ppm ammonia and nitrogen dioxide at 25°C, by steady state response times round 300–6000 s (Zhang et al., 2020a). Desorption process of the mentioned gases was incomplete in pure air; however, by heating a totally reversible sensor can be attained. The graphene sensors are moreover practical for detection of gas than sensors based on semiconducting metal oxide, which work at high temperature (>473K) (Li et al., 2019b, Yavari et al., 2010).

Carbon dioxide and methane are two main gases responsible for harmful effects like global warming. Carbon-based materials that can store and adsorb methane and carbon dioxide are gaining consideration because of their light weight, low-cost, high surface area and chemical inertness (Yavari et al., 2011). Graphene, carbon nanotube and carbon fibers would be predictable to be appropriate materials for this purpose. Activated carbon and graphene have high surface area to adsorb methane and carbon dioxide at 25°C and high pressure (Kumar et al., 2011). The amount of adsorbed methane and carbon dioxide increases with an increase in carbon-based material surface area. $B_xC_yN_z$ (borocarbonitrides), which have sequestering B- and N-doped and conducting graphene (and graphite) as limiting materials, have several possible uses (Kumar et al., 2011). $B_xC_yN_z$ provided with the reaction of urea, H_3BO_3 (boric acid) and activated carbon possess remarkable surface areas and they are possible to include borocarbonitride rings in addition to graphene and B- and N-doped domains, perform as well material storage for methane and CO_2 (Rao et al., 2014). The adsorption of carbon dioxide using the borocarbonitrides samples was calculated at –78°C and 1 bar. The highest carbon dioxide uptake was obtained for BC_2N that moreover had the highest surface area. The carbon dioxide uptake for BC_2N, obtained at 25°C and 50 bar, was 65 wt.%. Carbon dioxide is acidic and therefore the N_2 in the borocarbonitrides acts a key role in attaining high carbon dioxide absorbance (Keramatinia et al., 2021). The methane uptake contrasts from 7.6 to 18 wt.% at 25°C and at 20°C it was 16 wt.% for BC_2N. Uptake of methane and carbon dioxide furthermore is depending on borocarbonitrides materials surface area and the obtained results show the absorption of gases enhance exponentially by surface area. Adsorption of carbon dioxide using borocarbonitrides was found out to be nearly 9 times selective over nitrogen at 0.99 atm and 20°C (Ao et al., 2012).

10.3.1.2 Dye Adsorption

Natural microporous adsorbers like zeolites, expanded perlite, wool and sawdust have been applied for spill clean-up. However, the mentioned conventional substances have little oil loading and furthermore adsorb large making extraction values, water and the oil absorbents recycling impracticable. Microporous, hydrophobic polymers are moldable, able of adsorbing 4–26 times of own weight in organic solvents and oils (Liu et al., 2020). In place of a powerful substitute for cheap oil removal, graphite has indicated high loading, and up to 71% of composed oil can be eliminated. However, particulates do not show significant organic solvents adsorption, powder shape makes for complicated usage and substance can be used a few times (Chuah et al., 2020). The development of novel materials capable to reversibly and effectively remove oil spill and organic solvents pollutes is dangerous to the oil spill water remediation future and cleanup is necessary for protection of the environmental. Subsequently, graphene sponges by highly oleophilic and hydrophobic surfaces, chemical stability into organic solvents, high surface area, ability to withstand high temperature and uniform structure, have been established to be possible for adsorbers (Bi et al., 2012). For instance, graphene/carbon nano tube and graphene have been established as an adsorber for an extensive organic solvents variety, water soluble alcohols, alkanes and oils by special performance as given in Table 10.7.

Bi et al. (2012) applied 453K autoclaved reduction to persuade self-assembly a restore GO structure sponge with 4.3×10^8 cm^2/g surface area that indicated restore GO sponge was capable to adsorb high volume of substance in comparison with its own weight, i.e., 21–86 kg/kg. Zhao et al. (2012) informed a 400×10^8 cm^2/g restore GO sponge provided using a similar thermal reduction however, by addition of urea to crosslinking and functionalize the substance, and they indicated this restore GO sponge had higher capacities of adsorption about 74–55 kg/kg. Furthermore, absorption of dye in H_2O was also test indicating high dyes adsorption to color production, like methylene blue and rhodamine B (Chabot et al., 2014). Higher adsorption of dye was detected for Cu_2O/graphene (Wu et al., 2013). The composite graphene/Cu_2O sponges were tested using methylene blue, methyl orange and rhodamine B.

10.3.2 Energy Conversion and Storage

Graphene sponges treated or doped by metal oxides, indicating high electroactive surface area and limited sheet restacking, have been established to be appreciate components for conversion and energy storage devices (Chabot et al., 2014). Continuing attention for graphene sponges usage in energy storage is an additional reason to promise of mechanical stability and high elasticity, high electrochemical stability, 3D porous microstructure and high conductivity of interconnected networks (Liu et al., 2019). The mentioned devices contain the fuel cells, batteries, supercapacitors and solar cells. For polymer electrolyte membrane fuel cells, graphene-based materials are usually applied by catalyst supports and catalysts; for batteries like lithium batteries, graphene-based materials are applied as anode and cathode substances; for supercapacitors, fuel cells are applied as electrode substances for pseudo-capacitors and two-layer capacitors; also for solar cells, graphene-based materials are applied as dye-sensitizer (Chang et al., 2020).

10.3.2.1 Fuel Cell

A fuel cell is an electrochemical device that produces electricity over the fuel oxidation at the anode electrode and reduction of oxygen at the cathode electrode (Berktas et al., 2020). To get an important current density, oxygen reduction reaction kinetics must be increased with reduction reaction activation energy (Yu et al., 2010). General catalyst applied in oxygen reduction reaction is Pt that is not available and it is costly in large quantity on the crust of earth. But catalysts based on Pt suffer of poor tolerance in contradiction of fuel crossover and CO poisoning. Carbon-based catalysts have indicated improved activity of catalytic equal to that of Pt by good durability, low-cost and low fuel crossover. The most widely traveled fuel cell usages of materials based on graphene are in field of

TABLE 10.7
Several Non-Aqueous Liquids Absorption Capacity Using G-Sponges

Absorbing Components	Capacity of Adsorption (kg/kg)
Oils	
Gasoline	2.7×10^5
Crude oil	2.89×10^5
Pump oil	6.8×10^4, 8.5×10^4
Kerosene/paraffin oil	4.5×10^4
Castor oil	7.5×10^4
Soybean oil	5.5×10^4
Diesel oil	1.24×10^5
Motor oil	3.45×10^5
Olive oil	4.60×10^5
Vegetable oil	9.5×10^4, 1.03×10^4, 4.18×10^5, 81×10^4
Alkane	
C_6H_{14}	4.3×10^4, 2.15×10^5
C_7H_{16}	2.1×10^4, 7.5×10^4
C_8H_{18}	45×10^4
$C_{10}H_{22}$	35×10^4
Dioxane	4.89×10^5
C_6H_{12}	3.20×10^5
Dodecane	5.0×10^4
Aromatic compound	
Ethylbenzene	3.3×10^4
1,2-Dichlorobenzene	4.5×10^4, 1.27×10^5, 4.50×10^5
Toluene	5.5×10^4, 1.25×10^5, 2.0×10^5, 3.50×10^5
Nitrobenzene	6.3×10^4, 3.70×10^5
Organic solvent	
Acetic ester/ethyl acetate	9.2×10^4
Dimethyl sulfoxide	6.2×10^4, 4.30×10^5
Tetrahydrofuran	5.5×10^4, 2.50×10^5
Acetone	5.1×10^4, 3.50×10^5
Chloroform	8.6×10^4, 1.08×10^5, 1.54×10^5
Dimethylformamide	1.05×10^5
Phenoxin	6.0×10^5, 7.43×10^5
Ethylene glycol	1.33×10^5
Alcohols and aldehyde	
MeOH	4.9×10^4, 2.90×10^5
C_2H_5OH	5.1×10^4, 8.0×10^4, 3.0×10^5, 3.50×10^5
Ionic liquid	
1-Butyl-3-methylimidazolium tetrafluoroborate ([Bmim][BF_4])	5.27×10^5

Source: Modified After Chabot et al. (2014).

nitrogen coordinated metal catalysts and nonprecious metal oxide that have been followed as alternative catalyst to substitute expensive catalyst based on Pt (Liang et al., 2011). Though conventional nanostructured carbon support materials like carbon black (active carbons) that have high surface area could significantly recover performance of catalyst, insufficient electrochemical/chemical constancy of the mentioned supports is their limitation (Wang et al., 2011a). Due to graphene-based materials electrochemical/chemical stability, good surface area, promising conductivity, they were recognized as a good choice for catalyst usages. In addition to their high adhesion to catalyst particle. Moreover, the functional groups on GO can prepare extra opportunity for catalyst nanoparticle anchoring and nucleation, besides electron transport (Wu et al., 2013). Though Pt-based electrocatalyst indicated enhanced oxygen reduction reaction activity, its higher price hinders applications of platinum in fuel cells. A few researches are according to nonprecious transition metals (cobalt, iron) applied as cathode catalyst in fuel cells. N_2-doped G composites by transition metal oxides indicate unique improvement in electro-catalytic activity (Wang et al., 2014). Peng et al. (2013) presented Co-NO_x doped GO (N-rGO/Co_3O_4) based bifunctional catalysts for the oxygen evolution reaction and oxygen reduction reaction. The mentioned catalysts indicated positive onset potential and higher cathodic currents when associated to Co_3O_4 and N_2-doped reduced graphene (N-rGO).

The strong features of graphene as a catalyst support for Pt was lately investigated (Maiyalagan et al., 2012). Freestanding chemical vapor deposition graphene sponges supporting for Pt nanoparticles created better durability of catalytic and efficiency in comparison with carbon-based fiber supports. Wu et al. (2012) established Fe_3O_4 nanoparticles similar to deposition on a graphene sponge. The sponge was provided using autoclaving a GO solution including iron acetate and poly-pyrrole by different concentrations at 453K, followed using freeze-drying process. The process permitted simultaneous self-assembly for iron oxide growth on graphene site and N_2 source incorporation in graphene sponge lattice. N_2-doped restore GO structure sponges including crystalline iron oxide nanoparticles were formed with 873K thermal behavior under N_2 for 180 min. In comparison with nanoparticles that grown on carbon black and N_2-doped graphene sheets, N_2-doped restore GO structure had positive onset potential for O_2 reduction reaction and higher O_2 reduction reaction current density in alkaline solutions (Wang et al., 2012). Moreover, graphene-based materials cannot be applied as catalyst supports for nonprecious metal catalyst and platinum; however, have moreover been discovered as metal-free catalysts in the fuel cell usages (Parvez et al., 2012).

10.3.2.2 Dye Solar Cell

A usual solar cell based on dye sensitized has three key substances counting an iodide electrolyte, counter electrode and dye adsorbed titanium dioxide photo-anode (Grätzel, 2003). In practice, in the light beam on the titanium dioxide photo-anode, iodide can be oxidized to the triiodide, releasing an electron in titanium dioxide. Subsequently, to make trioxide reduction reaction fast sufficient, the catalyst like Pt catalyst is typically required (Xue et al., 2012). Like to fuel cells mentioned above, many energy has been put on substituting costly platinum-based electrodes by nonprecious catalysts, especially by heteroatom doped carbons like graphene-based materials though their catalytic activities and conductivity still underperform those of Pt catalysts (Zhang et al., 2011a). As mentioned above, G-sponges can prepare a great surface area for catalytic conversion and acceptable porosity for increased electrolyte diffusion phenomenon. To establish the mentioned application, Xue et al. (2012) performed a N_2-doped restore the GO structure sponge covered on the doped tin oxide glass top by way of counter electrode to attain efficiency of power conversion about 7% that was equal to 7.45% performance attained by the Pt counter electrode and also, better than applying un-doped restore the graphene sponge or a spin covered N_2-doped restore the GO film (Roy-Mayhew et al., 2010). It was detected that the N_2-doped restore the GO sponge could show higher current density and lower resistance than Pt. It was supposed by additional optimization; N_2-doped G-sponges applied as an efficient replacement for Pt in the dye sensitized solar cells (Ahn et al., 2013).

10.3.2.3 Lithium-Ion Battery

Lithium-ion battery has become main in battery technology because of its excellent storage characteristics, high reversible capacity, good cycle life, high power density and high energy density. The carbon electrode acts a significant role in battery efficiency (Ramanathan et al., 2019). Graphitic material usually is applied as anode components in commercial lithium-ion batteries. But, to improve efficiency, a novel anode material generation is required to prevail the capacity of intercalation limitations of graphite materials. Graphene shows a great reversible capacity compared to commercial graphite as anode material due to its high chemical stability, significant electrical conductivity and surface area. Doping boron or N_2 in graphene framework increases graphene sheets reversible capacity (Sun et al., 2018). Reddy et al. (2010) deposited N_2-doped graphene on copper foil using the chemical vapor deposition process and indicated its capacity for two reversible discharges. Enhancement in the capacity of reversible discharge can be ascribed from defects created using N_2-doped. N_2-doped graphene attained of thermal annealing of GO in NH_3 indicated an increase in the specific capacity with an increase in the charge/discharge cycles by stability of superior cyclic (Niu et al., 2013). Iron oxide/N_2-doped graphene composites indicated capacity of 1.012 Ah/g afterward 101 cycles that was higher than those of iron oxide/G and pure iron oxide itself (Du et al., 2012). Tin (IV) oxide/N_2-G indicated very well reversible rate of capability and capacity and well cyclic efficiency. Improvement in the electrochemical efficiency was attributed because of synergistic consequence among N_2-doped graphene and metal oxides (Vinayan and Ramaprabhu, 2013).

Graphene sheet has a big aspect ratio, is highly conductive and flexible, proposing possibility as the structure of confining to avoid pulverization and also, increase stability of cycle and improve retention capacity at beneficial discharge rate (Ren et al., 2013). It was presented that efficiency gains can be appreciated when coagulation was impeded using appropriate anchoring or mixing nanoparticles to graphene through growth (Huang et al., 2012a). Zhou et al. (2010) established the anode assembly by graphene/iron oxide, which gave 0.7 A.h/g afterward 101 series at 7×10^{-4} A/g, about two times upper than reversible commercial iron oxide capacity afterward 31 series at 0.35 A/g. Furthermore, the developed graphene composites increase cathode efficiency. Hu et al. (2013) increased commercial lithium iron phosphate material capacity 22%–25% and considerably increased the cycle life using coating components by a simple dropwise graphene solution. *S* (sulfur) has a great theoretical capacity when sulfur is applied as anode substance for battery that is numerous times greater than that of conventional cathode substances; however, like many high capacity anode substances, it suffers of little electrical conductivity, volume expansion issues and dissolution that result in a poor cycle. Wang et al. (2011b) indicated usage of decreased GO sheets to wrap *S* particles result in capacity of 36 A.min/g at 0.830 A/g, by 12% decay after 151 series. These consequences show that 2D graphene has a unique potential for improvement of cathode and anode substances in Li-ion battery. As documented, increasing the electrode materials porosity in lithium-ion battery can meaningfully reduce the Li-ions diffusion time in the chaotic graphene layers, increasing cycling performance and efficient capacity. Graphene, as a composite, could improve anchoring sites number yielding porosity, result in better capacity. Inappropriately, porosity can decrease the capacity of materials and can related to the large electrolyte intercalation value, result in an enhancement in irreversible capacity such as in the graphite-based materials (Liang and Zhi, 2009). For instance, Zhou et al. (2011) presented a restore the GO sponge ready using thermally annealing of a freeze-dried GO sponge. Original capacity of mentioned materials was more than 63.54 A.min/g (the low rate, 0.50 A/g). But large material surface area caused in an improvement in irreversible capacity, showed using retaining capacity only 2.4 A.min/g afterward a few capacity dropping and cycles to 13.80 A.min/g at higher discharge rates. In total, the Li diffusion time; furthermore, acts an important role in lithium-ion battery efficacy that can be improved by advent of nano-architectures. For instance, Yang et al. (2013) lately presented one pot synthesis of the freestanding sponge involving interconnecting V_2O_5 nanoribbons and restore the GO sheet, where the mixture of GO and vanadium oxide were decreased in the autoclave at 453K. GO composites were dried, tested and pressed as a freestanding electrode. Consequently, the graphene sponges can be applied as the cathode and anode materials for lithium-ion battery by high rates of discharge and improved efficiency using expressively decreasing inactive electrode mass. Further, combining chemical vapor deposition technique and the resulting graphene sponge by higher energy density substances like silicon and metal oxides can lead to a in much higher volume energy density (Gong et al., 2013).

10.3.2.4 Supercapacitor

Electrochemical capacitors are uses in many areas like power supply devices, mobile electronics and hybrid vehicles because of their high energy and power density, cost effectiveness and long cycle life (Kötz and Carlen, 2000). Carbon-based supercapacitors have best capacitance performance due to their high electrical conductivity and high surface area (An et al., 2001). The motivation why graphene is an appropriate base-material for supercapacitors is due to electron mobility at 298K and moreover graphene high surface area. First graphene-based supercapacitor was presented with Rao et al. (2014); they indicated that the specific graphene-based supercapacitors capacitance can reach 7.5×10^4 and 1.17×10^5 F/kg in ionic liquid (IL) and aqueous electrolytes, respectively. Theoretical electrochemical two-layer graphene capacitance is 5.2×10^5 F/kg supported using intrinsic graphene capacitance (Chuang et al., 2016). It is supposed, which graphene assembly in porous 3D supercapacitor electrode layers have potential to increase boost network conductivity, create large macro-pore channels and electrode active surface area to improve and facilitate ionic transport in the electrolyte. Subsequently, graphene materials freestanding networks can moreover decrease inactive electrode mass base from current collectors, polymer binders and conductive additives (Nhlane et al., 2021). By optimized processing, the graphene aerogels can outperform with respect to 2D graphene. Lately, Xu et al. (2012) showed possibilities of high power density G-carbon nano-tube sponge composites in ILs, where ILs conventionally propose poor rate efficiency. Electrodes created by spongy substance showed 89%–97% capacity retention afterward 10^4 series and established efficiency calculated up to 120 V/s or 10^2 A/g (Vivekchand et al., 2008). In dissimilarity, freestanding sponges formed using Xu et al. (2010) with 453K autoclaved GO reduction without cross-linkers and reducing agents were capable to yield 1.75×10^5 F/g at 0.01 V/s, 51% improved than separate graphene produced using similar process. But, freestanding, which restores GO sponge conductivity was still not sufficient to validate remarkability efficiency retention at rate >0.050 V/s even in the aqueous media by high conductivity of ionic. By continued improvement, the freestanding restore the GO sponges experienced secondary processing in 363K solution including the strong reduction agent, which efficiency was enhanced significantly because of an increase in conductivity (Zhang et al., 2011b).

Additional approach is in-situ restore GO sponge growth with Ni foam using a self-assembly process reduction, which shapes an electrode composite. This procedure prepares highly interconnected restore GO sheets in porous sponge by high surface area and limited restacking (Zhang et al., 2012). In addition, the 3D Ni foam support makes the short path to very conductive metal collector and enhances durability compared to overall strength by freestanding restore the GO sponge alone. Chen et al. (2012) used the mentioned technique using impregnating Ni foam by a 2000–6000 g/m^3 GO, followed with reduction self-assembly at 333–373K in the $C_6H_8O_6$ solution. The electrode surface area reached 1.26×10^9 cm^2/kg and electron conduction and high ion diffusion in composite electrode allowed outstanding electrovalence retention by electrovalence reducing just 9% for density (Chen et al., 2012). For supercapacitors, by aqueous electrolytes hold numerous benefits like easy handling, lower toxicity, high ion mobility, in addition to safety related to organic electrolytes applied in profitable electrochemical two-layer capacitors. But, low aqueous cells voltage limits density of energy (Wu et al., 2010). These substances have low conductivity that limits application at discharge rates needed for applied super-capacitive usage. Consequently, chemical vapor deposition

GO sponges could prepare some important advantages to increase pseudo-capacitance (Sun et al., 2011). The high surface area prepared sufficient to efficiently anchor nanostructured materials growth that results in the short electron transport distances by low transport resistance (Olabi et al., 2021). Indeed, high graphene network conductivity permits for useful utilization of pseudo-capacitive substance even at the high rates, enhancement making cycle life and high power retention more competitive by electrochemical double layer capacitors devices (Conway et al., 1997).

10.3.2.4.1 Supercapacitor Fuel Cell

Fuel cell is an apparatus that is applied for direct conversion of the fuel chemical energy in electricity by high performance and had showed promising consequences in several applications (Nassef et al., 2019). The support for catalyst on the membrane for fuel cell particularly proton exchange membrane is typically made of graphene (Rezk et al., 2019). Nowadays, many studies are being performed to assess replacing Pt-catalyst possibility by other nonprecious metal oxides/metal in addition to N_2 coordinated metal catalyst (Khatib et al., 2019). The mentioned catalysts have issues in the terms of activity and stability in compassion with Pt-catalyst. The active carbon is chain of mitigating of these challenges; however, they be disposed to have their own limits as well (Ogungbemi et al., 2019). The active carbon has a tendency to have a great surface area; however, active carbon support being un-stable increases a main concern without coupled by another material (Bai et al., 2011). Graphene technological development has alteration exemplum as active carbons are nowadays being investigated as a stronger alternative to Pt due to their conductivity, high surface area and adhesion to the catalyst particles (Abdelkareem et al., 2020). GO with functional groups is appropriate for nucleation in addition to catalyst nanoparticles being devoted to the surface (Liu et al., 2011). The graphene is widely applied in fuel cell mostly for anode catalyst support material, in addition to even substitute cathode catalyst, standalone electrolyte membrane and composite. Moreover, it is applied in bi-polar plates (Wang et al., 2011c). Platinum and platinum alloys are traditional active catalyst into fuel cell electrode, whether at cathode and anode at the low temperature fuel cell fed using H_2 and other hydrocarbons with low M_w (molecular weight) like MeOH (Abdelkareem et al., 2014), and C_2H_5OH (Akhairi and Kamarudin, 2016). But platinum is expensive; however, limited in resources, in addition to affected using intermediate products through oxidation of various fuels (Feng et al., 2013). Several techniques have been carried out to decrease completely replace platinum catalyst or catalyst loading by nonprecious catalysts at the both anodes (Abdelkareem et al., 2018) and fuel cells cathode (Dombrovskis and Palmqvist, 2016).

10.3.3 Graphene Catalyst Material

Catalyst is a material, which can accelerate chemical reaction rate and increase reaction selectivity without consuming itself (Higgins et al., 2016). The catalyzed reactions carry out through high energy well-organized routes that feed are capable to be applied more professionally and pollutants and byproducts are to be decreased (Haag and Kung, 2014). For instance, catalysts have been applied in synthesizing novel drugs, which increase human health. Catalyst materials are extensively used in the systems associated to storing energies, converting and generating like fuel cell, battery, new fuel production and oil refinery (Li et al., 2016). Catalyst materials are furthermore necessary to monitoring and removing pollutants from water, soil and air, reducing modern cars emissions and protection of environment (Jariwala et al., 2011). Catalyst materials can be synthetic materials or natural like metal oxides, metals, organic compounds and enzymes. Carbon nanomaterials containing graphene, carbon nanotube, carbon black and other derivatives are main substances of many catalyst-based synthetics. Carbon nanomaterials have been applied as useful catalysts and as the other catalysts supports (El-Gendy et al., 2019). Among carbon-based materials mentioned previously, graphene has newly attracted intense consideration. This is mostly because of the fact that graphene has many genuine benefits over the other carbon-allotrope to develop novel catalysts (Zheng et al., 2016). At the first, the theoretical SSA of graphene is about 2.6×10^7 m^2/g and it is twice of single walled carbon nanotubes and much higher than activated carbon and carbon black. Consequently, this structure makes graphene desirable for potential usages as two-dimensional support for catalyst loading. Also, locally conjugated structure dedicates graphene by increased capacities of adsorption to materials in catalytic reaction (Wang et al., 2020). Second, the graphene-based materials, particularly chemically modified graphene, can be determined at comparatively low-cost on a large scale using graphite and graphite oxide. The graphene-based substances are free from almost current unavoidable metallic impurities in carbon nanotubes that would decrease the carbon nanotube performance in the catalytic reactions (Zhang et al., 2020b). Third, greater electron flexibility of the graphene facilitates electron transfer through catalytic reactions, increasing the activity of graphene catalytic. Lastly, the graphene furthermore has great electrochemical, thermal and optical chemical constancies that can possibly improve catalysts lifetime (Alam et al., 2020).

The graphene-based catalysts can be only categorized in two kinds: composite and inherent catalyst. The inherent catalyst is composed of pure graphene material and usual instances are GO, rGO and also, heteroatom doped graphene (Saleh and AL-Hammadi, 2021). The graphene-based catalyst is the graphene component mixed by one and more other catalyst. In graphene-based catalyst, graphene typically plays as the conductive component for immobilizing of second catalytic substrate (Vatandost et al., 2020). Activities of catalytic of intrinsic catalysts are mostly associated by graphene sheets structures and compositions (Bai et al., 2021). But the efficiency of the graphene composite catalyst depends on inherent interfacial interaction and features and synergistic influence of graphene-based composite components. Sometimes, intrinsic catalyst is dispersible in the reaction medium and behaves similar to homogenous polymer catalysts, though composite catalysts are typically insoluble and play as heterogeneous catalysts (Kuniyil et al., 2021). Intrinsic catalysts lifetime depend on graphene materials structural stability and morphological, though those of composite catalysts were mostly controlled with catalyst immobilized stability on graphene sheet (Li et al., 2021). The composite catalyst is regularly provided using in-situ and mixing growth. As mentioned above, the graphene material can be applied to synthesize intrinsic and composite graphene catalysts. Graphene-based catalysts have useful thermal, optical, mechanical and graphene materials electrical features by abundant functions of other catalysts. As a consequence, graphene-based catalysts have been discovered for photochemical, electrochemical and catalyzing organic reactions (Garg and Ling, 2013). The most catalysts graphene-based and their uses are listed in Table 10.8.

10.3.3.1 Organic Reaction

Pure graphene is a substance by rare functional groups and also zero bandgap. Consequently, the activity of this kind of catalyst is weak. But, chemically modified graphenes like GO and rGO are semiconductors or insulators (Sahoo et al., 2021). Therefore, graphene-based materials are as potential catalysts for several organic reactions. Catalysts extensively applied for oxygen incorporation in organic components are transition metal-based materials that are typically toxic, expensive, limited in their resources and difficult to remove (Coros et al., 2020). Indeed, GO catalyst not only indicated great catalytic activities for several chemical reactions, but also, they could furthermore be usefully reused or eliminated using filtering it of reaction systems (Fouda et al., 2021). For example, GO was verified to be catalyzing capable the C_7H_8O (benzyl alcohol) oxidation reaction to C_7H_6O (benzaldehyde) by a good selectivity and high yield (>93%). GO is moreover capable to catalyze *cis*-stilbene oxidation and several alkynes hydration (Ngameni et al., 2021). Aldehydes and methyl ketones are often applied for the chalcones synthesis via Claisen-Schmidt reaction and chalcones are of iso-flavonoids and flavonoids main precursors. For instance, GO was applied as a catalyst for polymerization of dehydrative to produce carbon reinforced poly (phenyl-enemethylene) composites via a step reaction (Mohammadkhani et al., 2020). Carboxyl-modified GO has inherent peroxidase such as activity that can catalyze reaction of 3,3,5,5-tetramethylbenzidine, peroxidase substrate in the presence of hydrogen peroxide. The GO activity as a mimic peroxidase depends on reagent concentration, T and pH value (Jariwala et al., 2011). Furthermore, the GO catalyst has various benefits counting preparation ease, high surface/volume ratio, low cost and high stability.

TABLE 10.8
Graphene-Based Catalysts and Their Applications

Intrinsic Catalyst	Application	Composite Catalyst	Application
Graphene oxide	Oxidation reaction C–C coupling reaction Polymerization Mimic peroxidase Photoreaction Glucose sensor	Palladium/Reduced graphene oxide	C–C coupling reaction
Graphene/Graphitic carbon nitride	Selective oxidation	Gold/Reduced graphene oxide	Suzuki coupling reaction
L-Cysteine-graphene oxide	Matrine sensor	Calcium oxide/RGO	Transesterification
RGO	Hydrogen peroxide sensor β-nicotinamide adenine dinucleotide sensor Acetaminophen sensor Dopamine sensor Ascorbic acid sensor Uric acid sensor	Hemoglobin/Graphene oxide	Peroxidatic reaction
Ionic liquid-Reduced graphene oxide	β-nicotinamide adenine	Platinum/Reduced graphene oxide	Methanol oxidation reaction
Sulfonated polyaniline-Reduced graphene oxide	Ascorbic acid sensor	Platinum-Ruthenium/Reduced graphene oxide	Methanol oxidation reaction
Sulfonated-RGO	H_2O_2 sensor Sodium nitrite sensor	Platinum/ poly (diallyldimethyl-ammonium chloride)-Reduced graphene oxide	Methanol oxidation reaction
N_2-doped G	O_2 reduction reaction in fuel cells Hydrogen peroxide sensor	Platinum-Gold/poly (diallyldimethyl-ammonium chloride)-Reduced graphene oxide	Formic acid oxidation reaction
N_2-doped Edge-functionalized graphene	O_2 reduction reaction in fuel cells	Platinum/N_2-Reduced graphene oxide	Methanol oxidation reaction Oxygen reduction reaction
Carbon nitrate modified graphene	O_2 reduction reaction in the fuel cell	Platinum- Ruthenium/N-doped carbon nanotube-graphene hybrid nanostructure	Methanol oxidation reaction
Graphitic carbon nitride modified graphene	O_2 reduction reaction in the fuel cell	Cobalt oxide/Reduced mildly oxidized GO or N_2-Reduced mildly oxidized graphene oxide	Oxygen reduction reaction Oxygen evolving reaction
Carbon nitride modified graphene	O_2 reduction reaction in the fuel cell	$MnCo_2O_4$/Reduced mildly oxidized graphene oxide or N_2-Reduced mildly oxidized graphene oxide	Oxygen reduction reaction Oxygen evolving reaction
Nitrogen-doped graphene quantum dots	O_2 reduction reaction in the fuel cell	Fe-N_2-Reduced graphene oxide	Oxygen reduction reaction
Polyelectrolyte/Graphene	O_2 reduction reaction in the fuel cell	Manganese oxide/RGO-Ionic liquid	O_2 reduction reaction

Source: Modified after Huang et al. (2012b).

Consequently, it is hopeful for usage in sensing glucose in the real systems, like in buffer solution of fruit juice and diluted blood (Awaludin et al., 2020). Long et al. (2011) presented that GO foam had high capacity of adsorption for sulfur dioxide and could convert sulfur dioxide to sulfur trioxide using reacting by oxygen. GO played not only as the oxidant in reaction, but also, as a catalyst for sulfur dioxide and oxygen reactions. GO has moreover been applied as the solid acid catalyst for calix-4-pyrroles and dipyrromethane preparation in several solvents at 298K.

10.3.3.2 Photoreaction

It is famous that fundamental graphene features and graphene derivatives like surface chemical and physical properties, very well electrical conductivities and strong capability in accepting electrons are fundamentally dependent on graphene unique surface, structure and interface properties, like molecular adsorption, charge transport, electronic structure, surface chemistry and atomic arrangements; therefore, considerably influence the photocatalytic efficiency of graphene-based photo-catalysts (Xiang et al., 2012). Significantly, the electrical and surface features of graphene can be furthermore tuned through suitable chemical controlling and modification graphene structural order, respectively. Therefore, to realistically implement graphene and graphene derivatives features in designing composite photo-catalysts, a deep understanding on graphene and features of graphene derivatives is necessary (Long et al., 2011). The first report on heterogeneous photocatalytic environmental pollutants remediation (cyanide in H_2O) on Ti (titania) using Frank and Bard (1977) heterogeneous photocatalysis has been extensively applied in the widespread environmental purification like purification of water and air. In specific, it was indicated that some usual properties like crystalline shape and size, surface area, crystallinity, exposed facet and phase structures are main for increasing of photocatalytic performance in pollutants degradation in air or water. Consequently, a variety of ways for enhancement of photo-decomposition pollutants efficiency over semiconductors have been exploited (Mamba et al., 2020). Between them, specific consideration has been paid to creation of graphene composite photocatalysts containing superior volume of adsorption of separation and transfer capability, charge and light-harvesting and pollutants (Li et al., 2015).

It is well established that the graphene co-catalyst has moreover been extensively applied to extract electrons of excited semiconductors and consequently accelerate oxygen-reduction reactions on it, which have been observed as determining of rate step in the solar photocatalytic mineralization of organic contaminants (Kuang et al., 2020). It is identified that silver, gold, copper and platinum clusters/nanoparticles have been established to be best co-catalysts for photocatalytic oxidative organic pollutants decomposition of over titanium dioxide nanophotocatalysts (Singh et al., 2020). Photocatalytic water splitting in O_2 and H_2 by solar semiconductors and energy has concerned many proven and attention to be a hopeful way to solve green low-cost and hydrogen fuel production and solar energy storage. But, from the Gibbs energy change viewpoint, thermodynamically uphill photocatalytic H_2O splitting is more complicated than downhill photocatalytic degradation reactions like organic compounds oxidation by O_2 molecules (Wang et al., 2019). Subsequently pioneering works with Fujishima and Honda in 1972, several heterogeneous photocatalysts (like metal-free SiC and C_3N_4, organic dyes, [oxy]nitrides, sulfides and metal oxides) have been extensively used in photocatalytic H_2 production from H_2O reduction in recent years (Mishra and Acharya, 2021). Usually, photocatalytic splitting of H_2O systems can be separated in half reaction splitting of H_2O (for O_2 and H_2) and overall splitting of H_2O systems. Fascinatingly though, graphene has been widely integrated in composite photocatalysts in the two mentioned systems to improve their photocatalytic activity for splitting of H_2O (Maham et al., 2017).

The fast increase in anthropogenic CO_2 level related to fossil fuels shortages and global warming have become one of most important worldwide problems. Consequently, from lookout of maintainable energy extension, the carbon dioxide conversion to appreciate energy-bearing substances (like CH_4, methanol and CO) by sunshine as source of energy would be one of best explanations to solve the mentioned difficulties. Consequently, based on the first photocatalytic carbon dioxide demonstration reduction using Inoue et al. (1979), unique developments have been made in extraction

TABLE 10.9
Photo-Catalytic Reduction of Carbon Dioxide over Nano-Carbon-Based Photocatalysts

Semi-Conductor	Co-Catalysts	By-Product
Graphene oxide	Copper	Methanol, Acetaldehyde
Reduced graphene oxide	Titanium dioxide	Methane
Reduced graphene oxide	$Ti_{0.91}O_2$	Carbon monoxide, Methane
Reduced graphene oxide	Titania nanosheet	Methane
B-doped reduced graphene oxide	Titanium dioxide	Sodium sulfite, Methane
Solvent-exfoliated graphene	Titanium dioxide	Methane
Reduced graphene oxide	Cadmium sulfide nanorod	Methane
Graphene oxide	Tungsten trioxide	Methane
RGO	Copper oxide	Carbon monoxide
	Copper oxide	Methanol
	Tantalum pentoxide	Methanol, hydrogen
	Zinc oxide	Methanol
	$Fe_2V_4O_{13}$-Cadmium sulfide	Methane
	Titanium dioxide-Cadmium sulfide	Methane
	Graphitic carbon nitride	Methane
Graphene oxide	Cobalt phthalocyanine	Methanol
Reduced graphene oxide	Multianthraquinone-substituted porphyrin	Formic acid

Source: Modified after Li et al. (2016).

performance and practical semiconductors for carbon dioxide with H_2O during the last two decades. Among the semiconductor components, combination of rGO/GO and semiconductors has become a common strategy to increase the semiconductor activity for photocatalytic CO_2 conversion to valuable hydrocarbons, because of its low price, innocuity and excellent stability (Li et al., 2016). Also, more details are given in Table 10.9.

First, though the adsorption capacity of carbon dioxide and electron nanocomposite photocatalysts conductivity increase by increasing graphene content, the light protecting caused using excess values of graphene could in turn related to the reduced CO_2 photo-reduction activity. Consequently, the optimized values of graphene in composite photocatalysts are very critical to increase CO_2 photo-reduction efficiency (Li et al., 2016).

10.3.4 Biomedical Applications of Graphene

There have been several conflicting presents in field of the antimicrobial and biocompatibility activity of GO. Recently, it is maybe becoming stronger that well-prepared and purified GO is not antibacterial (Kumar et al., 2014). In one experiment (Ruiz et al., 2011) Luria-Bertani nutrient broth containing 2.5×10^{-10} kg.m/L GO was injected using *Escherichia coli* bacterial cells to a concentration of 0.025 OD_{600} (optical density of a sample measured at a wavelength of 600 nm) and raised for 960 min at 310K. The obtained results indicated, which the GO-containing samples attained an average value absorbance was 1.71 during 960 min and the bacteria growing in Luria-Bertani broth only attained an absorbance of 1.31. These mentioned results indicate that bacteria grew quicker and to a higher optical density than in the case of without GO and revealed that GO is not a bacteriostatic or bactericidal material; however, in its place a general growth enhancer that plays as a scaffold for proliferation and cell surface attachment.

Growing biomedical area of graphene-based substances applications increases questions about their toxicity of long- and short-term. It is identified that graphene cytotoxicity flakes depend on size of flake. The smaller flakes are cytotoxic and indicate effect of cellular functionality and higher cellular internalization to the greater level (Kumar et al., 2014). The number of O_2 functional groups, which are devoted to surface furthermore act a significant role. For larger levels of carbon/oxygen, flakes are less toxic that can be correlated to the structures of partially rGO. Shi et al. (2012) have indicated that cell treatment is powerfully responsive to rGO structure. Indeed, Shi et al. (2012) formed out few layer rGO films and controlled surface O_2 content and reduction level. It was additional detected that performance of cell reduced considerably with an increase in thermal reduction level. Numerous investigations have discovered toxicology and biocompatibility of GO using several human and animal cell kinds and generally, GO has proven biocompatible and nontoxic at moderate dosages (Liu et al., 2013). Zhang et al. (2011a) evaluated GO biocompatibility using monitoring influence of concentration of GO and incubation time on morphological variations of human erythrocytes. It was detected almost all erythrocyte membranes were retained integration when they incubated using phosphate buffer saline for up to 240 min. Though GO flakes were adhered to red blood cells surface, GO suspension indicated little influence on membrane integrity and erythrocyte morphology at dosage of 10^{-8} kg.m/L for 60 and 240 min. But an erythrocyte membrane's part was separated and ghost cells were detected when erythrocytes exposed to 8×10^{-8} kg.m/L of GO for 240 min. Since GO is biocompatible by blood cells, this will pave way for additional development of GO for other biomedical applications like targeted drug delivery (Chung et al., 2013).

Graphene is very hydrophobic, whereas GO is decorated using O_2-including hydrophilic groups. This significant surface chemistry permits for π-π electrostatic and stacking interaction to happen by other molecules in the vicinity of graphene. This permits for chemical and physical binding of drugs to surface of graphene/GO for application of drug delivery (Tadyszak et al., 2018). Liu et al. (2008) used PEGylated GO to deliver a camptothecin analogue. GO and graphene were used as vehicles for the drug delivery, containing anticancer drug, poor-soluble drug, antibodies, peptide, genes and antibiotic (Ghosal and Sarkar, 2018). It stays challenging to attain proper anticancer behavior because of factors like poor targeting and low bioavailability of chemotherapeutics. Graphene and GO have been widely discovered as biomolecule sensors, drug carriers and cellular imaging agents in anticancer treatments. The simplest strategy is a hybrid/nanocomposite built fabrication from drugs and GO. In one case, chlorogenic acid and GO were given as a power of hydrogen sensitive platform for chlorogenic acid slow release from GO. In another instance, GO was functionalized, chemically, by amino groups and joined by carboxymethyl cellulose as an anticancer system by a Doxorubicin drug controlled and targeted release (Cheng et al., 2020). Recently, folic acid mixed by polyethyleneimine functionalized GO was applied as a carrier for new Cu complexes to nasopharyngeal carcinoma cell line. Developed folic acid/polyethyleneimine/GO showed good biocompatibility and H_2O solubility and in vitro cytotoxicity investigations revealed that Cu complexes have an appreciate deterrence influence on carcinoma cell line (Iravani and Varma, 2020).

10.4 CONCLUSIONS

Graphene is a unique material by a single carbon atoms sheet packed in a hexagonal structure. Several morphologies of graphene have also been detected, containing zero-dimensional GQD, one-dimensional graphene nanoribbons and two-dimensional graphene nanosheets. The properties of zero-dimensional and one-dimensional graphene can be adjusted using their size and edges. The significant graphene features, like strong Young's modulus, fast charged carrier mobility, high thermal conductivity and high surface area lead to increase its new applications of it. Due to the inimitable properties of graphene and its wide application in electronics, medical, automotive, energy, heat dissipation, composites, sensor and aerospace fields, the production of graphene and its derivatives from low-cost and renewable carbon sources is noteworthy.

Leaf wastes can be considered as promising source to produce graphene because of their abundance, significant carbon content and structure. The leaves of plants and trees contain hemicellulose, lignin and cellulose, which are important carbon source. The pyrolysis of dried leaves at high temperature and under inert atmosphere can be one of the facile methods to produce graphene from these valuable resources. In this synthesis approach, temperature has key role in the formation of graphene with high carbon and low oxygen contents, lump-like, porous and flake-like structure and sp^2 or sp^3 carbon content. Also, synthesis conditions influence the thickness of graphene sheets, porous structure, the number of graphene layers and distance between them, graphitization degree and crystalline structure of produced graphene. Therefore, a lot of effort is required to provide a facile synthesis method with optimal conditions. On the other hand, some derivatives of leaves such as tannic acid, alginic acid and humic acids are appropriate source for graphene production. According to the characterization analyses, bilayer and few layers graphene can be produced from pyrolysis of alginic and tannic acids. The FESEM, XPS, HRTEM, AFM images, SAED pattern and Raman spectroscopy approve high quality and sixfold crystalline symmetry structure of produced graphene from these acids. Humic acids and organic redox-active components have chemical structure similar to graphene. These complex molecules can be converted to graphene during carbonization at high temperature and oxidation-exfoliation-thermal reduction. According to the SEM and TEM images, the surface morphology of obtained graphene from humic acid is very similar to that of synthesized samples from natural graphite. Graphene derivatives also have wide applications in various fields and compensate some shortcomings of graphene. GQD is one of the important alternatives to graphene, which possesses remarkable magnetic features because of quantum confinement and edge effects. The synthesis of GQD from leaves is a novel achievement to produce GQD without any organic solvent and chemical agents. Like the synthesis of graphene from leaves, temperature plays a very important role in the production of GQD from leaves. The reviewed literature indicated that the carbonization of leaves does not occur at low temperature, while at high temperatures the structure and surface of product are destroyed. Also, the effect of GQD magnetization and GQD spin are changed with temperature. Leaf extracts can also play reducing and capping roles instead of chemical agents in the chemical reduction of GO to fabricate graphene. But some parameters must be considered such as reduction duration, temperature and the concentration of GO. The results of energy-dispersive X-ray and FTIR spectroscopy proved that the oxygen content of rGO was decreased after GO reduction by leaf extract. Therefore, leaf extract has the ability and competence to be used in production of bio-rGO. As a result, the leaf wastes and their derivatives can be used as eco-friendly sources and reduction agents to produce graphene and its derivatives. The produced graphene has the same structure and properties in comparison with produced samples from nonrenewable carbon sources. To expand this green graphene production, much effort must be made to improve synthesis conditions and provide facile methods.

REFERENCES

Abd Ali, Ziad T. "Green synthesis of graphene-coated sand (GCS) using low-grade dates for evaluation and modeling of the pH-dependent permeable barrier for remediation of groundwater contaminated with copper." *Separation Science and Technology* 56, no. 1 (2021): 14–25. https://doi.org/10.1080/01496395.2019.1708937

Abdelkareem, Mohammad A., Masdar M. Shahbudin, Takuya Tsujiguchi, Nobuyoshi Nakagawa, Enas Taha Sayed, and Nasser A. M. Barakat. "Elimination of toxic products formation in vapor-feed passive DMFC operated by absolute methanol using air cathode filter." *Chemical Engineering Journal* 240 (2014): 38–44. https://doi.org/10.1016/j.cej.2013.11.043

Abdelkareem, Mohammad A., Yazan Al Haj, Mohannad Alajami, Hussain Alawadhi, and Nasser A. M. Barakat. "Ni-Cd carbon nanofibers as an effective catalyst for urea fuel cell." *Journal of Environmental Chemical Engineering* 6, no. 1 (2018): 332–337. https://doi.org/10.1016/j.jece.2017.12.007

Abdelkareem, Mohammad A., Enas T. Sayed, Hussain Alawadhi, and Abdul H. Alami. "Synthesis and testing of cobalt leaf-like nanomaterials as an active catalyst for ethanol oxidation." *International Journal of Hydrogen Energy* 45, no. 35 (2020): 17311–17319. https://doi.org/10.1016/j.ijhydene.2020.04.156

Ahn, Youngkun, Hyein Kim, Young Hwan Kim, Yeonjin Yi, Seong Il Kim. "Procedure of removing polymer residues and its influences on electronic and structural characteristics of graphene." *Applied Physics Letters* 102, no. 9 (2013): 091602. https://doi.org/10.1063/1.4794900

Akhairi, Maf, and Siti Kartom Kamarudin. "Catalysts in direct ethanol fuel cell (DEFC): an overview." *International Journal of Hydrogen Energy* 41, no. 7 (2016): 4214–4228. https://doi.org/10.1016/j.ijhydene.2015.12.145

Alam, Khurshed, Yelyn Sim, Ji-Hun Yu, Janani Gnanaprakasam, Hyeonuk Choi, Yujin Chae, Uk Sim, and Hoonsung Cho. "In-situ deposition of graphene oxide catalyst for efficient photoelectrochemical hydrogen evolution reaction using atmospheric plasma." *Materials* 13, no. 1 (2020): 12. https://doi.org/10.3390/ma13010012

An, Kay H., Won S. Kim, Young S. Park, Young C. Choi, Seung M. Lee, Dong C. Chung, Dong J. Bae, Seong C. Lim, and Young H. Lee. "Supercapacitors using single-walled carbon nanotube electrodes." *Advanced Materials* 13, no. 7 (2001): 497–500. https://doi.org/10.1002/1521-4095(200104)13:7%3C497::AID-ADMA497%3E3.0.CO;2-H

Ao, Zhimin, Alexander David Hernández-Nieves, F.M. Peeters, and S. Li. "The electric field as a novel switch for uptake/release of hydrogen for storage in nitrogen doped graphene." *Physical Chemistry Chemical Physics* 14, no. 4 (2012): 1463–1467. https://doi.org/10.1039/C1CP23153G

Awaludin, Norhafniza, Jaafar Abdullah, Faridah Salam, Kogeethavani Ramachandran, Nor A. Yusof, and Helmi Wasoh. "Fluorescence-based immunoassay for the detection of Xanthomonas oryzae pv. oryzae in rice leaf." *Analytical Biochemistry* 610 (2020): 113876. https://doi.org/10.1016/j.ab.2020.113876

Bai, Hua, Chun Li, and Gaoquan Shi. "Functional composite materials based on chemically converted graphene." *Advanced Materials* 23, no. 9 (2011): 1089–1115. https://doi.org/10.1002/adma.201003753

Bai, Liqun, Abdolreza Tajikfar, Sajad Tamjidi, Rauf Foroutan, and Hossein Esmaeili. "Synthesis of $MnFe_2O_4$@ graphene oxide catalyst for biodiesel production from waste edible oil." *Renewable Energy* 170 (2021): 426–437. https://doi.org/10.1016/j.renene.2021.01.139

Basu, Sukumar, and Partha Bhattacharyya. "Recent developments on graphene and graphene oxide based solid state gas sensors." *Sensors and Actuators B: Chemical* 173 (2012): 1–21. https://doi.org/10.1016/j.snb.2012.07.092

Berktas, Ilayda, Marjan Hezarkhani, Leila Haghighi Poudeh, and Burcu Saner Okan. "Recent developments in the synthesis of graphene and graphene-like structures from waste sources by recycling and upcycling technologies: a review." *Graphene Technology* 5, no. 3 (2020): 59–73. https://doi.org/10.1007/s41127-020-00033-1

Bi, Hengchang, Xiao Xie, Kuibo Yin, Yilong Zhou, Shu Wan, Longbing He, Feng Xu, Florian Banhart, Litao Sun, and Rodney S. Ruoff. "Spongy graphene as a highly efficient and recyclable sorbent for oils and organic solvents." *Advanced Functional Materials* 22, no. 21 (2012): 4421–4425. https://doi.org/10.1002/adfm.201200888

Biswal, Mandakini, Abhik Banerjee, Meenal Deo, and Satishchandra Ogale. "From dead leaves to high energy density supercapacitors." *Energy and Environmental Science* 6, no. 4 (2013): 1249–1259. https://doi.org/10.1039/c3ee22325f

Bonaccorso, Francesco, Antonio Lombardo, Tawfique Hasan, Zhipei Sun, Luigi Colombo, and Andrea C. Ferrari. "Production and processing of graphene and 2D crystals." *Materials Today* 15, no. 12 (2012): 564–589. https://doi.org/10.1016/S1369-7021(13)70014-2

Chabot, Victor, Drew Higgins, Aiping Yu, Xingcheng Xiao, Zhongwei Chen, and Jiujun Zhang. "A review of graphene and graphene oxide sponge: material synthesis and applications to energy and the environment." *Energy and Environmental Science* 7, no. 5 (2014): 1564–1596. https://doi.org/10.1039/c3ee43385d

Chandu, Basavaiah, Jalaja Chittajallu, Raghu R. Mukkavilli, Asha B. Pilli, and Hari B. Bollikolla. "Synthesis and antimicrobial studies of graphene-silver nanocomposite through a highly environmentally benign reduction methodology." *Materials Technology* 33, no. 11 (2018): 730–736. https://doi.org/10.1080/10667857.2018.1498608

Chang, Wei, Xiao-Ying Zhang, Jin Qu, Zhe Chen, Yu-Jiao Zhang, Yanqiu Sui, Xiu-Feng Ma, and Zhong-Zhen Yu. "Freestanding $Na_3V_2O_2$ (PO_4) 2F/graphene aerogels as high-performance cathodes of sodium-ion full batteries." *ACS Applied Materials and Interfaces* 12, no. 37 (2020): 41419–41428. https://doi.org/10.1021/acsami.0c11074

Chen, Ji, Kaixuan Sheng, Peihui Luo, Chun Li, and Gaoquan Shi. "Graphene hydrogels deposited in nickel foams for high-rate electrochemical capacitors." *Advanced Materials* 24, no. 33 (2012): 4569–4573. https://doi.org/10.1002/adma.201201978

Chen, Feng, Juan Yang, Tao Bai, Bo Long, and Xiangyang Zhou. "Facile synthesis of few-layer graphene from biomass waste and its application in lithium ion batteries." *Journal of Electroanalytical Chemistry* 768 (2016): 18–26. https://doi.org/10.1016/j.jelechem.2016.02.035

Chen, Jian, Yingda Ma, Lichun Wang, Wenyan Han, Yamin Chai, Tingting Wang, Jian Li, and Lailiang Ou. "Preparation of chitosan/SiO_2-loaded graphene composite beads for efficient removal of bilirubin." *Carbon* 143 (2019): 352–361. https://doi.org/10.1016/j.carbon.2018.11.045

Cheng, Hongyang, Jianrong Lin, Yanning Su, Denglong Chen, Xuelin Zheng, and Hu Zhu. "Green synthesis of soluble graphene in organic solvent via simultaneous functionalization and reduction of graphene oxide with urushiol." *Materials Today Communications* 23 (2020): 100938. https://doi.org/10.1016/j.mtcomm.2020.100938

Choi, Wonbong, Indranil Lahiri, Raghunandan Seelaboyina, and Yong S. Kang. "Synthesis of graphene and its applications: a review." *Critical Reviews in Solid State and Materials Sciences* 35, no. 1 (2010): 52–71. https://doi.org/10.1080/10408430903505036

Chuah, Regnant, Subash C. B. Gopinath, Periasamy Anbu, Midhat Nabil Salimi, Ahmad R.W. Yaakub, and Thangavel Lakshmipriya. "Synthesis and characterization of reduced graphene oxide using the aqueous extract of Eclipta prostrata." *3 Biotech* 10, no. 8 (2020): 1–10. https://doi.org/10.1007/s13205-020-02365-4

Chung, Chul, Young-Kwan Kim, Dolly Shin, Soo-Ryoon Ryoo, Byung H. Hong, and Dal-Hee Min. "Biomedical applications of graphene and graphene oxide." *Accounts of Chemical Research* 46, no. 10 (2013): 2211–2224. https://doi.org/10.1021/ar300159f

Chuang, Chiashain, Prathik Roy, Rini Ravindranath, Arun P. Periasamy, Huan-Tsung Chang, and Chi-Te Liang. "Intrinsic magnetic properties of plant leaf-derived graphene quantum dots." *Materials Letters* 170 (2016): 110–113. https://doi.org/10.1016/j.matlet.2016.01.130

Conway, Brian E., Viola Ingrid Birss, and John Wojtowicz. "The role and utilization of pseudocapacitance for energy storage by supercapacitors." *Journal of Power Sources* 66, no. 1–2 (1997): 1–14. https://doi.org/10.1016/S0378-7753(96)02474-3

Coros, Maria, Florina Pogacean, Alexandru Turza, Monica Dan, Camelia Berghian-Grosan, Ioan-Ovidiu Pana, and Stela Pruneanu. "Green synthesis, characterization and potential application of reduced graphene oxide." *Physica E: Low-dimensional Systems and Nanostructures* 119 (2020): 113971. https://doi.org/10.1016/j.physe.2020.113971

Dombrovskis, Johanna Katharina, and Anders E. C. Palmqvist. "Recent progress in synthesis, characterization and evaluation of non-precious metal catalysts for the oxygen reduction reaction." *Fuel Cells* 16, no. 1 (2016): 4–22. https://doi.org/10.1002/fuce.201500123

Du, Meng, Chaohe Xu, Jing Sun, and Lian Gao. "One step synthesis of Fe_2O_3/nitrogen-doped graphene composite as anode materials for lithium ion batteries." *Electrochimica Acta* 80 (2012): 302–307. https://doi.org/10.1016/j.electacta.2012.07.029

Eigler, Siegfried. "Graphene. An introduction to the fundamentals and industrial applications edited by Madhuri Sharon and Maheshwar Sharon." (2016): 5122–5122. Wiley Online Library. https://doi.org/10.1002/anie.201602067

Ekennia, Anthony C., Dickson N. Uduagwu, Njemuwa N. Nwaji, Obinna O. Oje, Chimerem O. Emma-Uba, Sandra I. Mgbii, Olawale J. Olowo, and Obianuju L. Nwanji. "Green synthesis of biogenic zinc oxide nanoflower as dual agent for photodegradation of an organic dye and tyrosinase inhibitor." *Journal of Inorganic and Organometallic Polymers and Materials* 31, no. 2 (2020): 1–12. https://doi.org/10.1007/s10904-020-01729-w

Elemike, Elias E., Damian C. Onwudiwe, Lei Wei, Chaogang Lou, and Zhiwei Zhao. "Synthesis of nanostructured ZnO, AgZnO and the composites with reduced graphene oxide (rGO-AgZnO) using leaf extract of Stigmaphyllon ovatum." *Journal of Environmental Chemical Engineering* 7, no. 3 (2019): 103190. https://doi.org/10.1016/j.jece.2019.103190

El-Gendy, Dalia M., Nabil A. Abdel Ghany, and Nageh K. Allam. "Green, single-pot synthesis of functionalized Na/N/P co-doped graphene nanosheets for high-performance supercapacitors." *Journal of Electroanalytical Chemistry* 837 (2019): 30–38. https://doi.org/10.1016/j.jelechem.2019.02.009

Elsen, Renold. S, Ramalingam Bharanidaran, and Sundaramoorthy Surendarnath. "Biodegradable composites from leaf wastes for packing applications." In *Advances in Manufacturing Technology*, pp. 233–241. Springer, Singapore, 2019. https://doi.org/10.1007/978-981-13-6374-0

Evanoff, Kara, Alexandre Magasinski, Junbing Yang, and Gleb Yushin. "Nanosilicon-coated graphene granules as anodes for Li-ion batteries." *Advanced Energy Materials* 1, no. 4 (2011): 495–498. https://doi.org/10.1002/aenm.201100071

Faiz, Amir. M. S., Che Abdullah Che Azurahanim, Syahidah Azis Raba'ah, and Mohd Zawawi Ruzniza. "Low cost and green approach in the reduction of graphene oxide (GO) using palm oil leaves extract for potential in industrial applications." *Results in Physics* 16 (2020): 102954. https://doi.org/10.1016/j.rinp.2020.102954

Feng, Chen, Taizo Takeuchi, Mohammad A. Abdelkareem, Takuya Tsujiguchi, and Nobuyoshi Nakagawa. "Carbon-CeO_2 composite nanofibers as a promising support for a PtRu anode catalyst in a direct methanol fuel cell." *Journal of Power Sources* 242 (2013): 57–64. https://doi.org/10.1016/j.jpowsour.2013.04.157

Ferrari, Andrea C., Francesco Bonaccorso, Vladimir Fal'Ko, Konstantin S. Novoselov, Stephan Roche, Peter Bøggild, Stefano Borini et al. "Science and technology roadmap for graphene, related two-dimensional crystals, and hybrid systems." *Nanoscale* 7, no. 11 (2015): 4598–4810. https://doi.org/10.1039/C4NR01600A

Feriancikova, Lucia, and Shangping Xu. "Deposition and remobilization of graphene oxide within saturated sand packs." *Journal of Hazardous Materials* 235 (2012): 194–200. https://doi.org/10.1016/j.jhazmat.2012.07.041

Fouda, Aly Nabeih, Duraia. M. El Shazly, and Ali A. Almaqwashi. "Facile and scalable green synthesis of N-doped graphene/CNTs nanocomposites via ball milling." *Ain Shams Engineering Journal* 12, no. 1 (2021): 1017–1024. https://doi.org/10.1016/j.asej.2020.04.011

Frank, Steven N., and Allen J. Bard. "Heterogeneous photocatalytic oxidation of cyanide and sulfite in aqueous solutions at semiconductor powders." *The Journal of Physical Chemistry* 81, no. 15 (1977): 1484–1488. https://doi.org/10.1021/j100530a011

Garg, Bhaskar, and Yong-Chien Ling. "Versatilities of graphene-based catalysts in organic transformations." *Green Materials* 1, no. 1 (2013): 47–61. http://doi.org/10.1680/gmat.12.00008

Georgakilas, Vasilios, Michal Otyepka, Athanasios B. Bourlinos, Vimlesh Chandra, Namdong Kim, Christian K. Kemp, Pavel Hobza, Radek Zboril, and Kwang S. Kim. "Functionalization of graphene: covalent and non-covalent approaches, derivatives and applications." *Chemical reviews* 112, no. 11 (2012): 6156–6214. https://doi.org/10.1021/cr3000412

Ghosal, Krishanu, and Kishor Sarkar. "Biomedical applications of graphene nanomaterials and beyond." *ACS Biomaterials Science and Engineering* 4, no. 8 (2018): 2653–2703. https://doi.org/10.1021/acsbiomaterials.8b00376

Gokulkumar, Sivanantham, PR Thyla, Loganathan Prabhu, Selvaraj Sathish, and N Karthi. "A comparative study on epoxy based composites filled with pineapple/areca/ramie hybridized with industrial tea leaf wastes/GFRP." *Materials Today: Proceedings* 27 (2020): 2474–2476. https://doi.org/10.1016/j.matpr.2019.09.221

Gong, Yongji, Shubin Yang, Zheng Liu, Lulu Ma, Robert Vajtai, and Pulickel M. Ajayan. "Graphene-network-backboned architectures for high-performance lithium storage." *Advanced Materials* 25, no. 29 (2013): 3979–3984. https://doi.org/10.1002/adma.201301051

González, William A., Diana López, and Juan F. Pérez. "Biofuel quality analysis of fallen leaf pellets: effect of moisture and glycerol contents as binders." *Renewable Energy* 147 (2020): 1139–1150. https://doi.org/10.1016/j.renene.2019.09.094

Gopinath, Subash C. B., Periasamy Anbu, Thirugnanasambandan Theivasanthi, Mohd Khairuddin Arshad, Thangavel Lakshmipriya, Chun H. Voon, Kannaiyan Pandian, Palaniyandi Velusamy, and Suresh V. Chinni. "Characterization of reduced graphene oxide obtained from vacuum-assisted low-temperature exfoliated graphite." *Microsystem Technologies* 24, no. 12 (2018): 5007–5016. https://doi.org/10.1007/s00542-018-3921-3

Grätzel, Michael. "Dye-sensitized solar cells." *Journal of Photochemistry and Photobiology C: Photochemistry Reviews* 4, no. 2 (2003): 145–153. https://doi.org/10.1016/S1389-5567(03)00026-1

Haag, DelRae, and Harold H. Kung. "Metal free graphene based catalysts: a review." *Topics in Catalysis* 57, no. 6–9 (2014): 762–773. https://doi.org/10.1007/s11244-013-0233-9

Higgins, Drew, Pouyan Zamani, Aiping Yu, and Zhongwei Chen. "The application of graphene and its composites in oxygen reduction electrocatalysis: a perspective and review of recent progress." *Energy and Environmental Science* 9, no. 2 (2016): 357–390. https://doi.org/10.1039/c5ee02474a

Hu, Lung-Hao, Feng-Yu Wu, Cheng-Te Lin, Andrei N. Khlobystov, and Lain-Jong Li. "Graphene-modified $LiFePO_4$ cathode for lithium ion battery beyond theoretical capacity." *Nature Communications* 4, no. 1 (2013): 1–7. https://doi.org/10.1038/ncomms2705

Huang, Xiao, Zongyou Yin, Shixin Wu, Xiaoying Qi, Qiyuan He, Qichun Zhang, Qingyu Yan, Freddy Boey, and Hua Zhang. "Graphene-based materials: synthesis, characterization, properties, and applications." *Small* 7, no. 14 (2011): 1876–1902. https://doi.org/10.1002/smll.201002009

Huang, Xiao, Xiaoying Qi, Freddy Boey, and Hua Zhang. "Graphene-based composites." *Chemical Society Reviews* 41, no. 2 (2012a): 666–686. https://doi.org/10.1039/c1cs15078b

Huang, Cancan, Chun Li, and Gaoquan Shi. "Graphene based catalysts." *Energy and Environmental Science* 5, no. 10 (2012b): 8848–8868. https://doi.org/10.1039/c2ee22238h

Huang, Binyan, Yunguo Liu, Bin Li, Shaobo Liu, Guangming Zeng, Zhiwei Zeng, Xiaohua Wang, Qimeng Ning, Bohong Zheng, and Chunping Yang. "Effect of Cu (II) ions on the enhancement of tetracycline adsorption by Fe_3O_4@SiO_2-Chitosan/graphene oxide nanocomposite." *Carbohydrate Polymers* 157 (2017): 576–585. https://doi.org/10.1016/j.carbpol.2016.10.025

Huang, Dan, Jizi Wu, Lu Wang, Xingmei Liu, Jun Meng, Xianjin Tang, Caixian Tang, and Jianming Xu. "Novel insight into adsorption and co-adsorption of heavy metal ions and an organic pollutant by magnetic graphene nanomaterials in water." *Chemical Engineering Journal* 358 (2019): 1399–1409. https://doi.org/10.1016/j.cej.2018.10.138

Inoue, Tooru, Akira Fujishima, Satoshi Konishi, and Kenichi Honda. "Photoelectrocatalytic reduction of carbon dioxide in aqueous suspensions of semiconductor powders." *Nature* 277, no. 5698 (1979): 637–638. https://doi.org/10.1038/277637a0

Iravani, Siavash, and Rajender S. Varma. "Green synthesis, biomedical and biotechnological applications of carbon and graphene quantum dots. A review." *Environmental Chemistry Letters* 18, no. 3 (2020): 703–727. https://doi.org/10.1007/s10311-020-00984-0

Ismail, Zulhelmi. "Green reduction of graphene oxide by plant extracts: a short review." *Ceramics International* 45, no. 18 (2019): 23857–23868. https://doi.org/10.1016/j.ceramint.2019.08.114

Jariwala, Deep, Anchal Srivastava, and Pulickel M. Ajayan. "Graphene synthesis and band gap opening." *Journal of Nanoscience and Nanotechnology* 11, no. 8 (2011): 6621–6641. https://doi.org/10.1166/jnn.2011.5001

Jayarathne, Ayomi, Prasanna Egodawatta, Godwin A. Ayoko, and Ashantha Goonetilleke. "Geochemical phase and particle size relationships of metals in urban road dust." *Environmental Pollution* 230 (2017): 218–226. https://doi.org/10.1016/j.envpol.2017.06.059

Keramatinia, Motahhare, Mohammad Ramezanzadeh, Ghasem Bahlakeh, and Bahram Ramezanzadeh. "Synthesis of a multi-functional zinc-centered nitrogen-rich graphene-like thin film from natural sources on the steel surface for achieving superior anti-corrosion properties." *Corrosion Science* 178 (2021): 109077. https://doi.org/10.1016/j.corsci.2020.109077

Khatib, Fawwad.Nisar, Tabbi Wilberforce, Oluwatosin Ijaodola, Emmanuel Ogungbemi, Zaki El-Hassan, Andy Durrant, James Thompson, and Abdul Ghani Olabi. "Material degradation of components in polymer electrolyte membrane (PEM) electrolytic cell and mitigation mechanisms: a review." *Renewable and Sustainable Energy Reviews* 111 (2019): 1–14. https://doi.org/10.1016/j.rser.2019.05.007

Khose, Rahul V., Goutam Chakraborty, Mahesh P. Bondarde, Pravin H. Wadekar, Alok K. Ray, and Surajit Some. "Red-fluorescent graphene quantum dots from guava leaf as a turn-off probe for sensing aqueous Hg (II)." *New Journal of Chemistry* 45, no. 10 (2021): 4617–4625.

Kötz, Ruediger, and Martin W Carlen. Principles and applications of electrochemical capacitors. *Electrochimica acta*. 2000 May 3;45(15–16):2483–2498. https://doi.org/10.1016/S0013-4686(00)00354-6

Kuang, Panyong, Mahmoud Sayed, Jiajie Fan, Bei Cheng, and Jiaguo Yu. "3D graphene-based H2-production photocatalyst and electrocatalyst." *Advanced Energy Materials* 10, no. 14 (2020): 1903802. https://doi.org/10.1002/aenm.201903802

Kumar, Nitesh, Kota Surya Subrahmanyam, Piyush Chaturbedy, Kalyan Raidongia, Achutharao Govindaraj, Kailash P. S. S. Hembram, Abhishek K. Mishra, Umesh V. Waghmare, and Chintamani Nagesa Ramachandra Rao. "Remarkable uptake of CO_2 and CH_4 by graphene-like borocarbonitrides, $B_xC_yN_z$." *ChemSusChem* 4, no. 11 (2011): 1662–1670. https://doi.org/10.1002/cssc.201100197

Kumar, Ram, Venkata M. Suresh, Tapas K. Maji, and Chintamani Nagesa Ramachandra Rao. "Porous graphene frameworks pillared by organic linkers with tunable surface area and gas storage properties." *Chemical Communications* 50, no. 16 (2014): 2015–2017. https://doi.org/10.1039/c3cc46907g

Kumawat, Mukesh K., Mukeshchand Thakur, Raju B. Gurung, and Rohit Srivastava. "Graphene quantum dots from *Mangifera indica*: application in near-infrared bioimaging and intracellular nanothermometry." *ACS Sustainable Chemistry and Engineering* 5, no. 2 (2017): 1382–1391. https://doi.org/10.1021/acssuschemeng.6b01893

Kuniyil, Mufsir, Jagarlapudi Venkata Shanmukha Kumar, Syed F. Adil, Mohamed E. Assal, Mohammed R. Shaik, Mujeeb Khan, Abdulrahman Al-Warthan, and Mohammed R. H. Siddiqui. "Production of biodiesel from waste cooking oil using ZnCuO/N-doped graphene nanocomposite as an efficient heterogeneous catalyst." *Arabian Journal of Chemistry* 14, no. 3 (2021): 102982. https://doi.org/10.1016/j.arabjc.2020.102982

Kurmarayuni, Chandra M., Swarnalatha Kurapati, Syed Akhil, Basavaiah Chandu, Bala M. K. Khandapu, Prabhakara R. Koya, and Hari B. Bollikolla. "Synthesis of multifunctional graphene exhibiting excellent sonochemical dye removal activity, green and regioselective reduction of cinnamaldehyde." *Materials Letters* 263 (2020): 127224. https://doi.org/10.1016/j.matlet.2019.127224

Lazar, Petr, Frantisek Karlicky, Petr Jurecka, Mikuláš Kocman, Eva Otyepková, Klára Šafářová, and Michal Otyepka. "Adsorption of small organic molecules on graphene." *Journal of the American Chemical Society* 135, no. 16 (2013): 6372–6377. https://doi.org/10.1021/ja403162r

Lee, Changgu, Xiaoding Wei, Jeffrey W. Kysar, and James Hone. "Measurement of the elastic properties and intrinsic strength of monolayer graphene." *Science* 321, no. 5887 (2008): 385–388. https://doi.org/10.1126/science.1157996

Li, Qin, Xin Li, S. Wageh, Ahmed A. Al-Ghamdi, and Jiaguo Yu. "CdS/graphene nanocomposite photocatalysts." *Advanced Energy Materials* 5, no. 14 (2015): 1500010. https://doi.org/10.1002/aenm.201500010

Li, Xin, Jiaguo Yu, S. Wageh, Ahmed A. Al-Ghamdi, and Jun Xie. "Graphene in photocatalysis: a review." *Small* 12, no. 48 (2016): 6640–6696. https://doi.org/10.1002/smll.201600382

Li, Chengyang, Zechao Zhuang, Xiaoying Jin, and Zuliang Chen. "A facile and green preparation of reduced graphene oxide using Eucalyptus leaf extract." *Applied Surface Science* 422 (2017): 469–474. https://doi.org/10.1016/j.apsusc.2017.06.032

Li, Min, Jian Feng, Kun Huang, Si Tang, Ruihua Liu, Huan Li, Fuye Ma, and Xiaojing Meng. "Amino group functionalized SiO_2@graphene oxide for efficient removal of Cu (II) from aqueous solutions." *Chemical Engineering Research and Design* 145 (2019a): 235–244. https://doi.org/10.1016/j.cherd.2019.03.028

Li, Songwei, Peipei Yang, Xianhu Liu, Jiaoxia Zhang, Wei Xie, Chao Wang, Chuntai Liu, and Zhanhu Guo. "Graphene oxide based dopamine mussel-like cross-linked polyethylene imine nanocomposite coating with enhanced hexavalent uranium adsorption." *Journal of Materials Chemistry A* 7, no. 28 (2019b): 16902–16911. https://doi.org/10.1039/c9ta04562g

Li, Jiacheng, Miao Li, Jing Li, Sai Wang, Gongbo Li, and Xiang Liu. "Hydrodechlorination and deep hydrogenation on single-palladium-atom-based heterogeneous catalysts." *Applied Catalysis B: Environmental* 282 (2021): 119518. https://doi.org/10.1016/j.apcatb.2020.119518

Liang, Minghui, and Linjie Zhi. "Graphene-based electrode materials for rechargeable lithium batteries." *Journal of Materials Chemistry* 19, no. 33 (2009): 5871–5878. https://doi.org/10.1039/B901551E

Liang, Yongye, Yanguang Li, Hailiang Wang, Jigang Zhou, Jian Wang, Tom Regier, and Hongjie Dai. "CO_3O_4 nanocrystals on graphene as a synergistic catalyst for oxygen reduction reaction." *Nature Materials* 10, no. 10 (2011): 780–786. https://doi.org/10.1038/nmat3087

Liu, Zhuang, Joshua T. Robinson, Xiaoming Sun, and Hongjie Dai. "PEGylated nanographene oxide for delivery of water-insoluble cancer drugs." *Journal of the American Chemical Society* 130, no. 33 (2008): 10876–10877. https://doi.org/10.1021/ja803688x

Liu, Ruili, Christian von Malotki, Lena Arnold, Nobuyoshi Koshino, Hideyuki Higashimura, Martin Baumgarten, and Klaus Müllen. "Triangular trinuclear metal-N4 complexes with high electrocatalytic activity for oxygen reduction." *Journal of the American Chemical Society* 133, no. 27 (2011): 10372–10375. https://doi.org/10.1021/ja203776f

Liu, Jingquan, Liang Cui, and Dusan Losic. "Graphene and graphene oxide as new nanocarriers for drug delivery applications." *Acta Biomaterialia* 9, no. 12 (2013): 9243–9257. https://doi.org/10.1016/j.actbio.2013.08.016

Liu, Mingkai, Peng Zhang, Zehua Qu, Yan Yan, Chao Lai, Tianxi Liu, and Shanqing Zhang. "Conductive carbon nanofiber interpenetrated graphene architecture for ultra-stable sodium ion battery." *Nature Communications* 10, no. 1 (2019): 1–11. 917. https://doi.org/10.1038/s41467-019-11

Liu, Yue, Fengrui Zhang, Wenxia Zhu, Dong Su, Zhiyuan Sang, Xiao Yan, Sheng Li, Ji Liang, and Shi X. Dou. "A multifunctional hierarchical porous SiO_2/GO membrane for high efficiency oil/water separation and dye removal." *Carbon* 160 (2020): 88–97. https://doi.org/10.1016/j.carbon.2020.01.002

Liu, Chenrui, Yun Liu, Zhi Dang, Shuai Zeng, and Chengcheng Li. "Enhancement of heterogeneous photo-Fenton performance of core-shell structured boron-doped reduced graphene oxide wrapped magnetical Fe_3O_4 nanoparticles: Fe (II)/Fe (III) redox and mechanism." *Applied Surface Science* 544 (2021): 148886. https://doi.org/10.1016/j.apsusc.2020.148886

Long, Ying, Congcong Zhang, Xingxin Wang, Jianping Gao, Wei Wang, and Yu Liu. "Oxidation of SO_2 to SO_3 catalyzed by graphene oxide foams." *Journal of Materials Chemistry* 21, no. 36 (2011): 13934–13941. https://doi.org/10.1039/c1jm12031j

Maham, Mehdi, Mahmoud Nasrollahzadeh, S. Mohammad Sajadi, and Mehdi Nekoei. "Biosynthesis of Ag/reduced graphene oxide/Fe_3O_4 using Lotus garcinii leaf extract and its application as a recyclable nanocatalyst for the reduction of 4-nitrophenol and organic dyes." *Journal of Colloid and Interface Science* 497 (2017): 33–42. https://doi.org/10.1016/j.jcis.2017.02.064

Maiyalagan, Thandavarayan, Xiaochen Dong, Peng Chen, and Xin Wang. "Electrodeposited Pt on three-dimensional interconnected graphene as a free-standing electrode for fuel cell application." *Journal of Materials Chemistry* 22, no. 12 (2012): 5286–5290. https://doi.org/10.1039/c2jm16541d

Mamba, Gcina, Gumani Gangashe, Lerato Moss, Hari S. Ganesh, Sourbh Thakur, Sethumathavan Vadivel, Ajay Kumar Mishra, Gcina Doctor Vilakati, Velluchamy Muthuraj, and Thabo T.I. Nkambule. "State of the art on the photocatalytic applications of graphene based nanostructures: from elimination of hazardous pollutants to disinfection and fuel generation." *Journal of Environmental Chemical Engineering* 8, no. 2 (2020): 103505. https://doi.org/10.1016/j.jece.2019.103505

Mao, Hong Y., Sophie Laurent, Wei Chen, Omid Akhavan, Mohammad Imani, Ali A. Ashkarran, and Morteza Mahmoudi. "Graphene: promises, facts, opportunities, and challenges in nanomedicine." *Chemical Reviews* 113, no. 5 (2013): 3407–3424. https://doi.org/10.1021/cr300335p

Mermin, Nathaniel David. "Crystalline order in two dimensions." *Physical Review* 176, no. 1 (1968): 250. https://doi.org/10.1103/PhysRev.176.250

Mindivan, Ferda, and Meryem Göktaş. "Rosehip-extract-assisted green synthesis and characterization of reduced graphene oxide." *ChemistrySelect* 5, no. 29 (2020): 8980–8985. https://doi.org/10.1002/slct.202001656

Mishra, Subhasish, and Rashmi Acharya. "Photocatalytic applications of graphene based semiconductor composites: a review." *Materials Today: Proceedings* 35 (2021): 164–169. https://doi.org/10.1016/j.matpr.2020.04.066

Mohammadkhani, Rahman, Mohammad Ramezanzadeh, Sajjad Akbarzadeh, Ghasem Bahlakeh, and Bahram Ramezanzadeh. "Graphene oxide nanoplatforms reduction by green plant-sourced organic compounds for construction of an active anti-corrosion coating; experimental/electronic-scale DFT-D modeling studies." *Chemical Engineering Journal* 397 (2020): 125433. https://doi.org/10.1016/j.cej.2020.125433

Mohan, Velram B., Kin-tak Lau, David Hui, and Debes Bhattacharyya. "Graphene-based materials and their composites: a review on production, applications and product limitations." *Composites Part B: Engineering* 142 (2018): 200–220. https://doi.org/10.1016/j.compositesb.2018.01.013

Morgado, Tiago A., and Mário G. Silveirinha. "Drift-induced unidirectional graphene plasmons." *ACS Photonics* 5, no. 11 (2018): 4253–4258. https://doi.org/10.1021/acsphotonics.8b00987

Nassef, Ahmed M., Ahmed Fathy, Enas T. Sayed, Mohammad A. Abdelkareem, Hegazy Rezk, Waqas Hassan Tanveer, and Abdul Ghani Olabi. "Maximizing SOFC performance through optimal parameters identification by modern optimization algorithms." *Renewable Energy* 138 (2019): 458–464. https://doi.org/10.1016/j.renene.2019.01.072

Ngameni, Bathélémy, Kamdoum Cedric, Armelle T. Mbaveng, Musa Erdoğan, Ingrid Simo, Victor Kuete, and Arif Daştan. "Design, synthesis, characterization, and anticancer activity of a novel series of O-substituted chalcone derivatives." *Bioorganic and Medicinal Chemistry Letters* 35 (2021): 127827. https://doi.org/10.1016/j.bmcl.2021.127827

Nhlane, Dineo, Heidi Richards, and Anita Etale. "Facile and green synthesis of reduced graphene oxide for remediation of Hg (II)-contaminated water." *Materials Today: Proceedings* 38 (2021): 737–742. https://doi.org/10.1016/j.matpr.2020.04.163

Niu, Lengyuan, Zhangpeng Li, Wei Hong, Jinfeng Sun, Zhaofeng Wang, Limin Ma, Jinqing Wang, and Shengrong Yang. "Pyrolytic synthesis of boron-doped graphene and its application as electrode material for supercapacitors." *Electrochimica Acta* 108 (2013): 666–673. https://doi.org/10.1016/j.electacta.2013.07.025

Ogungbemi, Emmanuel, Oluwatosin Ijaodola, Fawwad Nisar Khatib, Tabbi Wilberforce, Zaki El Hassan, James Thompson, Mohamad Ramadan, and Abdul Ghani Olabi. "Fuel cell membranes–Pros and cons." *Energy* 172 (2019): 155–172. https://doi.org/10.1016/j.energy.2019.01.034

Olabi, Abdul Ghani, Mohammad A. Abdelkareem, Tabbi Wilberforce, and Enas T. Sayed. "Application of graphene in energy storage device – a review." *Renewable and Sustainable Energy Reviews* 135 (2021): 110026. https://doi.org/10.1016/j.rser.2020.110026

Parvez, Khaled, Shubin Yang, Yenny Hernandez, Andreas Winter, Andrey Turchanin, Xinliang Feng, and Klaus Müllen. "Nitrogen-doped graphene and its iron-based composite as efficient electrocatalysts for oxygen reduction reaction." *ACS Nano* 6, no. 11 (2012): 9541–9550. https://doi.org/10.1021/nn302674k

Peng, Hongliang, Zaiyong Mo, Shijun Liao, Huagen Liang, Lijun Yang, Fan Luo, Huiyu Song, Yiliang Zhong, and Bingqing Zhang. "High performance Fe-and N-doped carbon catalyst with graphene structure for oxygen reduction." *Scientific Reports* 3, no. 1 (2013): 1–7. https://doi.org/10.1038/srep0176

Peres, Nuno M. R. "Colloquium: the transport properties of graphene: an introduction." *Reviews of Modern Physics* 82, no. 3 (2010): 2673. https://doi.org/10.1103/RevModPhys.82.2673

Phukan, Palash, Rewrewa Narzary, and Partha P. Sahu. "A green approach to fast synthesis of reduced graphene oxide using alcohol for tuning semiconductor property." *Materials Science in Semiconductor Processing* 104 (2019): 104670. https://doi.org/10.1016/j.mssp.2019.104670

Phukan, Palash, and Partha P. Sahu. "High performance UV photodetector based on metal-semiconductor-metal structure using TiO_2-rGO composite." *Optical Materials* 109 (2020): 110330. https://doi.org/10.1016/j.optmat.2020.110330
Pikal, Michael J., and Saroj Shah. "The collapse temperature in freeze drying: dependence on measurement methodology and rate of water removal from the glassy phase." *International Journal of Pharmaceutics* 62, no. 2–3 (1990): 165–186. https://doi.org/10.1016/0378-5173(90)90231-R
Qi, Jian, Shanshan Zhang, Chun Xie, Qiufen Liu, and Shuxia Yang. "Fabrication of Erythrina senegalensis leaf extract mediated reduced graphene oxide for cardiac repair applications in the nursing care." *Inorganic and Nano-Metal Chemistry* 51, no. 1 (2021): 143–149. https://doi.org/10.1080/24701556.2020.1769663
Ramanathan, Subramanian, Steplin P. Selvin, Asir Obadiah, Arulappan Durairaj, Palanisamy Santhoshkumar, Sharmila Lydia, Subramaian Ramasundaram, and Samuel Vasanthkumar. "Synthesis of reduced graphene oxide/ZnO nanocomposites using grape fruit extract and Eichhornia crassipes leaf extract and a comparative study of their photocatalytic property in degrading Rhodamine B dye." *Journal of Environmental Health Science and Engineering* 17, no. 1 (2019): 195–207. https://doi.org/10.1007/s40201-019-00340-7
Rao, Chintamani Nagesa Ramachandra, Kailash Gopalakrishnan, and Govinda Raj. "Synthesis, properties and applications of graphene doped with boron, nitrogen and other elements." *Nano Today* 9, no. 3 (2014): 324–343. https://doi.org/10.1016/j.nantod.2014.04.010
Reddy, Arava L. M., Anchal Srivastava, Sanketh R. Gowda, Hemtej Gullapalli, Madan Dubey, and Pulickel M. Ajayan. "Synthesis of nitrogen-doped graphene films for lithium battery application." *ACS Nano* 4, no. 11 (2010): 6337–6342. https://doi.org/10.1021/nn101926g
Ren, Jian-Guo, Qi-Hui Wu, Guo Hong, Wen-Jun Zhang, Huiming Wu, Khalil Amine, Junbing Yang, and Shuit-Tong Lee. "Silicon-graphene composite anodes for high-energy lithium batteries." *Energy Technology* 1, no. 1 (2013): 77–84. https://doi.org/10.1002/ente.201200038
Rezk, Hegazy, Enas T. Sayed, Mujahed Al-Dhaifallah, Mohamed Obaid, Abou Hashema, Mohammad A. Abdelkareem, and Abdul Ghani Olabi. "Fuel cell as an effective energy storage in reverse osmosis desalination plant powered by photovoltaic system." *Energy* 175 (2019): 423–433. https://doi.org/10.1016/j.energy.2019.02.167
Roy, Prathik, Arun P. Periasamy, Chiashain Chuang, Yi-Rou Liou, Yang-Fang Chen, Joseph Joly, Chi-Te Liang, and Huan-Tsung Chang. "Plant leaf-derived graphene quantum dots and applications for white LEDs." *New Journal of Chemistry* 38, no. 10 (2014): 4946–4951. https://doi.org/10.1039/c4nj01185f
Roy, Amrita, Saptarshi Kar, Ranjan Ghosal, Kinsuk Naskar, and Anil K. Bhowmick. "Facile synthesis and characterization of few-layer multifunctional graphene from sustainable precursors by controlled pyrolysis, understanding of the graphitization pathway, and its potential application in polymer nanocomposites." *ACS Omega* 6, no. 3 (2021): 1809–1822. https://doi.org/10.1021/acsomega.0c03550
Roy-Mayhew, Joseph D., David J. Bozym, Christian Punckt, and Ilhan A. Aksay. "Functionalized graphene as a catalytic counter electrode in dye-sensitized solar cells." *ACS Nano* 4, no. 10 (2010): 6203–6211. https://doi.org/10.1021/nn1016428
Ruiz, Oscar N., Shiral K.A. Fernando, Baojiang Wang, Nicholas A. Brown, Pengju G. Luo, Nicholas D. McNamara, Marlin Vangsness, Ya-Ping Sun, and Christopher E. Bunker. "Graphene oxide: a nonspecific enhancer of cellular growth." *ACS Nano* 5, no. 10 (2011): 8100–8107. https://doi.org/10.1021/nn202699t
Saeed, Khalid, Idrees Khan, and Muhammad Sadiq. "Synthesis of graphene-supported bimetallic nanoparticles for the sunlight photodegradation of Basic Green 5 dye in aqueous medium." *Separation Science and Technology* 51, no. 8 (2016): 1421–1426. http://dx.doi.org/10.1080/01496395.2016.1154872
Sahoo, Lipipuspa, Sanjit Mondal, Cb Nayana, and Ujjal K. Gautam. "Facile d-band tailoring in Sub-10 nm Pd cubes by in-situ grafting on nitrogen-doped graphene for highly efficient organic transformations." *Journal of Colloid and Interface Science* 590 (2021): 175–185. https://doi.org/10.1016/j.jcis.2020.12.118
Saleh, Tawfik A., and Saddam A. AL-Hammadi. "A novel catalyst of nickel-loaded graphene decorated on molybdenum-alumina for the HDS of liquid fuels." *Chemical Engineering Journal* 406 (2021): 125167. https://doi.org/10.1016/j.cej.2020.125167
Schniepp, Hannes C., Je-Luen Li, Michael J. McAllister, Hiroaki Sai, Margarita Herrera-Alonso, Douglas H. Adamson, Robert K. Prud'homme, Roberto Car, Dudley A. Saville, and Ilhan A. Aksay. "Functionalized single graphene sheets derived from splitting graphite oxide." *The Journal of Physical Chemistry B* 110, no. 17 (2006): 8535–8539. https://doi.org/10.1021/jp060936f
Shakak, Mohammad, Reza Rezaee, Afshin Maleki, Ali Jafari, Mahdi Safari, Behzad Shahmoradi, Hiua Daraei, and Seung-Mok Lee. "Synthesis and characterization of nanocomposite ultrafiltration membrane (PSF/PVP/SiO_2) and performance evaluation for the removal of amoxicillin from aqueous solutions." *Environmental Technology & Innovation* 17 (2020): 100529. https://doi.org/10.1016/j.eti.2019.100529

Shams, Syed Saqib, Li S. Zhang, Renhao Hu, Ruoyu Zhang, and Jin Zhu. "Synthesis of graphene from biomass: a green chemistry approach." *Materials Letters* 161 (2015): 476–479. https://doi.org/10.1016/j.matlet.2015.09.022

Shi, Xuetao, Haixin Chang, Song Chen, Chen Lai, Ali Khademhosseini, and Hongkai Wu. "Regulating cellular behavior on few-layer reduced graphene oxide films with well-controlled reduction states." *Advanced Functional Materials* 22, no. 4 (2012): 751–759. https://doi.org/10.1002/adfm.201102305

Singh, Pardeep, Pooja Shandilya, Pankaj Raizada, Anita Sudhaik, Abolfazl Rahmani-Sani, and Ahmad Hosseini-Bandegharaei. "Review on various strategies for enhancing photocatalytic activity of graphene based nanocomposites for water purification." *Arabian Journal of Chemistry* 13, no. 1 (2020): 3498–3520. https://doi.org/10.1016/j.arabjc.2018.12.001

Siong, Valerie L. E., Kian M. Lee, Joon C. Juan, Chin W. Lai, Xin H. Tai, and Cheng S. Khe. "Removal of methylene blue dye by solvothermally reduced graphene oxide: a metal-free adsorption and photodegradation method." *RSC Advances* 9, no. 64 (2019): 37686–37695. https://doi.org/10.1039/c9ra05793e

Somanathan, Thirunavukkarasu, Karthika Prasad, Kostya Ken Ostrikov, Arumugam Saravanan, and Vemula Mohana Krishna. "Graphene oxide synthesis from agro waste." *Nanomaterials* 5, no. 2 (2015): 826–834. https://doi.org/10.3390/nano5020826

Subrahmanyam, Kota Surya, Prashant Kumar, Urmimala Maitra, Govinda Raj, K. P. S. S. Hembram, Umesh V. Waghmare, and Chintamani Nagesa Ramachandra Rao. "Chemical storage of hydrogen in few-layer graphene." *Proceedings of the National Academy of Sciences* 108, no. 7 (2011): 2674–2677. https://doi.org/10.1073/pnas.1019542108

Sun, Xiang, Gongkai Wang, Jiann-Yang Hwang, and Jie Lian. "Porous nickel oxide nano-sheets for high performance pseudocapacitance materials." *Journal of Materials Chemistry* 21, no. 41 (2011): 16581–16588. https://doi.org/10.1039/c1jm12734a

Sun, Haifeng, Yanli Nan, Ruijie Feng, and Ruiyao Ma. "Novel method for in situ investigation into graphene quantum dots effects on the adsorption of nitrated polycyclic aromatic hydrocarbons by crop leaf surfaces." *Ecotoxicology and Environmental Safety* 162 (2018): 10–16. https://doi.org/10.1016/j.ecoenv.2018.06.059

Sun, Zhuxing, Siyuan Fang, and Yun H. Hu. "3D graphene materials: from understanding to design and synthesis control." *Chemical Reviews* 120, no. 18 (2020): 10336–10453. https://doi.org/10.1021/acs.chemrev.0c00083

Suryawanshi, Anil, Mandakini Biswal, Dattakumar Mhamane, Rohan Gokhale, Shankar Patil, Debanjan Guin, and Satishchandra Ogale. "Large scale synthesis of graphene quantum dots (GQDs) from waste biomass and their use as an efficient and selective photoluminescence on-off-on probe for Ag^+ ions." *Nanoscale* 6, no. 20 (2014): 11664–11670. https://doi.org/10.1039/c4nr02494j

Tadyszak, Krzysztof, Jacek K. Wychowaniec, and Jagoda Litowczenko. "Biomedical applications of graphene-based structures." *Nanomaterials* 8, no. 11 (2018): 944. https://doi.org/10.3390/nano8110944

Tewari, Chetna, Gaurav Tatrari, Manoj Karakoti, Sandeep Pandey, Mintu Pal, Sravendra Rana, Boddepalli Santhi Bhushan, Anand B. Melkani, Anurag Srivastava, and Nanda G. Sahoo. "A simple, eco-friendly and green approach to synthesis of blue photoluminescent potassium-doped graphene oxide from agriculture waste for bio-imaging applications." *Materials Science and Engineering: C* 104 (2019): 109970. https://doi.org/10.1016/j.msec.2019.109970

Torres, Luis E. F. F., Stephan Roche, and Jean-Christophe Charlier. In *Introduction to Graphene-Based Nanomaterials: from Electronic Structure to Quantum Transport*. Cambridge University Press. 2020. https://doi.org/10.1557/mrs.2015.108

Trivedi, Shivam, Kenneth Lobo, and H. S. S Hss Ramakrishna Matte. "Synthesis, properties, and applications of graphene." In *Fundamentals and Sensing Applications of 2D Materials*, pp. 25–90. Woodhead Publishing, 2019. https://doi.org/10.1016/B978-0-08-102577-2.00003-8

Vatandost, Elham, Azade Ghorbani-HasanSaraei, Fereshteh Chekin, Shahram Naghizadeh Raeisi, and Seyed-Ahmad Shahidi. "Green tea extract assisted green synthesis of reduced graphene oxide: application for highly sensitive electrochemical detection of sunset yellow in food products." *Food Chemistry: X* 6 (2020): 100085. https://doi.org/10.1016/j.fochx.2020.100085

Vinayan, Bhaghavathi Parambath, and Sundara Ramaprabhu. "Facile synthesis of SnO_2 nanoparticles dispersed nitrogen doped graphene anode material for ultrahigh capacity lithium ion battery applications." *Journal of Materials Chemistry A* 1, no. 12 (2013): 3865–3871. https://doi.org/10.1039/c3ta01515g

Vivekchand, S. R. C., Chandra Sekhar Rout, K. S. Subrahmanyam, Achutharao Govindaraj, and Chintamani Nagesa Ramachandra Rao. "Graphene-based electrochemical supercapacitors." *Journal of Chemical Sciences* 120, no. 1 (2008): 9–13. https://doi.org/ 10.1007/s12039-008-0002-7

Voros, Virag, Enrico Drioli, Claudio Fonte, and Gyorgy Szekely. "Process intensification via continuous and simultaneous isolation of antioxidants: an upcycling approach for olive leaf waste." *ACS Sustainable Chemistry & Engineering* 7, no. 22 (2019): 18444–18452. https://doi.org/10.1021/acssuschemeng.9b04245

Wang, Yan, Zhiqiang Shi, Yi Huang, Yanfeng Ma, Chengyang Wang, Mingming Chen, and Yongsheng Chen. "Supercapacitor devices based on graphene materials." *The Journal of Physical Chemistry C* 113, no. 30 (2009): 13103–13107. https://doi.org/10.1021/jp902214f

Wang, Yan-Jie, David P. Wilkinson, and Jiujun Zhang. "Noncarbon support materials for polymer electrolyte membrane fuel cell electrocatalysts." *Chemical Reviews* 111, no. 12 (2011a): 7625–7651. https://doi.org/10.1021/cr100060r

Wang, Hailiang, Yuan Yang, Yongye Liang, Joshua T. Robinson, Yanguang Li, Ariel Jackson, Yi Cui, and Hongjie Dai. "Graphene-wrapped sulfur particles as a rechargeable lithium-sulfur battery cathode material with high capacity and cycling stability." *Nano Letters* 11, no. 7 (2011b): 2644–2647. https://doi.org/10.1021/nl200658a

Wang, Hailiang, Yongye Liang, Yanguang Li, and Hongjie Dai. "Co1-xS-graphene hybrid: a high-performance metal chalcogenide electrocatalyst for oxygen reduction." *Angewandte Chemie* 123, no. 46 (2011c): 11161–11164. https://doi.org/10.1002/ange.201104004

Wang, Haibo, Thandavarayan Maiyalagan, and Xin Wang. "Review on recent progress in nitrogen-doped graphene: synthesis, characterization, and its potential applications." *Acs Catalysis* 2, no. 5 (2012): 781–794. https://doi.org/10.1021/cs200652y

Wang, Xuewan, Gengzhi Sun, Parimal Routh, Dong-Hwan Kim, Wei Huang, and Peng Chen. "Heteroatom-doped graphene materials: syntheses, properties and applications." *Chemical Society Reviews* 43, no. 20 (2014): 7067–7098. https://doi.org/10.1039/c4cs00141a

Wang, Hou, Xingzhong Yuan, Guangming Zeng, Yan Wu, Yang Liu, Qian Jiang, and Shansi Gu. "Three dimensional graphene based materials: synthesis and applications from energy storage and conversion to electrochemical sensor and environmental remediation." *Advances in Colloid and Interface Science* 221 (2015): 41–59. https://doi.org/10.1016/j.cis.2015.04.005

Wang, Kaixi, Yong Liu, Xiaoying Jin, and Zuliang Chen. "Characterization of iron nanoparticles/reduced graphene oxide composites synthesized by one step eucalyptus leaf extract." *Environmental Pollution* 250 (2019): 8–13. https://doi.org/10.1016/j.envpol.2019.04.002

Wang, Jiashi, Qinhong Wei, Qingxiang Ma, Zhongya Guo, Fangfang Qin, Zinfer R. Ismagilov, and Wenzhong Shen. "Constructing Co@N-doped graphene shell catalyst via Mott-Schottky effect for selective hydrogenation of 5-hydroxylmethylfurfural." *Applied Catalysis B: Environmental* 263 (2020): 118339. https://doi.org/10.1016/j.apcatb.2019.118339

Wanno, Banchob, and Chanukorn Tabtimsai. "A DFT investigation of CO adsorption on VIIIB transition metal-doped graphene sheets." *Superlattices and Microstructures* 67 (2014): 110–117. https://doi.org/10.1016/j.spmi.2013.12.025

Wu, Zhong-Shuai, Da-Wei Wang, Wencai Ren, Jinping Zhao, Guangmin Zhou, Feng Li, and Hui-Ming Cheng. "Anchoring hydrous RuO_2 on graphene sheets for high-performance electrochemical capacitors." *Advanced Functional Materials* 20, no. 20 (2010): 3595–3602. https://doi.org/10.1002/adfm.201001054

Wu, Zhong-Shuai, Shubin Yang, Yi Sun, Khaled Parvez, Xinliang Feng, and Klaus Müllen. "3D nitrogen-doped graphene aerogel-supported Fe_3O_4 nanoparticles as efficient electrocatalysts for the oxygen reduction reaction." *Journal of the American Chemical Society* 134, no. 22 (2012): 9082–9085. https://doi.org/10.1021/ja3030565

Wu, Tao, Mingxi Chen, Lei Zhang, Xiaoyang Xu, Yu Liu, Jing Yan, Wei Wang, and Jianping Gao. "Three-dimensional graphene-based aerogels prepared by a self-assembly process and its excellent catalytic and absorbing performance." *Journal of Materials Chemistry A* 1, no. 26 (2013): 7612–7621. https://doi.org/10.1039/c3ta10989e

Wu, Xi-Lin, Yanpeng Shi, Shuxian Zhong, Hongjun Lin, and Jian-Rong Chen. "Facile synthesis of Fe_3O_4-graphene@ mesoporous SiO_2 nanocomposites for efficient removal of methylene blue." *Applied Surface Science* 378 (2016): 80–86. https://doi.org/10.1016/j.apsusc.2016.03.226

Xiang, Quanjun, Jiaguo Yu, and Mietek Jaroniec. "Graphene-based semiconductor photocatalysts." *Chemical Society Reviews* 41, no. 2 (2012): 782–796. https://doi.org/10.1039/c1cs15172j

Xing, Baolin, Ruifu Yuan, Chuanxiang Zhang, Guangxu Huang, Hui Guo, Zhengfei Chen, Lunjian Chen, Guiyun Yi, Yude Zhang, and Jianglong Yu. "Facile synthesis of graphene nanosheets from humic acid for supercapacitors." *Fuel Processing Technology* 165 (2017): 112–122. https://doi.org/10.1016/j.fuproc.2017.05.021

Xu, Yuxi, Kaixuan Sheng, Chun Li, and Gaoquan Shi. "Self-assembled graphene hydrogel via a one-step hydrothermal process." *ACS Nano* 4, no. 7 (2010): 4324–4330. https://doi.org/10.1021/nn101187z

Xu, Zhanwei, Zhi Li, Chris M. B. Holt, Xuehai Tan, Huanlei Wang, Babak S. Amirkhiz, Tyler Stephenson, and David Mitlin. "Electrochemical supercapacitor electrodes from sponge-like graphene nanoarchitectures with ultrahigh power density." *The Journal of Physical Chemistry Letters* 3, no. 20 (2012): 2928–2933. https://doi.org/10.1021/jz301207g

Xue, Yuhua, Jun Liu, Hao Chen, Ruigang Wang, Dingqiang Li, Jia Qu, and Liming Dai. "Nitrogen-doped graphene foams as metal-free counter electrodes in high-performance dye-sensitized solar cells." *Angewandte Chemie International Edition* 51, no. 48 (2012): 12124–12127. https://doi.org/10.1002/anie.201207277

Yang, Shubin, Yongji Gong, Zheng Liu, Liang Zhan, Daniel P. Hashim, Lulu Ma, Robert Vajtai, and Pulickel M. Ajayan. "Bottom-up approach toward single-crystalline VO_2-graphene ribbons as cathodes for ultrafast lithium storage." *Nano Letters* 13, no. 4 (2013): 1596–1601. https://doi.org/10.1021/nl400001u

Yavari, Fazel, Zongping Chen, Abhay V. Thomas, Wencai Ren, Hui-Ming Cheng, and Nikhil Koratkar. "High sensitivity gas detection using a macroscopic three-dimensional graphene foam network." *Scientific Reports* 1, no. 1 (2011): 1–5. https://doi.org/10.1038/srep0016

Yavari, Fazel, Christo Kritzinger, Churamani Gaire, Li Song, Hemtej Gulapalli, Theodorian Borca-Tasciuc, Pulickel M. Ajayan, and Nikhil Koratkar. "Tunable bandgap in graphene by the controlled adsorption of water molecules." *Small* 6, no. 22 (2010): 2535–2538. https://doi.org/10.1002/smll.201001384

Yu, Dingshan, Enoch Nagelli, Feng Du, and Liming Dai. "Metal-free carbon nanomaterials become more active than metal catalysts and last longer." *The Journal of Physical Chemistry Letters* 1, no. 14 (2010): 2165–2173. https://doi.org/10.1021/jz100533t

Zhang, Xiaoyong, Jilei Yin, Cheng Peng, Weiqing Hu, Zhiyong Zhu, Wenxin Li, Chunhai Fan, and Qing Huang. "Distribution and biocompatibility studies of graphene oxide in mice after intravenous administration." *Carbon* 49, no. 3 (2011a): 986–995. https://doi.org/10.1016/j.carbon.2010.11.005

Zhang, Dingwen, Xiaodong Li, Haibo Li, Si Chen, Zhuo Z. Sun, Xijiang Yin, and Sumei Huang. "Graphene-based counter electrode for dye-sensitized solar cells." *Carbon* 49, no. 15 (2011b): 5382–5388. https://doi.org/10.1016/j.carbon.2011.08.005

Zhang, Li L., Xin Zhao, Hengxing Ji, Meryl D. Stoller, Linfei Lai, Shanthi Murali, Stephen Mcdonnell, Brandon Cleveger, Robert M. Wallace, and Rodney S. Ruoff. "Nitrogen doping of graphene and its effect on quantum capacitance, and a new insight on the enhanced capacitance of N-doped carbon." *Energy and Environmental Science* 5, no. 11 (2012): 9618–9625. https://doi.org/10.1016/0.1039/C2EE23442D

Zhang, Long, Fan Zhang, Xi Yang, Guankui Long, Yingpeng Wu, Tengfei Zhang, Kai Leng et al. "Porous 3D graphene-based bulk materials with exceptional high surface area and excellent conductivity for supercapacitors." *Scientific Reports* 3, no. 1 (2013): 1–9. https://doi.org/10.1038/srep0140

Zhang, Yan, Zhe Jiao, Yongyou Hu, Sihao Lv, Hongbo Fan, Yanyan Zeng, Jun Hu, and Mianmian Wang. "Removal of tetracycline and oxytetracycline from water by magnetic Fe_3O_4@graphene." *Environmental Science and Pollution Research* 24, no. 3 (2017): 2987–2995. https://doi.org/10.1007/s11356-016-7964-7

Zhang, Quan, Qinxuan Hou, Guanxing Huang, and Qi Fan. "Removal of heavy metals in aquatic environment by graphene oxide composites: a review." *Environmental Science and Pollution Research* 27, no. 1 (2020a): 190–209. https://doi.org/10.1007/s11356-019-06683-w

Zhang, Huan, Qin-Qin Gu, Yi-Wen Zhou, Shou-Qing Liu, Wen-Xiao Liu, Li Luo, and Ze-Da Meng. "Direct Z-scheme photocatalytic removal of ammonia via the narrow band gap MoS_2/N-doped graphene hybrid catalyst upon near-infrared irradiation." *Applied Surface Science* 504 (2020b): 144065. https://doi.org/10.1016/j.apsusc.2019.144065

Zhao, Jinping, Wencai Ren, and Hui-Ming Cheng. "Graphene sponge for efficient and repeatable adsorption and desorption of water contaminations." *Journal of Materials Chemistry* 22, no. 38 (2012): 20197–20202. https://doi.org/10.1039/c2jm34128j

Zheng, Yuhong, Aiwu Wang, Wen Cai, Zhong Wang, Feng Peng, Zhong Liu, and Li Fu. "Hydrothermal preparation of reduced graphene oxide–silver nanocomposite using Plectranthus amboinicus leaf extract and its electrochemical performance." *Enzyme and Microbial Technology* 95 (2016): 112–117. https://doi.org/10.1016/j.enzmictec.2016.05.010

Zhou, Guangmin, Da-Wei Wang, Feng Li, Lili Zhang, Na Li, Zhong-Shuai Wu, Lei Wen, Gao Qing Lu, and Hui-Ming Cheng. "Graphene-wrapped Fe_3O_4 anode material with improved reversible capacity and cyclic stability for lithium ion batteries." *Chemistry of Materials* 22, no. 18 (2010): 5306–5313. https://doi.org/10.1021/cm101532x

Zhou, Xufeng, and Zhaoping Liu. "Graphene foam as an anode for high-rate Li-ion batteries." In IOP Conference Series: Materials Science and Engineering, vol. 18, no. 6, p. 062006. IOP Publishing, 2011. https://doi.org/10.1088/1757-899X/18/6/062006

Zhu, Yanwu, Shanthi Murali, Meryl D. Stoller, Kameswaran J Ganesh, Weiwei Cai, Paulo J. Ferreira, Adam Pirkle. "Carbon-based supercapacitors produced by activation of graphene." *Science* 332, no. 6037 (2011): 1537–1541. https://doi.org/10.1126/science.1200770

11 Graphene from Leaf Wastes

Devadas Bhat Panemangalore,
Spandana Bhat Kuruveri, and Udaya Bhat Kuruveri

CONTENTS

11.1 INTRODUCTION

Graphene is a wonder material as it finds several energy-based applications in the fabrication of sensors, supercapacitors, solar cells, fuel cells, etc. It also possesses greater thermal conductivity (Balandin et al. 2008) values, exhibits magnetic properties, and also claims to be the strongest material (Lee et al. 2008) ever tested. Graphene has ultrahigh intrinsic mobility of about 200,000 $cm^2/v/s$ (Bolotin et al. 2008). There are several methods of obtaining graphene, most notably the Scotch Tape method by Novoselov et al. (2004) and Geim in 2004. Other conventional methods involve chemical vapor deposition (CVD) (Naghdi, Rhee, and Park 2018), epitaxial growth (Tetlow et al. 2014), arc discharge (Borand, Akçamlı, and Uzunsoy 2020), exfoliation (Ma and Shen 2020), etc. A schematic of different synthesis methods is shown in Figure 11.1.

Today, there is a need for sustainable materials processing and recycling and hence our focus should be on green synthesis methods. Therefore, adapting inexpensive starting materials like wastes to produce carbon nanotubes (CNTs) and graphene is very beneficial. Figure 11.2 gives information regarding different carbon precursors such as biomass, waste plastics, and conventional precursors required for the synthesis of graphene and its derivatives.

Graphene can also be synthesized using carbon wastes via pyrolysis (Shams et al. 2015). Several innovative methods also have been explored, such as burning Mg metal in dry ice (Chakrabarti et al. 2011) to produce few-layer graphene (FLG). Several precursors have been identified for graphene production such as glucose (Li et al. 2014), rice husk (Muramatsu et al. 2014), hemp (Wang et al. 2013), disposable paper cups (Zhao and Zhao 2013), coal (Ye et al. 2013), chitosan (Terzopoulou, Kyzas, and Bikiaris 2015), etc. Ruan et al. (2011) used different raw low-valued carbon-containing materials such as cookies, chocolate, grass, etc. to synthesize high-quality monolayer graphene. Using exfoliated GO as a precursor, Zhu et al. (2010) synthesized graphene nanosheets using glucose, fructose, and sucrose as reducing agents.

Using flash joule heating, Algozeeb et al. (2020) converted plastic waste into flash graphene. Several other light hydrocarbons were also formed in addition to flash graphene. For every ton of

DOI: 10.1201/9781003169741-11

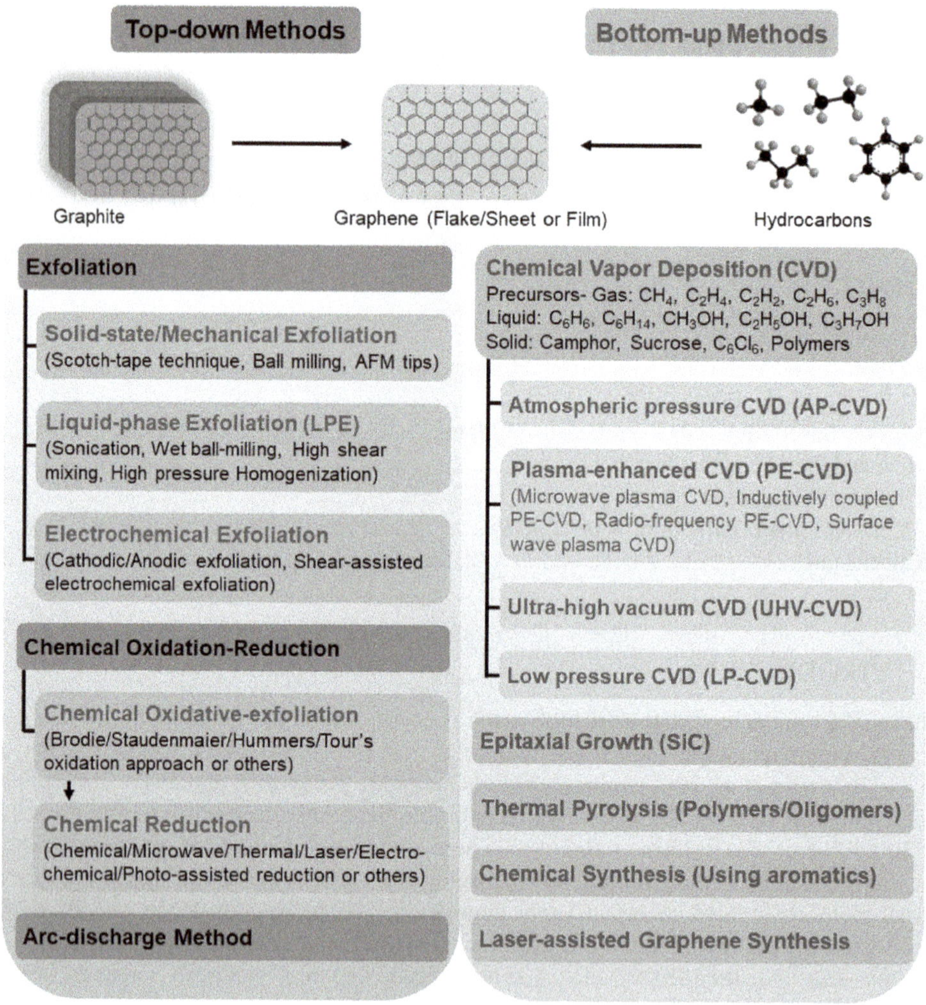

FIGURE 11.1 A diagrammatic representation of synthesizing graphene using bottom-up method and top-down method. (Reproduced with permission from Kumar et al. 2021, Elsevier.)

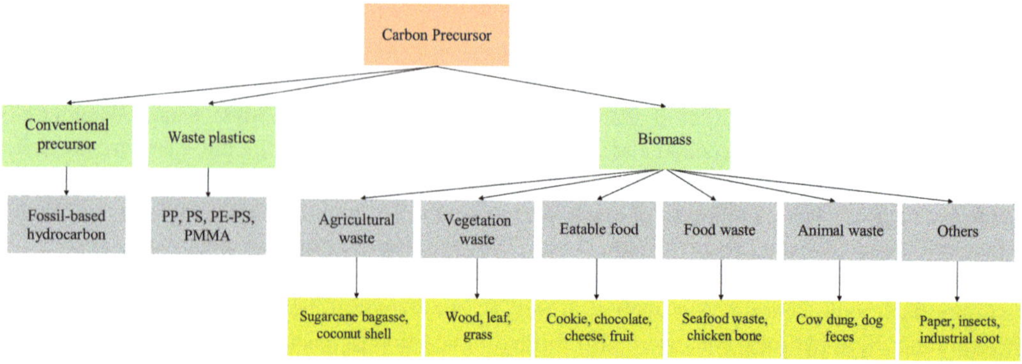

FIGURE 11.2 Different carbon precursors for the synthesis of graphene and its derivatives). (Reproduced with permission from Lee et al. 2019, Elsevier.)

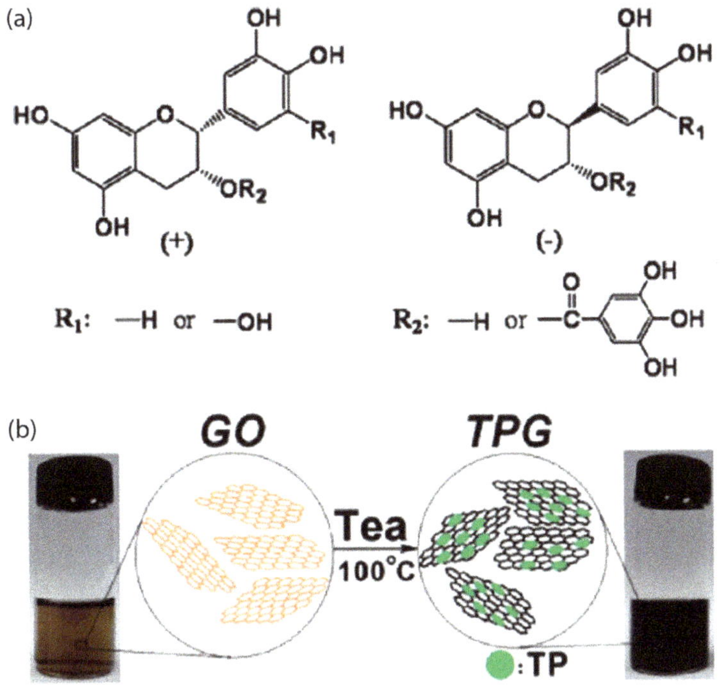

FIGURE 11.3 (a) Tea polyphenols (TPs) and its chemical structure. (b) Schematic of the synthesis of TP reduced graphene. (Reproduced with permission from Wang, Shi, and Yin 2011. Copyright 2011, American Chemical Society.)

plastic waste, 125 USD (23 J/g) worth of electricity was required for flash joule heating, and hence it was very economical. Transmission electron microscopy (TEM) analysis showed the interlayer spacing as 3.45 Å. Gao et al. (2010) employed another environment-friendly technique to synthesize graphene using amino acid and vitamin C as a stabilizer and reductant, respectively. Wang, Shi, and Yin (2011) used tea polyphenols for graphene oxide reduction that was able to remove the O-containing groups very effectively. A schematic of the preparation method is given in Figure 11.3. It can be observed that the color of graphene oxide has changed to black from brownish yellow and it was attributed to the electronic conjugation restoration.

Moosa and Noori Jaafar (2017) used black tea leaves extract for the environment friendly conversion of graphene oxide. These were synthesized targeting removal of lead ions and the usage of tea leaves extract served as a natural reductant. Vatandost et al. (2020) used green tea extract to prepare rGO from GO precursor via the refluxing method for 6 h at 60°C. This material was targeted toward application in the detection of coloring agents added to food products. A schematic of the preparation methodology is represented in Figure 11.4. Purkait et al. (2017) have produced graphene of a couple of layers from peanut shell.

11.2 CARBON-BASED MATERIALS FROM DIFFERENT BIOMASS

Among different categories of biomass, oil palm wastes including shells, fiber, fruit bunches, etc. contribute to energy production (Abdullah, Sulaiman, and Aliasak 2013). Subramanian et al. (2007) used banana fibers to synthesize carbon materials with enhanced surface area via effective pore generations by treating them with $ZnCl_2$ and KOH. Increased specific capacitance and current density values were obtained targeting supercapacitor applications.

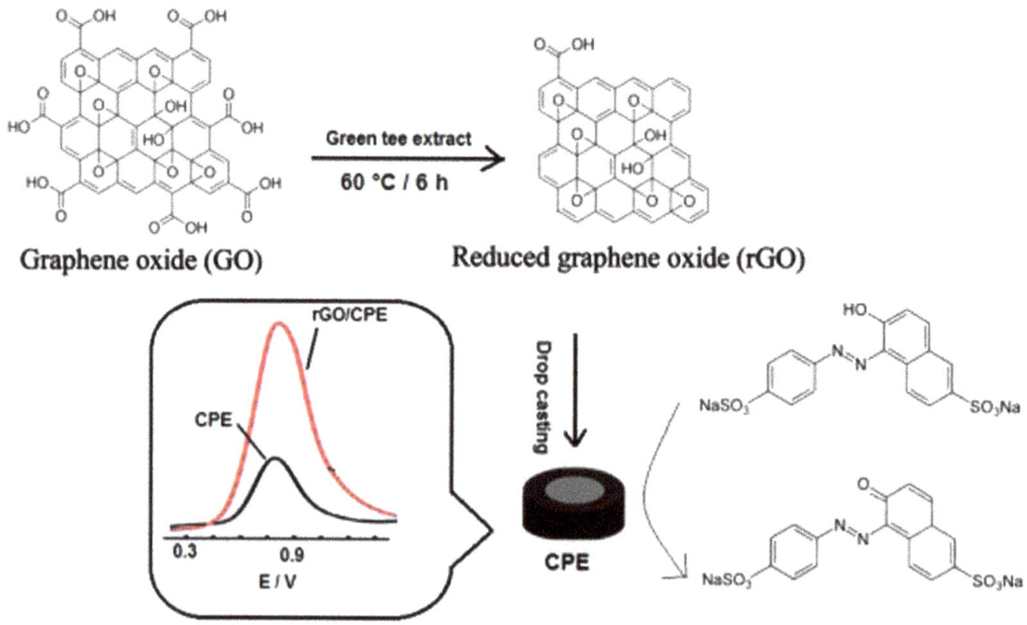

FIGURE 11.4 Schematic presentation of formation of rGO and use of rGO in the electrochemical sensor for the oxidation of colors. (Reproduced with permission from Vatandost et al. 2020, Elsevier.)

Ismanto et al. (2010) prepared carbon-based electrodes from cassava peel. Oxidative chemical agents were used for modification of the surface of the activated carbon and HNO_3 led to increased values of specific capacitance targeting electric double layer capacitor (EDLC) applications.

Li et al. (2011) used sunflower seed shells and prepared a series of nanoporous carbon for EDLC applications. Guo et al. (2003) used rice husk to synthesize porous carbon and the carbons constituted capacitance values greater than its commercial-grade counterpart. Yang et al. (2008) explored walnut shells as starting material for developing activated carbon electrode materials. Kim et al. (2006) studied activated carbons used for EDLC applications using bamboo activation. $ZnCl_2$ activation of bagasse produced from sugarcane for supercapacitor electrode applications was studied by Rufford et al. (2010). Steam activation method was used by Wu et al. (2004) to synthesize activated carbons from firwoods. Adewumi et al. (2018) used coconut fiber to synthesize CNTs and carbon nanospheres. Sujiono et al. (2020) used shell waste of coconut and produced GO using the modified Hummers method. Waste ground coffee treated with $ZnCl_2$ was studied by Rufford et al. (2008) for activated carbon synthesis. Yadav et al. (2021) used waste rice to produce porous graphene oxide nanosheets. Somanathan et al. (2015) used sugarcane bagasse agricultural waste to produce graphene oxide. Using pulping black liquor, a biomass waste, Ding et al. (2020) adapted a one-pot activation-synthesis method to fabricate graphene sheets.

11.2.1 Quantum Dots

Suryawanshi et al. (2014) prepared graphene quantum dots (GQDs) from bio-waste using H_2SO_4 and HNO_3. Turbostratic carbon is generated during incomplete graphitization when pyrolysis of the biomass is carried out below 1000°C. This carbon is refluxed with strong acids for a long time to produce GQDs. Wang et al. (2016) used rice husk biomass to produce high-quality GQDs that not only exhibited bright and tunable photoluminescence but also showed excellent biocompatibility characteristics.

11.3 CARBON-BASED MATERIALS FROM LEAVES OF PLANT

Biswal et al. (2013) used dead Neem and Ashoka leaves to fabricate functional microporous conducting carbon targeting supercapacitor applications. The enhanced conductivity and microporosity values along with the carbonaceous materials of high surface area stressing the importance of natural constituents for energy applications.

Ray et al. (2012) produced graphene by heating lotus and hibiscus flower petals for 30 min under Ar atmosphere. With increased preparation temperature from 800°C to 1200°C, the graphene quality was improved.

11.4 REDUCED GRAPHENE OXIDE

Graphene oxide has immense biomimetic potential due to its similar features to that of plant cell walls w.r.t. applications such as water transport, and hence, GO can also be termed – synthetic leaf (Lamb et al. 2015). rGO is prepared by first employing oxidative-exfoliation reaction of graphite and then reduction to form rGO. The schematic presentation is shown in Figure 11.5 (Yang et al. 2010). Brodie in 1859 (Botas et al. 2013) who is an Oxford Chemist first prepared graphene oxide by treating graphite with $KClO_3$ and fuming HNO_3 (Brodie 1859) at 60°C for 3–4 days. Another method devised by Staudenmaier performs the above reaction at 0°C that involves an intermediary suspension method. Hofmann used non-fuming acids to synthesize GO. It was observed that graphene prepared following conventional Hummers method exhibited enhanced heterogeneous electron transfer compared to other methods as the former employs potassium permanganate as oxidant (Poh et al. 2012). Moreover, it does not produce toxic and explosive ClO_2, hence this method is also termed eco-friendly. Today, usage of the improved Hummers method by not using $NaNO_3$ is gaining importance, which eases the disposal of water (Chen et al. 2013). Partial replacement of $KMnO_4$ with K_2FeO_4 and $NaNO_3$-free Hummers method was also studied by Yu et al. (2016).

Li et al. (2017) synthesized rGO using Eucalyptus leaf extract. Similar biosynthesis was carried out by Lin et al. (2020). Similarly, rGO nanosheets were synthesized using *Annona squamosa* leaves by Chandu et al. (2017). Chettri et al. (2016) used an extract from Artemisia vulgaris dry leaves for the conversion of graphene oxide. Nasir et al. (2017) used wastes from palm oil to produce rGO. Palm oil leaves were used by Amir Faiz, Che Azurahanim, Raba'ah, et al. (2020) in the reduction of GO. They used modified Hummer's method to produce GO and later it was used to reflux with palm oil leaves extract to form rGO. Several countries like Malaysia generate plenty of lignocellulosic biomass and this could lead to waste disposal issues. These studies lead to the generation of wealth from waste and also the unwanted palm oil leaves from Malaysia.

Omid Akhavan et al. (2014) used ginseng to prepare rGO from chemically exfoliated graphene oxide. They also used green tea polyphenols for the reduction of GO and to increase their antioxidant

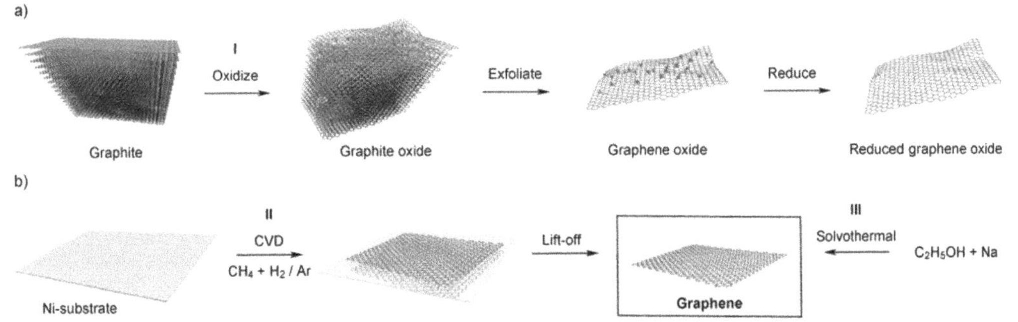

FIGURE 11.5 (a) Production of individual sheets of rGO from graphite using oxidation, exfoliation and reduction route (b) schematic presentation of graphene production through chemical vapor deposition (CVD) process (left or method II) or solvothermal route (right or method III). (Reproduced with permission from Yang et al. 2010, Wiley.)

activity it was carried out in the presence of iron (Akhavan et al. 2012). Lingaraju et al. (2019) used *Euphorbia heterophylla* (L.) for the biocompatible synthesis of rGO. Mahmoud (2020) used aquatic macrophytes or agricultural byproducts to avail plant extracts, by using which graphene oxide was reduced in an eco-friendly manner. Lee and Kim (2014) used extracts from seven different types of leaves to study the biological reduction of graphene oxide. Different plants such as pine, cherry, persimmon, etc. were chosen to prepare plant extracts. Neem oil, also known as margosa oil is a very rich source of carbon and Kumar, Tiwari, and Srivastava (2011) used them to synthesize bundles of aligned CNTs. Fenugreek also contains fatty acid distributions with an increased amount of hydrocarbons and nitrogen-containing groups (Ciftci et al. 2011). In an alternate study, four different plant leaf extracts such as Malabar nut, Neem, tea, and drumsticks were considered to prepare reduced graphene oxide (Roy, Jing, and Basu 2017).

11.5 GRAPHENE USING LEAF WASTES

Fathy et al. (2019) used oil palm leaves to prepare graphene oxide using the catalytic acid spray method. The first step is hydrolysis where oil palm leaves were reacted with H_2SO_4 at 120°C for an hour. Amorphous microcrystalline cellulose that remained as residue after this step was subjected to an alkaline peroxide delignification process. Silica and cobalt silicate were used in later steps for the preparation of GO sheets. Fourier transform infrared spectroscopy (FTIR) results showed three characteristic peaks for the functional groups. Using XRD analysis, it was observed in reflection mode the presence of a strong (0 0 1) peak. The results from Raman spectroscopy showed the presence of low defects and disorders in the synthesized GO. p–p* corresponding to aromatic C–C bonds was observed in UV–V is spectroscopy corresponding to two bands at 270 nm. A larger distance of 1 nm between GO layers was observed in HRTEM, which indicated its growth on the surface of the nanoparticles of cobalt silicate. This study mainly targeted the removal of copper ions from the effluents in the industry. Amir Faiz, Che Azurahanim, Yazid, et al. (2020) used tea waste for the synthesis of graphene oxide. At first, they powdered the dried tea waste and it was then graphitized via carbonization under an inert atmosphere at 650°C–850°C. Using Hummer's method this carbonized sample was oxidized to synthesize GO and rGO was formed using a bed reactor furnace when it is heated at 250°C for 30 min. XRD and Raman peaks indicate shifts with respect to the commercial graphite and it was assigned to the functional units in it (Amir Faiz, Che Azurahanim, Yazid, et al. 2020).

Shams et al. (2015) used dead camphor leaves to synthesize a few (about seven) layered graphene (FLG) without any catalytic reaction. The synthesis scheme is represented in Figure 11.6. First, acetone was used to wash the collected camphor leaves to remove the dirt and using a vacuum oven at 60°C. Second, the leaves were dried for 4 h to remove the presence of any solvent. Third, leaves were cut uniformly into small pieces and it was introduced inside a ceramic boat with a lid in an N_2 atmosphere-filled vacuum oven at 1200°C. The sample after pyrolysis was sonicated in an ice bath with chloroform and D-Tyrosine for 15 min. High rpm of centrifugation was set such that amorphous carbons were precipitated out. After this process, the suspended material was few-layered graphene that was filtered and the remaining forms of carbon got retained in the sediment.

A bio-based dispersion agent, called as D-Tyrosine was used to separate the char and FLG. Thermogravimetric analysis (TGA) shown in Figure 11.7 indicated rapid degradation between 220°C and 315°C, which resembled that of hemicellulose (one of the primary constituents of leaves) pyrolysis curve (Yang et al. 2007). The scanning electron microscope (SEM) cross-sectional microstructure shown in Figure 11.7 revealed porous morphology that is responsible for the thin layer of graphene due to rapid heat transfer during the pyrolysis process.

The purified product was confirmed via TEM analysis as shown in Figure 11.8 and the porous structure of the leaves is responsible for graphene production. Atomic force microscopy (AFM) analysis is shown in Figure 11.8 estimated the thickness of a few (seven) layered graphene to be 2.37 nm. Raman spectra of FLG led to weak D, G, and 2D peaks and this was attributed to the defect structure (Shen et al. 2013). Brunauer-Emmett-Teller (BET) surface area analysis of the product resulted in a surface area of 296 m^2/g.

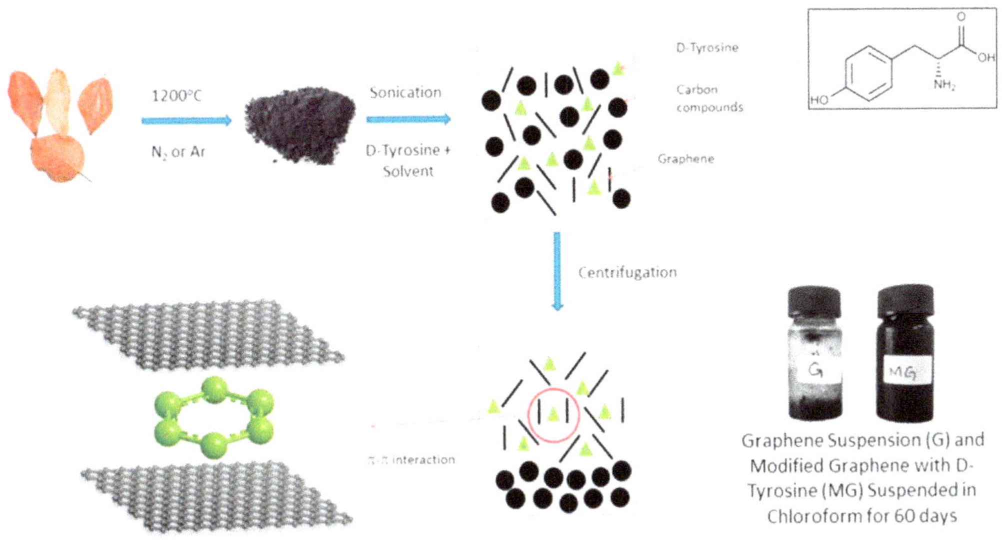

FIGURE 11.6 Synthetic scheme for the preparation of graphene. (Reproduced with permission from Shams et al. 2015, Elsevier.)

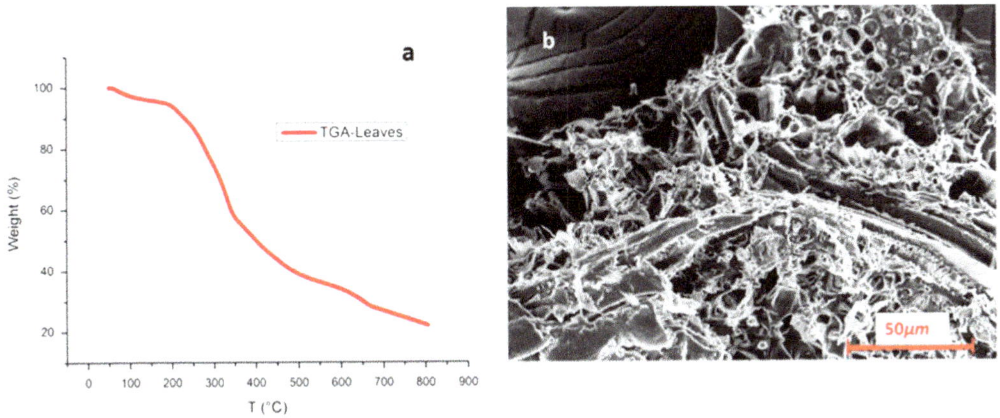

FIGURE 11.7 (a) Thermogravimetric analysis of leaves; and (b) scanning electron microscopy of the cross-sectional area of a leaf presenting its porous nature. (Reproduced with permission from Shams et al. 2015, Elsevier.)

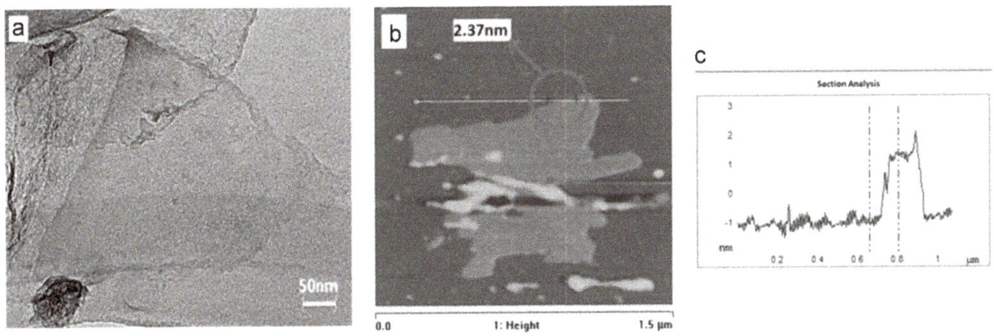

FIGURE 11.8 (a) TEM micrographs of graphene produced from leaves; (b) AFM micrograph of graphene (produced from leaves) presenting a layer thickness of 2.37 nm; and (c) spectrum presenting thickness of the graphene layer. (Reproduced with permission from Shams et al. 2015, Elsevier.)

Wang et al. (2019) transformed waste leaf into 3D graphene with enhanced surface area for applications in water treatment. It was successfully employed in the removal of residues of tetracycline antibiotics from the aqueous solution. This 3D graphene showed excellent adsorption characteristics – 909 mg/g as calculated via the Langmuir model.

Ravani et al. (2013) produced graphene via dissociation of camphor molecules on Ni substrate. Kavitha et al. used camphor precursors and used low-temperature CVD to grow graphene layers on Cu substrate. Chaliyawala et al. (2019) studied the controlled island formation of camphor-based mono-/bilayer graphene (MLG) sheets via CVD. Pyrolysis of camphor via CVD to yield planer nano-graphenes was studied by Somani, Somani, and Umeno (2006). Kalita et al. (2010) also reported graphene sheets from solid botanical derivative camphor.

11.5.1 Graphene Quantum Dots (GQDs)

Using raw Neem and fenugreek plant extracts, Roy et al. (2014) synthesized GQDs for white LED applications. The quantum yields obtained using Neem (*Azadirachta indica*) and fenugreek (*Trigonella foenum-graecum*) are 41.2% and 38.9%, respectively. Figure 11.9 depicts the formation of GQDs from a Neem leaf extract via a hydrothermal method. The leaf extract was prepared by grinding fresh leaves into a fine powder and boiling at 80°C for an hour. As Neem exhibited higher quantum yield, it was chosen to prepare white UV-LED. To remove residual solids, the mixture was subjected to centrifugation and the supernatant was further filtered. Then it is sonicated for some time, which was then subjected inside an autoclave for 8 h at 300°C. With a yield of 25.2%, the GQDs obtained consisted of brown transparent suspension, which was subjected to further treatments to obtain pure GQDs.

The TEM analysis is shown in Figure 11.10, where GO sheets formed via hydrothermal treatment could be seen in Figure 11.10(a). The spherical GQDs could be seen in Figure 11.10(b) along with the histogram. Pyrolysis temperature was chosen such that above 300°C, the resultant material would not provide adequate photoluminescent properties due to its structure, and below 300°C temperature is not enough to carbonize Neem leaves to GO. The high-resolution TEM micrograph of GQDs as presented in Figure 11.10(c) is used to calculate the d-spacing, which is 0.34 nm. The value is similar to that presented by Tang et al. (2012) and Peng et al. (2012). The bottom inset of Figure 11.10(c) shows the fast-Fourier transform pattern that indicates the GQD's spherical nature. X-ray photoelectron spectrometry (XPS) and FTIR measurements confirm the functionalization of GQDs with hydroxyl, carbonyl, and carboxylic groups. Raman spectroscopy showed G band at 1546 cm^{-1} and D band at 1306 cm^{-1}.

The TEM analysis of GQDs obtained from fenugreek leaves is shown in Figure 11.11. The results are similar to Figure 11.10, but the size distribution showed an average value of 7 nm, as compared

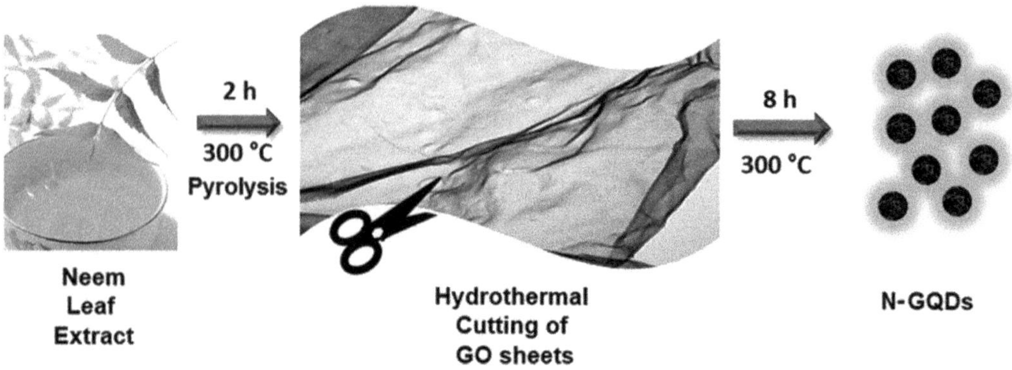

FIGURE 11.9 Schematic presentation for the preparation of the GQDs from the Neem leaf extract using hydrothermal method. (Reproduced with permission from Roy et al. 2014, RSC.)

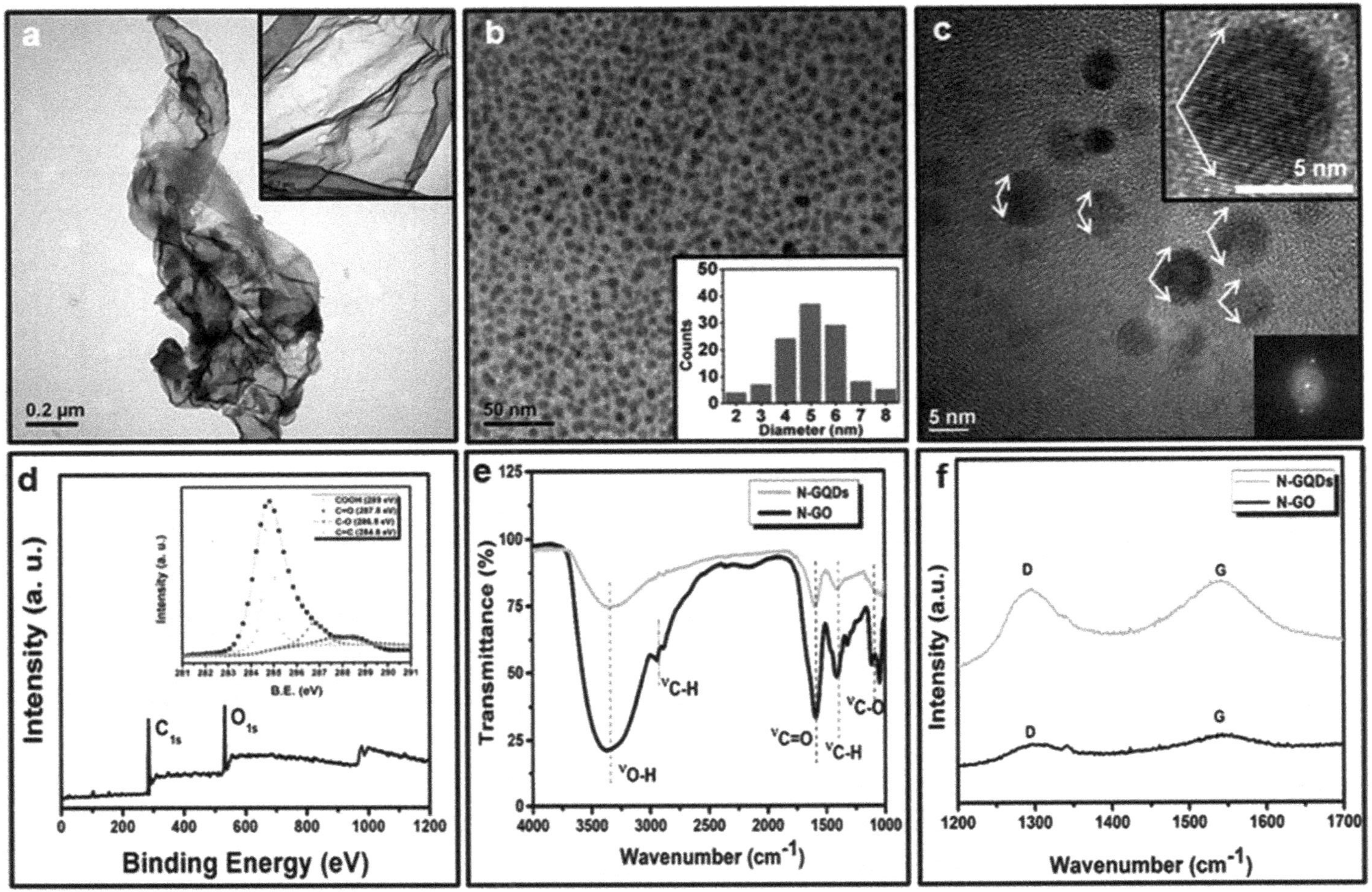

FIGURE 11.10 TEM image (a) of the N-GO layer (a high magnification micrograph is presented as inset); (b) a low magnification TEM micrograph presenting diameter of the N-GQD; (c) HRTEM micrograph of a single N-GQD and a FFT pattern of a single N-GQD is presented as a inset; (d) XPS spectrum of Cls presented with a de-convoluted XPS spectra of the N-GQD is presented inside; (e) FTIR spectra of N-GO and N-GQDs is presented; and (f) Raman spectra of the N-GO and N-GQDs. (Reproduced with permission from Roy et al. 2014, RSC.)

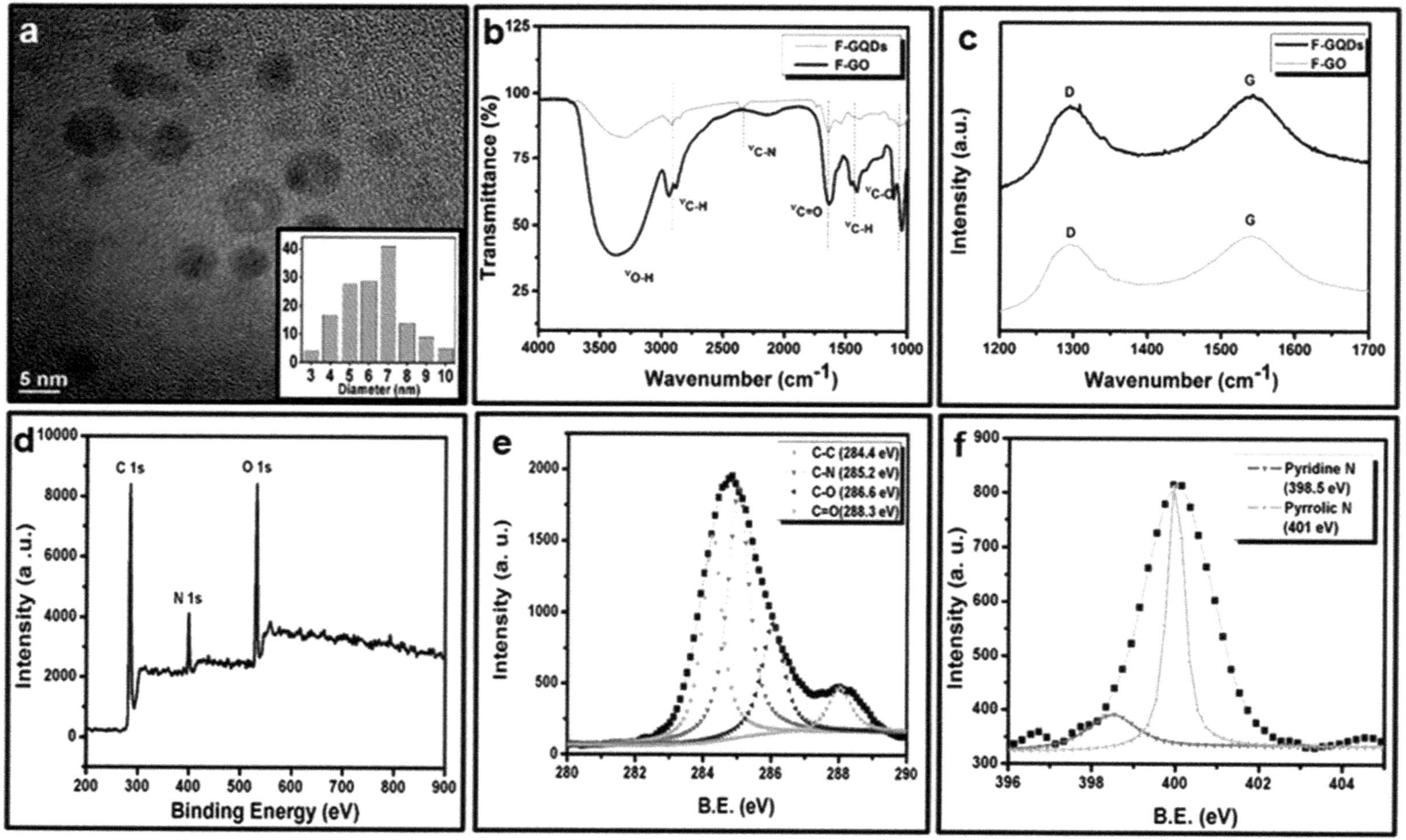

FIGURE 11.11 (a) High-resolution TEM micrograph of the F-GQDs (diameter distribution is presented inside); (b) FTIR spectra of the F-GO and F-GQDs; (c) Raman spectra of the F-GO and F-GQDs; (d) the XPS spectrum C 1s species; (e) XPS spectra after deconvolution; and (f) XPS spectra of the F-GQDs for N1s. (Reproduced with permission from Roy et al. 2014, RSC.)

to 5 nm for Neem leaves. Also, the presence of nitrogen-related peaks was observed in FTIR spectra that were absent in the former, and hence such N-doped GQDs find energy-based applications. This is due to the intrinsic modulation of electron donors, which were studied by Sun et al. (2012) and Hu et al. (2013).

11.6 CONCLUSION

We have explored the bio-based fabrication of graphene and graphene-related materials in this chapter, with emphasis given to leaf wastes. Waste disposal these days typically involves burning plant wastes including dry leaves. These are unused waste materials that can yield significant results w.r.t. graphene production owing to applications in the field of energy management. Many synthesis methods avoid the usage of toxic chemicals and do not generate any hazardous wastes. These leaves and their extracts also play a major role as an alternative reducing agent that also plays a major role in minimizing the reduction time. Hence, these bio-based materials also can pave way for the cheaper and large quantity production of graphene sheets. Future research work has to be in a direction that accounts for production scalability and sustainability.

REFERENCES

Abdullah, Nurhayati, Fauziah Sulaiman, and Zalila Aliasak. 2013. "A Case Study of Pyrolysis of Oil Palm Wastes in Malaysia." *AIP Conference Proceedings* 1528: 331–6. https://doi.org/10.1063/1.4803619

Adewumi, Gloria A., Freddie Inambao, Andrew Eloka-Eboka, and Neerish Revaprasadu. 2018. "Synthesis of Carbon Nanotubes and Nanospheres from Coconut Fibre and the Role of Synthesis Temperature on Their Growth." *Journal of Electronic Materials* 47 (7): 3788–94. https://doi.org/10.1007/s11664-018-6248-z

Akhavan, Omid., Mohammadreza Reza. Kalaee, Z. S. Alavi, Seyyed Mohammad Amin Ghiasi, and Ali Esfandiar. 2012. "Increasing the Antioxidant Activity of Green Tea Polyphenols in the Presence of Iron for the Reduction of Graphene Oxide." *Carbon* 50 (8): 3015–25. https://doi.org/10.1016/j.carbon.2012.02.087

Akhavan, Omid, Elham Ghaderi, Elham Abouei, Shadie Hatamie, and Effat Ghasemi. 2014. "Accelerated Differentiation of Neural Stem Cells into Neurons on Ginseng-Reduced Graphene Oxide Sheets." *Carbon* 66. https://doi.org/10.1016/j.carbon.2013.09.015

Algozeeb, Wala A., Paul E. Savas, Duy Xuan Luong, Weiyin Chen, Carter Kittrell, Mahesh Bhat, Rouzbeh Shahsavari, and James M. Tour. 2020. "Flash Graphene from Plastic Waste." *ACS Nano* 14 (11): 15595–604. https://doi.org/10.1021/acsnano.0c06328

Amir Faiz, M. S., Che. Abdullah. Che Azurahanim, S. A. Raba'ah, and Mohd. Zawawi. Ruzniza. 2020. "Low Cost and Green Approach in the Reduction of Graphene Oxide (GO) Using Palm Oil Leaves Extract for Potential in Industrial Applications." *Results in Physics* 16 (October 2019): 102954. https://doi.org/10.1016/j.rinp.2020.102954

Amir Faiz, M. S., Che. Abdullah. Che Azurahanim, Yaakob. Yazid, Abu. Bakar. Suriani, and Md Jamil Siti Nurul Ain. 2020. "Preparation and Characterization of Graphene Oxide from Tea Waste and It's Photocatalytic Application of TiO_2/Graphene Nanocomposite." *Materials Research Express* 7 (1). https://doi.org/10.1088/2053-1591/ab689d

Balandin, Alexander A., Suchismita Ghosh, Wenzhong Bao, Irene Calizo, Desalegne Teweldebrhan, Feng Miao, and Chun Ning Lau. 2008. "Superior Thermal Conductivity of Single-Layer Graphene." *Nano Letters* 8 (3): 902–7. https://doi.org/10.1021/nl0731872

Biswal, Mandakini, Abhik Banerjee, Meenal Deo, and Satishchandra Ogale. 2013. "From Dead Leaves to High Energy Density Supercapacitors." *Energy and Environmental Science* 6 (4): 1249–59. https://doi.org/10.1039/c3ee22325f

Bolotin, Kirill. I., K. J. Sikes, Zhigang. Jiang, M. Klima, G. Fudenberg, James. Hone, Philip. Kim, and Horst. L. Stormer. 2008. "Ultrahigh Electron Mobility in Suspended Graphene." *Solid State Communications* 146 (9–10): 351–5. https://doi.org/10.1016/j.ssc.2008.02.024

Borand, Gökçe, Nazlı Akçamlı, and Deniz Uzunsoy. 2020. "Structural Characterization of Graphene Nanostructures Produced via Arc Discharge Method." *Ceramics International* 47 (6): 8044–52. https://doi.org/10.1016/j.ceramint.2020.11.158

Botas, Cristina, Patricia Álvarez, Patricia Blanco, Marcos Granda, Clara Blanco, Ricardo Santamaría, Laura J. Romasanta, Raquel Verdejo, Miguel A. López-Manchado, and Rosa Menéndez. 2013. "Graphene Materials with Different Structures Prepared from the Same Graphite by the Hummers and Brodie Methods." *Carbon* 65 (December): 156–64. https://doi.org/10.1016/j.carbon.2013.08.009

Brodie, B. C. 1859. "XIII. On the Atomic Weight of Graphite." *Philosophical Transactions of the Royal Society of London* 149 (December): 249–59. https://doi.org/10.1098/rstl.1859.0013

Chakrabarti, Amartya, Jun Lu, Jennifer C. Skrabutenas, Tao Xu, Zhili Xiao, John A. Maguire, and Narayan S. Hosmane. 2011. "Conversion of Carbon Dioxide to Few-Layer Graphene." *Journal of Materials Chemistry* 21 (26): 9491–3. https://doi.org/10.1039/c1jm11227a

Chaliyawala, Harsh A., Narasimman Rajaram, Roma Patel, Abhijit Ray, and Indrajit Mukhopadhyay. 2019. "Controlled Island Formation of Large-Area Graphene Sheets by Atmospheric Chemical Vapor Deposition: Role of Natural Camphor." *ACS Omega* 4 (5): 8758–66. https://doi.org/10.1021/acsomega.9b00051

Chandu, Basavaiah, Venkata Sai Sriram Mosali, Bhanu Mullamuri, and Hari Babu Bollikolla. 2017. "A Facile Green Reduction of Graphene Oxide Using *Annona squamosa* Leaf Extract." *Carbon Letters* 21 (1): 74–80. https://doi.org/10.5714/CL.2017.21.074

Chen, Ji, Bowen Yao, Chun Li, and Gaoquan Shi. 2013. "An Improved Hummers Method for Eco-Friendly Synthesis of Graphene Oxide." *Carbon* 64 (November): 225–9. https://doi.org/10.1016/j.carbon.2013.07.055

Chettri, Prajwal, V. S. Vendamani, Ajay Tripathi, Anand P. Pathak, and Archana Tiwari. 2016. "Self Assembly of Functionalised Graphene Nanostructures by One Step Reduction of Graphene Oxide Using Aqueous Extract of Artemisia Vulgaris." *Applied Surface Science* 362 (January): 221–9. https://doi.org/10.1016/j.apsusc.2015.11.231

Ciftci, Ozan Nazim, Roman Przybylski, Magdalena Rudzinska, and Surya Acharya. 2011. "Characterization of Fenugreek (*Trigonella foenum-graecum*) Seed Lipids." *Journal of the American Oil Chemists' Society* 88 (10): 1603–10. https://doi.org/10.1007/s11746-011-1823-y

Ding, Zheyuan, Tongqi Yuan, Jialong Wen, Xuefei Cao, Shaoni Sun, Ling Ping Xiao, Quentin Shi, Xiluan Wang, and Runcang Sun. 2020. "Green Synthesis of Chemical Converted Graphene Sheets Derived from Pulping Black Liquor." *Carbon* 158: 690–7. https://doi.org/10.1016/j.carbon.2019.11.041

Fathy, Mahmoud, Rasha Hosny, Mohamed Keshawy, and Amany Gaffer. 2019. "Green Synthesis of Graphene Oxide from Oil Palm Leaves as Novel Adsorbent for Removal of Cu(II) Ions from Synthetic Wastewater." *Graphene Technology* 4 (1–2): 33–40. https://doi.org/10.1007/s41127-019-00025-w

Gao, Jian, Fang Liu, Yiliu Liu, Ning Ma, Zhiqiang Wang, and Xi Zhang. 2010. "Environment-Friendly Method to Produce Graphene That Employs Vitamin C and Amino Acid." *Chemistry of Materials* 22 (7): 2213–8. https://doi.org/10.1021/cm902635j

Guo, Yupeng, Jurui Qi, Yanqiu Jiang, Shaofeng Yang, Zichen Wang, and Hongding Xu. 2003. "Performance of Electrical Double Layer Capacitors with Porous Carbons Derived from Rice Husk." *Materials Chemistry and Physics* 80 (3): 704–9. https://doi.org/10.1016/S0254-0584(03)00105-6

Hu, Chaofan, Yingliang Liu, Yunhua Yang, Jianghu Cui, Zirong Huang, Yaling Wang, Lufeng Yang, Haibo Wang, Yong Xiao, and Jianhua Rong. 2013. "One-Step Preparation of Nitrogen-Doped Graphene Quantum Dots from Oxidized Debris of Graphene Oxide." *Journal of Materials Chemistry B* 1 (1): 39–42. https://doi.org/10.1039/c2tb00189f

Ismanto, Andrian Evan, Steven Wang, Felycia Edi Soetaredjo, and Suryadi Ismadji. 2010. "Preparation of Capacitor's Electrode from Cassava Peel Waste." *Bioresource Technology* 101 (10): 3534–40. https://doi.org/10.1016/j.biortech.2009.12.123

Kalita, Golap, Matsushima Masahiro, Hideo Uchida, Koichi Wakita, and Masayoshi Umeno. 2010. "Few Layers of Graphene as Transparent Electrode from Botanical Derivative Camphor." *Materials Letters* 64 (20): 2180–3. https://doi.org/10.1016/j.matlet.2010.07.005

Kim, Yong Jung, Byoung Ju Lee, Hiroaki Suezaki, Teruaki Chino, Yusuke Abe, Takashi Yanagiura, Ki Chul Park, and Morinobu Endo. 2006. "Preparation and Characterization of Bamboo-Based Activated Carbons as Electrode Materials for Electric Double Layer Capacitors." *Carbon*. https://doi.org/10.1016/j.carbon.2006.02.011

Kumar, Neeraj, Reza Salehiyan, Vongani Chauke, Orebotse Joseph Botlhoko, Katlego Setshedi, Manfred Scriba, Mike Masukume, and Suprakas Sinha Ray. 2021. "Top-Down Synthesis of Graphene: A Comprehensive Review." *FlatChem* 27 (February): 100224. https://doi.org/10.1016/j.flatc.2021.100224

Kumar, Rajesh, Radhey Shyam Tiwari, and Onkar Nath Srivastava. 2011. "Scalable Synthesis of Aligned Carbon Nanotubes Bundles Using Green Natural Precursor: Neem Oil." *Nanoscale Research Letters* 6 (1): 92. https://doi.org/10.1186/1556-276X-6-92

Lamb, Marilla, George W. Koch, Eric R. Morgan, and Michael W. Shafer. 2015. "A Synthetic Leaf: The Biomimetic Potential of Graphene Oxide." In *Bioinspiration, Biomimetics, and Bioreplication 2015*, edited by Akhlesh Lakhtakia, Mato Knez, and Raúl J. Martín-Palma, 9429:183–92. SPIE. https://doi.org/10.1117/12.2086567

Lee, Changgu, Xiaoding Wei, Jeffrey W. Kysar, and James Hone. 2008. "Measurement of the Elastic Properties and Intrinsic Strength of Monolayer Graphene." *Science* 321 (5887): 385–8. https://doi.org/10.1126/science.1157996

Lee, Geummi, and Beom Soo Kim. 2014. "Biological Reduction of Graphene Oxide Using Plant Leaf Extracts." *Biotechnology Progress* 30 (2): 463–9. https://doi.org/10.1002/btpr.1862

Lee, Xin Jiat, Billie Yan Zhang Hiew, Kar Chiew Lai, Lai Yee Lee, Suyin Gan, Suchithra Thangalazhy-Gopakumar, and Sean Rigby. 2019. "Review on Graphene and Its Derivatives: Synthesis Methods and Potential Industrial Implementation." *Journal of the Taiwan Institute of Chemical Engineers* 98 (May): 163–80. https://doi.org/10.1016/j.jtice.2018.10.028

Li, Chengyang, Zechao Zhuang, Xiaoying Jin, and Zuliang Chen. 2017. "A Facile and Green Preparation of Reduced Graphene Oxide Using Eucalyptus Leaf Extract." *Applied Surface Science* 422: 469–74. https://doi.org/10.1016/j.apsusc.2017.06.032

Li, Qing, Chuang Zhang, Jian Yao Zheng, Yong Sheng Zhao, and Jiannian Yao. 2014. "Large-Scale Production of High-Quality Graphene Using Glucose and Ferric Chloride." *Chemical Science* 5 (12): 4656–60. https://doi.org/10.1039/c4sc01950d

Li, Xiao, Wei Xing, Shuping Zhuo, Jin Zhou, Feng Li, Shi Zhang Qiao, and Gao Qing Lu. 2011. "Preparation of Capacitor's Electrode from Sunflower Seed Shell." *Bioresource Technology* 102 (2): 1118–23. https://doi.org/10.1016/j.biortech.2010.08.110

Lin, Jiajiang, Chao Xue, Shen Guo, Gary Owens, and Zuliang Chen. 2020. "Impact of Green Reduced Graphene Oxide on Sewage Sludge Bioleaching with *Acidithiobacillus ferrooxidanse*." *Environmental Pollution* 267: 115455. https://doi.org/10.1016/j.envpol.2020.115455

Lingaraju, K., Hanumanaika Raja Naika, Ganganagappa Nagaraju, and H. Nagabhushana. 2019. "Biocompatible Synthesis of Reduced Graphene Oxide from *Euphorbia heterophylla* (L.) and Their In-Vitro Cytotoxicity against Human Cancer Cell Lines." *Biotechnology Reports* 24: e00376. https://doi.org/10.1016/j.btre.2019.e00376

Ma, Han, and Zhigang Shen. 2020. "Exfoliation of Graphene Nanosheets in Aqueous Media." *Ceramics International*. https://doi.org/10.1016/j.ceramint.2020.05.314

Mahmoud, Alaa El Din. 2020. "Eco-Friendly Reduction of Graphene Oxide via Agricultural Byproducts or Aquatic Macrophytes." *Materials Chemistry and Physics* 253 (May): 123336. https://doi.org/10.1016/j.matchemphys.2020.123336

Moosa, Ahmed, and Jaafar Noori Jaafar. 2017. "Green Reduction of Graphene Oxide Using Tea Leaves Extract with Applications to Lead Ions Removal from Water." *Nanoscience and Nanotechnology* 7 (2): 38–47. https://doi.org/10.5923/j.nn.20170702.03

Muramatsu, Hiroyuki, Yoong Ahm Kim, Kap-Seung Yang, Rodolfo Cruz-Silva, Ikumi Toda, Takumi Yamada, Mauricio Terrones, Morinobu Endo, Takuya Hayashi, and Hidetoshi Saitoh. 2014. "Rice Husk-Derived Graphene with Nano-Sized Domains and Clean Edges." *Small* 10 (14): 2766–70. https://doi.org/10.1002/smll.201400017

Naghdi, Samira, Kyong Yop Rhee, and Soo Jin Park. 2018. "A Catalytic, Catalyst-Free, and Roll-to-Roll Production of Graphene via Chemical Vapor Deposition: Low Temperature Growth." *Carbon*. https://doi.org/10.1016/j.carbon.2017.10.065

Nasir, Salisu, Mohd Zobir Hussein, Nor Azah Yusof, and Zulkarnain Zainal. 2017. "Oil Palm Waste-Based Precursors as a Renewable and Economical Carbon Sources for the Preparation of Reduced Graphene Oxide from Graphene Oxide." *Nanomaterials* 7 (7): 1–18. https://doi.org/10.3390/nano7070182

Novoselov, Konstantin Sergeevich, Andre K. Geim, S. V. Morozov, D. Jiang, Y. Zhang, S. V. Dubonos, Irina V. Grigorieva, and Alexander A. Firsov. 2004. "Electric Field in Atomically Thin Carbon Films." *Science* 306 (5696): 666–9. https://doi.org/10.1126/science.1102896

Peng, Juan, Wei Gao, Bipin Kumar Gupta, Zheng Liu, Rebeca Romero-Aburto, Liehui Ge, Li Song, et al. 2012. "Graphene Quantum Dots Derived from Carbon Fibers." *Nano Letters* 12 (2): 844–9. https://doi.org/10.1021/nl2038979

Poh, Hwee Ling, Filip Šaněk, Adriano Ambrosi, Guanjia Zhao, Zdeněk Sofer, and Martin Pumera. 2012. "Graphenes Prepared by Staudenmaier, Hofmann and Hummers Methods with Consequent Thermal Exfoliation Exhibit Very Different Electrochemical Properties." *Nanoscale* 4 (11): 3515–22. https://doi.org/10.1039/c2nr30490b

Purkait, Taniya, Guneet Singh, Mandeep Singh, Dinesh Kumar, and Ramendra Sundar Dey. 2017. "Large Area Few-Layer Graphene with Scalable Preparation from Waste Biomass for High-Performance Supercapacitor." *Scientific Reports* 7 (1): 1–14. https://doi.org/10.1038/s41598-017-15463-w

Ravani, Fotini, Konstantinos Papagelis, Vassileios Dracopoulos, John Parthenios, Konstantinos G. Dassios, Angeliki Siokou, and Costas Galiotis. 2013. "Graphene Production by Dissociation of Camphor Molecules on Nickel Substrate." *Thin Solid Films* 527 (January): 31–7. https://doi.org/10.1016/j.tsf.2012.12.029

Ray, Ajoy K., Ranjan K. Sahu, V. Rajinikanth, Himangshu Bapari, Mainak Ghosh, and Parimal Paul. 2012. "Preparation and Characterization of Graphene and Ni-Decorated Graphene Using Flower Petals as the Precursor Material." *Carbon* 50 (11): 4123–9. https://doi.org/10.1016/j.carbon.2012.04.060

Roy, Babli, Yunke Jing, and Basudeb Basu. 2017. "Reduced Graphene Oxides (RGOs) Using Nature-Based Reducing Sources: Detailed Studies on Properties, Morphologies and Catalytic Activity." *Current Graphene Science* 1 (1). https://doi.org/10.2174/2452273201666170519155915

Roy, Prathik, Arun Prakash Periasamy, Chiashain Chuang, Yi Rou Liou, Yang Fang Chen, Joseph Joly, Chi Te Liang, and Huan Tsung Chang. 2014. "Plant Leaf-Derived Graphene Quantum Dots and Applications for White LEDs." *New Journal of Chemistry* 38 (10): 4946–51. https://doi.org/10.1039/c4nj01185f

Ruan, Gedeng, Zhengzong Sun, Zhiwei Peng, and James M. Tour. 2011. "Growth of Graphene from Food, Insects, and Waste." *ACS Nano* 5 (9): 7601–7. https://doi.org/10.1021/nn202625c

Rufford, Thomas E., Denisa Hulicova-Jurcakova, Kiran Khosla, Zhonghua Zhu, and Gao Qing Lu. 2010. "Microstructure and Electrochemical Double-Layer Capacitance of Carbon Electrodes Prepared by Zinc Chloride Activation of Sugar Cane Bagasse." *Journal of Power Sources* 195 (3): 912–8. https://doi.org/10.1016/j.jpowsour.2009.08.048

Rufford, Thomas E., Denisa Hulicova-Jurcakova, Zhonghua Zhu, and Gao Qing Lu. 2008. "Nanoporous Carbon Electrode from Waste Coffee Beans for High Performance Supercapacitors." *Electrochemistry Communications* 10 (10): 1594–7. https://doi.org/10.1016/j.elecom.2008.08.022

Shams, S. Saqib, Li Sheng Zhang, Renhao Hu, Ruoyu Zhang, and Jin Zhu. 2015. "Synthesis of Graphene from Biomass: A Green Chemistry Approach." *Materials Letters* 161 (September): 476–9. https://doi.org/10.1016/j.matlet.2015.09.022

Shen, Bin, Dingding Lu, Wentao Zhai, and Wenge Zheng. 2013. "Synthesis of Graphene by Low-Temperature Exfoliation and Reduction of Graphite Oxide under Ambient Atmosphere." *Journal of Materials Chemistry C* 1 (1): 50–3. https://doi.org/10.1039/c2tc00044j

Somanathan, Thirunavukkarasu, Karthika Prasad, Kostya (Ken) Ostrikov, Arumugam Saravanan, and Vemula Mohana Krishna. 2015. "Graphene Oxide Synthesis from Agro Waste." *Nanomaterials* 5: 826–34. https://doi.org/10.3390/nano5020826

Somani, Prakash R., Savita P. Somani, and Masayoshi Umeno. 2006. "Planer Nano-Graphenes from Camphor by CVD." *Chemical Physics Letters* 430 (1–3): 56–9. https://doi.org/10.1016/j.cplett.2006.06.081

Subramanian, V., Cheng Luo, A. M. Stephan, K. S. Nahm, Sabu Thomas, and Bingqing Wei. 2007. "Supercapacitors from Activated Carbon Derived from Banana Fibers." *Journal of Physical Chemistry C* 111 (20): 7527–31. https://doi.org/10.1021/jp067009t

Sujiono, Eko. Hadi, Zurnansyah, Dirfan Zabrian, Muhammad Yusriadi Dahlan, B. D. Amin, Samnur, and Jasdar Agus. 2020. "Graphene Oxide Based Coconut Shell Waste: Synthesis by Modified Hummers Method and Characterization." *Heliyon* 6 (8). https://doi.org/10.1016/j.heliyon.2020.e04568

Sun, Li, Lei Wang, Chungui Tian, Taixing Tan, Ying Xie, Keying Shi, Meitong Li, and Honggang Fu. 2012. "Nitrogen-Doped Graphene with High Nitrogen Level via a One-Step Hydrothermal Reaction of Graphene Oxide with Urea for Superior Capacitive Energy Storage." *RSC Advances* 2 (10): 4498–506. https://doi.org/10.1039/c2ra01367c

Suryawanshi, Anil, Mandakini Biswal, Dattakumar Mhamane, Rohan Gokhale, Shankar Patil, Debanjan Guin, and Satishchandra Ogale. 2014. "Large Scale Synthesis of Graphene Quantum Dots (GQDs) from Waste Biomass and Their Use as an Efficient and Selective Photoluminescence on-off-on Probe for Ag+ Ions." *Nanoscale* 6 (20): 11664–70. https://doi.org/10.1039/c4nr02494j

Tang, Libin, Rongbin Ji, Xiangke Cao, Jingyu Lin, Hongxing Jiang, Xueming Li, Kar Seng Teng, et al. 2012. "Deep Ultraviolet Photoluminescence of Water-Soluble Self-Passivated Graphene Quantum Dots." *ACS Nano* 6 (6): 5102–10. https://doi.org/10.1021/nn300760g

Terzopoulou, Zoi, George Z. Kyzas, and Dimitrios N. Bikiaris. 2015. "Recent Advances in Nanocomposite Materials of Graphene Derivatives with Polysaccharides." *Materials*. https://doi.org/10.3390/ma8020652

Tetlow, Holly, Joel Posthuma de Boer, Ian J. Ford, Dimitri Dimitrievich Vvedensky, Johann Coraux, and Lev N. Kantorovich. 2014. "Growth of Epitaxial Graphene: Theory and Experiment." *Physics Reports*. https://doi.org/10.1016/j.physrep.2014.03.003

Vatandost, Elham, Azade Ghorbani-HasanSaraei, Fereshteh Chekin, Shahram Naghizadeh Raeisi, and Seyed Ahmad Shahidi. 2020. "Green Tea Extract Assisted Green Synthesis of Reduced Graphene Oxide: Application for Highly Sensitive Electrochemical Detection of Sunset Yellow in Food Products." *Food Chemistry: X* 6 (March): 100085. https://doi.org/10.1016/j.fochx.2020.100085

Wang, Huanlei, Zhanwei Xu, Alireza Kohandehghan, Zhi Li, Kai Cui, Xuehai Tan, Tyler James Stephenson, et al. 2013. "Interconnected Carbon Nanosheets Derived from Hemp for Ultrafast Supercapacitors with High Energy." *ACS Nano* 7 (6): 5131–41. https://doi.org/10.1021/nn400731g

Wang, Xuechun, Baosen Liu, Baoyou Shi, Laizhou Song, and Yuan Zhuang. 2019. "Transformation of Leaf Waste into 3D Graphene for Water Treatment." *Desalination and Water Treatment* 168: 348–56. https://doi.org/10.5004/dwt.2019.24654

Wang, Yan, Zi Xing Shi, and Jie Yin. 2011. "Facile Synthesis of Soluble Graphene via a Green Reduction of Graphene Oxide in Tea Solution and Its Biocomposites." *ACS Applied Materials and Interfaces* 3 (4): 1127–33. https://doi.org/10.1021/am1012613

Wang, Zhaofeng, Jingfang Yu, Xin Zhang, Na Li, Bin Liu, Yanyan Li, Yuhua Wang, et al. 2016. "Large-Scale and Controllable Synthesis of Graphene Quantum Dots from Rice Husk Biomass: A Comprehensive Utilization Strategy." *ACS Applied Materials and Interfaces* 8 (2): 1434–9. https://doi.org/10.1021/acsami.5b10660

Wu, Feng Chin, Ru Ling Tseng, Chi Chang Hu, and Chen Ching Wang. 2004. "Physical and Electrochemical Characterization of Activated Carbons Prepared from Firwoods for Supercapacitors." *Journal of Power Sources* 138 (1–2): 351–9. https://doi.org/10.1016/j.jpowsour.2004.06.023

Yadav, Krishna K., Ritika Wadhwa, Nausad Khan, and Menaka Jha. 2021. "Efficient Metal-Free Supercapacitor Based on Graphene Oxide Derived from Waste Rice." *Current Research in Green and Sustainable Chemistry* 4 (January): 100075. https://doi.org/10.1016/j.crgsc.2021.100075

Yang, Haiping, Rong Yan, Hanping Chen, Dong Ho Lee, and Chuguang Zheng. 2007. "Characteristics of Hemicellulose, Cellulose and Lignin Pyrolysis." *Fuel* 86 (12–13): 1781–8. https://doi.org/10.1016/j.fuel.2006.12.013

Yang, Jing, Yafei Liu, Xiaomei Chen, Zhonghua Hu, and Guohua Zhao. 2008. "Carbon Electrode Material with High Densities of Energy and Power." *Acta Physico - Chimica Sinica* 24 (1): 13–9. https://doi.org/10.1016/S1872-1508(08)60002-9

Yang, Wenrong, Kyle R. Ratinac, Simon P. Ringer, Pall Thordarson, J. Justin Gooding, and Filip Braet. 2010. "Carbon Nanomaterials in Biosensors: Should You Use Nanotubes or Graphene?" *Angewandte Chemie International Edition* 49 (12): 2114–38. https://doi.org/10.1002/anie.200903463

Ye, Ruquan, Changsheng Xiang, Jian Lin, Zhiwei Peng, Kewei Huang, Zheng Yan, Nathan P. Cook, et al. 2013. "Coal as an Abundant Source of Graphene Quantum Dots." *Nature Communications* 4 (1): 1–7. https://doi.org/10.1038/ncomms3943

Yu, Huitao, Bangwen Zhang, Chaoke Bulin, Ruihong Li, and Ruiguang Xing. 2016. "High-Efficient Synthesis of Graphene Oxide Based on Improved Hummers Method." *Scientific Reports* 6 (1): 1–7. https://doi.org/10.1038/srep36143

Zhao, Hong, and Tianshou Zhao. 2013. "Graphene Sheets Fabricated from Disposable Paper Cups as a Catalyst Support Material for Fuel Cells." *Journal of Materials Chemistry A* 1 (2): 183–7. https://doi.org/10.1039/c2ta00018k

Zhu, Chengzhou, Shaojun Guo, Youxing Fang, and Shaojun Dong. 2010. "Reducing Sugar: New Functional Molecules for the Green Synthesis of Graphene Nanosheets." *ACS Nano* 4 (4): 2429–37. https://doi.org/10.1021/nn1002387

12 Biosynthesis of Reduced Graphene Oxide and Its Functionality as an Antibacterial Template

Aruna Jyothi Kora

CONTENTS

12.1 INTRODUCTION

In recent years, graphene has gained importance due to its outstanding physical, mechanical, thermal, electrical, optical, chemical and biological properties attributed to distinctive single atom thick two-dimensional structure. The graphene-based materials are extensively utilized in the fields of electronics, optics, photovoltaics, catalysis, green chemistry, fuel cells, wastewater treatment, sensing, tissue engineering, cancer therapy, drug delivery, imaging etc. (Maddinedi et al. 2014, Maddinedi et al. 2015, Xu et al. 2018, Nguyen et al. 2019, Noorunnisa Khanam and Hasan 2019, Vargas et al. 2019, Dat et al. 2020). Thus, studies have been concentrated on the preparation, characterization and applications of nearly graphene equivalent material, reduced graphene oxide (rGO). The electrically insulating, oxygen containing functional group (epoxy, hydroxyl, carbonyl, carboxyl, ether) rich GO is obtained from the oxidation of graphite, during which no of layers of graphite are decreased. The electrically conductive, rGO with a higher surface area is obtained by reduction/deoxygenation of GO using various methods.

The current review briefly focuses on the biosynthesis of rGO employing various biological reductants and its characterization. The applications of rGO as an antibacterial template for the functionalization of different agents, fabrication of rGO nanocomposites and antibacterial action of the nanocomposites on a vast number of bacteria are detailed. The mode of action of rGO nanocomposites on bacteria is touched upon.

12.2 BIOSYNTHESIS

Various methodologies have been standardized for the bulk production of rGO, such as micromechanical exfoliation, chemical vapor deposition, epitaxial growth, pyrolysis, chemical reduction etc. Among the methods, the chemical reduction is widely practiced for obtaining the rGO by deoxygenation of GO using an array of reductants, including hydrazine, dimethyl hydrazine, phenyl hydrazine,

DOI: 10.1201/9781003169741-12

hydroxylamine, borohydride, hydroquinone, pyrrole etc. Some of the chemical reduction methods are limited by low yield and irreversible agglomeration of rGO. Moreover, the chemical reductants are explosive, highly reactive, harmful and pose a threat to flora and fauna of the environment (Maddinedi et al. 2015, Luo, Xu, and Zhao 2017, Xu et al. 2018, Noorunnisa Khanam and Hasan 2019, Ikram, Jan, and Ahmad 2020, Narayanan, Park, and Han 2021). Thus, the trust has been shifted to nontoxic, environment friendly, renewable, biodegradable, alternative dual functional biological reductant and stabilizing agents to synthesize rGO from GO.

Generally, the synthesis of rGO is carried at an alkaline pH of 12 from the GO precursor, which is produced by Hummers, modified Hummers or Offenman process (Luo, Xu, and Zhao 2017, Narayanan, Park, and Han 2021). An array of reductant and stabilizing agents such as biomolecules (pyrogallol, glucose), biopolymers (casein, silk fibroin, starch), antibiotics (vancomycin), extracts of seed, fruits, leaves, bulbs etc. and bacteria are extensively utilized for the biosynthesis of rGO by heating at different temperatures for a variable duration (Table 12.1).

TABLE 12.1
A Comparison of Biosynthesis of Reduced Graphene Oxide (rGO), in Terms of Reductant, Precursors, Reaction Conditions and rGO Morphology

Reductant	Precursors	Reaction Conditions	rGO Morphology	Reference
Pyrogallol	Pyrogallol (1 mg/mL), GO (0.1 mg/mL)	pH 12, 100°C, 8 h	Transparent silk like sheet	Luo, Xu, and Zhao (2017)
Glucose	Glucose (9 mg/mL), GO (1 mg/mL)	pH 8–9, 95°C, 3 h	Few-layered sheet	Sadhukhan et al. (2019)
Casein	Casein (10 mg/mL), GO (1 mg/mL)	pH 12, 90°C, 7 h	Few-layered nano sheets, 419 nm	Maddinedi et al. (2014)
Silk fibroin	Silk fibroin (100 mg/mL), GO (1 mg/mL)	125°C, 5 h	4–5-layered nano sheets	Zhang, Wang, Fan et al. (2021)
Potato soluble starch	Starch (0.2%), GO (0.5 mg/mL)	pH 12, 120°C, 15 min	Few-layered sheet	Narayanan, Park, and Han (2021)
Vancomycin	Vancomycin (2 mg/mL), GO (1 mg/mL)	pH 8.5, 60°C, 24 h	Lamellar structured, 1.45 nm	Xu et al. (2018)
Terminalia chebula seed extract	Seed extract (0.8%), GO (0.5 mg/mL)	pH 12, 90°C, 24 h	Silk like ultrathin 3-layered sheets, 1014 nm	Maddinedi et al. (2015)
Platanus orientalis leaf extract	Leaf extract (0.7%), GO (0.5 mg/mL)	pH 12, 100°C, 10 h	Silk like thin, transparent curled sheet	Xing et al. (2016)
Tagetes erecta flower extract	Flower extract (13.3%), GO (3.3 mg/mL)	90°C, 24 h	Curve and wave like sheets	Navya Rani et al. (2019)
Allium cepa bulb extract	Bulb extract (3.3%), GO (5 mg/mL)	Room temperature, 6h	Thin, multilayered sheets, 140 nm	Noorunnisa Khanam and Hasan (2019)
Gram-negative (*Escherichia coli, Enterobacter cloacae, Shewanella baltica*) and Gram-positive (*Bacillus* sp.) bacteria, extremophilic consortium	Bacterial biomass (30 mg/mL), GO (0.4 mg/mL)	20–25°C, 72 h	Single sheets	Vargas et al. (2019)

The biomolecule pyrogallol (1 mg/mL) was utilized as a dual functional reducing and stabilizing agent for the biosynthesis of rGO by heating the alkaline (pH 12) solution of GO (0.1 mg/mL) at 100°C for 8 h. The hydroxyl groups of the pyrogallol mediated the deoxygenation of GO (Luo, Xu, and Zhao 2017). The monosaccharide, glucose (9 mg/mL) was employed for the alkaline (pH 8–9) reduction of GO (1 mg/mL) at 95°C for 3 h (Sadhukhan et al. 2019). The milk protein, casein (10 mg/mL) was used for the green synthesis of rGO by refluxing the pH-12-maintained solution of GO (1 mg/mL) at 90°C for 7 h. The aspartic and glutamic acid residues of the casein molecules were involved in the reduction of GO (Maddinedi et al. 2014). The silk fibroin (100 mg/mL) was employed for the reduction of GO (1 mg/mL) at 125°C, 5 h and preparation of regenerated silk fibroin (RSF)/rGO fibrous mats (Zhang, Wang, Fan et al. 2021). The rGO was hydrothermally synthesized from GO (0.5 mg/mL) within 15 min at 120°C, pH 12 by the biopolymer soluble starch (0.2%) (Narayanan, Park, and Han 2021). The seed extract of *Terminalia chebula* (0.8%) was employed for facile biosynthesis of rGO from GO (0.5 mg/mL) and the polyphenols, including gallic acid, methyl gallate, ascorbic acid and pyrogallol were involved in the reduction and stabilization of rGO (Maddinedi et al. 2015). In another study, polyphenols in the leaf extract (0.7%) of *Platanus orientalis* were employed for the biosynthesis of rGO from GO (0.5 mg/mL) at 100°C for 10 h at a solution pH of 12 (Xing et al. 2016). The flower extract of *Tagetes erecta* (13.3%) was employed for the reduction of GO (3.3 mg/mL) at 90°C for 24 h (Navya Rani et al. 2019). The bulb extract of onion (*Allium cepa*) was used for the biological reduction of GO into rGO (Noorunnisa Khanam and Hasan 2019). The Gram-negative (*Escherichia coli, Enterobacter cloacae, Shewanella baltica*) and Gram-positive (*Bacillus* sp.) bacterial strains isolated from the natural environment and the extremophilic consortium (30 mg/mL) were used for the biological reduction of GO (0.4 mg/mL) under aerobic conditions without addition of nutrient or carbon source at 20–25°C for 72 h (Vargas et al. 2019).

The nanocomposites of rGO either loaded, functionalized or cosynthesized along with antibiotics, streptomycin (Zhao et al. 2020), vancomycin (Xu et al. 2018), ciprofloxacin (Bhattacharya et al. 2021); anticancer drug, sorafenib (Xu et al. 2019), citrus flavonoid naringenin (Shanmuganathan et al. 2020), nanoparticles (NPs) of silver (Ag) (Zheng et al. 2016, Chen, Li, and Chen 2019, Song and Shi 2019, Veisi et al. 2019, Dat et al. 2020, Esmaeili et al. 2020, Moghayedi et al. 2020, Pan et al. 2020, Tan et al. 2020, Gu et al. 2021), silver sulfide (Ag_2S) (Huo et al. 2018), copper (Cu) (Zhang et al. 2020), copper oxide (CuO) (Alayande, Obaid, and Kim 2020), cuprous oxide (Cu_2O) (Navya Rani et al. 2019), gold (Au) (Saikia et al. 2016), iron (Fe) (Wang et al. 2019), magnetite (Fe_3O_4) (Sadeghi Rad et al. 2021), iron oxide (Fe_2O_3) (Naseer et al. 2020), titanium dioxide (TiO_2) (Wanag et al. 2018,Zhang, Wang, Liu et al. 2021,), cerium doped TiO_2 (Behera, Barik, and Mohapatra 2021), cadmium oxide (CdO) (Sadhukhan et al. 2019), bioactive glass (Ashok raja et al. 2018), polyaniline (Mirmohseni, Azizi, and Dorraji 2020); Schiff's base (Omidi, Kakanejadifard, and Azarbani 2017) etc. were synthesized by biological, chemical, hydrothermal, UV, microwave assisted reduction; and co-precipitation methods.

12.3 CHARACTERIZATION

The rGO and rGO nanocomposites synthesized by various methods are characterized by different analytical instrumentation techniques, including UV-Vis absorption spectroscopy (UV-Vis), X-ray diffraction (XRD), transmission electron microscopy (TEM), Fourier-transform infrared spectroscopy (FTIR) etc. The reduction of GO into rGO is evident from the solution's color change, i.e., yellowish brown to black (Narayanan, Park, and Han 2021). The rGO biosynthesized using potato soluble starch via autoclaving at 121°C for 30 min is depicted in Figure 12.1. The GO exhibited a characteristic absorption peak at 235 nm in the UV-Vis absorption spectra, attributed to π-π^* transitions of the aromatic C-C bonds. Upon reduction, the peak was shifted

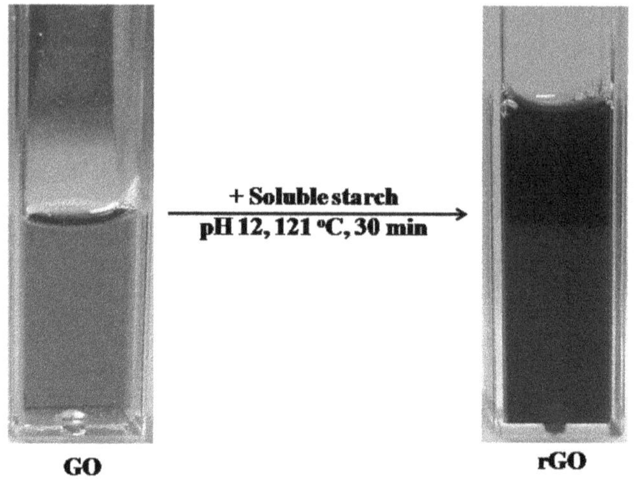

FIGURE 12.1 The autoclave mediated biosynthesis of reduced graphene oxide (rGO) from graphene oxide (GO) by starch at reaction conditions of pH 12, 121°C, 30 min.

to 260 nm confirming the reduction of GO by starch (Figure 12.2) (Maddinedi et al. 2015, Luo, Xu, and Zhao 2017, Narayanan, Park, and Han 2021). Also, the biosynthesis of rGO by starch was confirmed from the broad diffraction peak in the XRD pattern at a 2θ angle of 21.6° from (002) crystal plane of rGO (Figure 12.3(a))(Maddinedi et al. 2014, Narayanan, Park, and Han 2021). The TEM image of Ag/rGO nanocomposite indicated the homogeneous distribution of silver NPs (Ag NPs) on the surface of thin, transparent, silky and wavy rGO nanosheets (Figure 12.3(b)) (Veisi et al. 2019, Shanmuganathan et al. 2020).

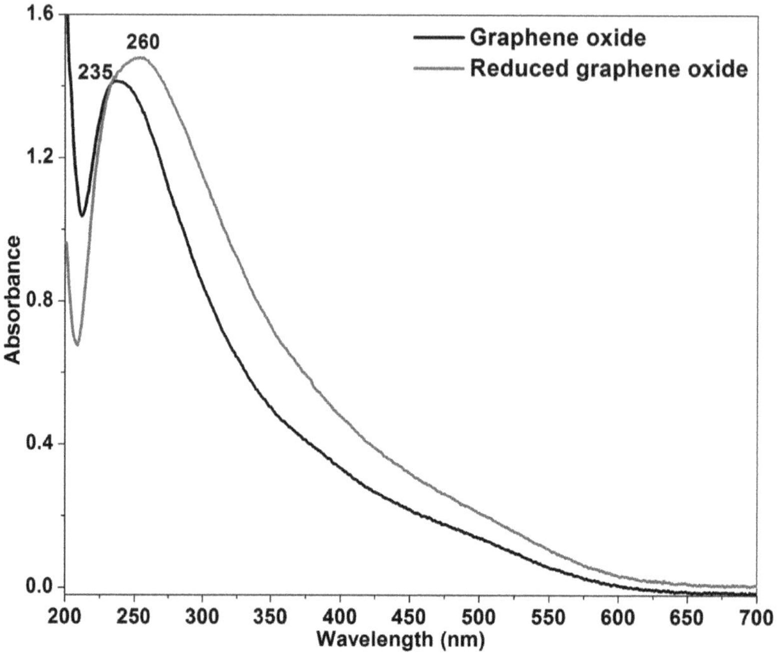

FIGURE 12.2 The UV-visible absorption spectra of graphene oxide and starch reduced graphene oxide.

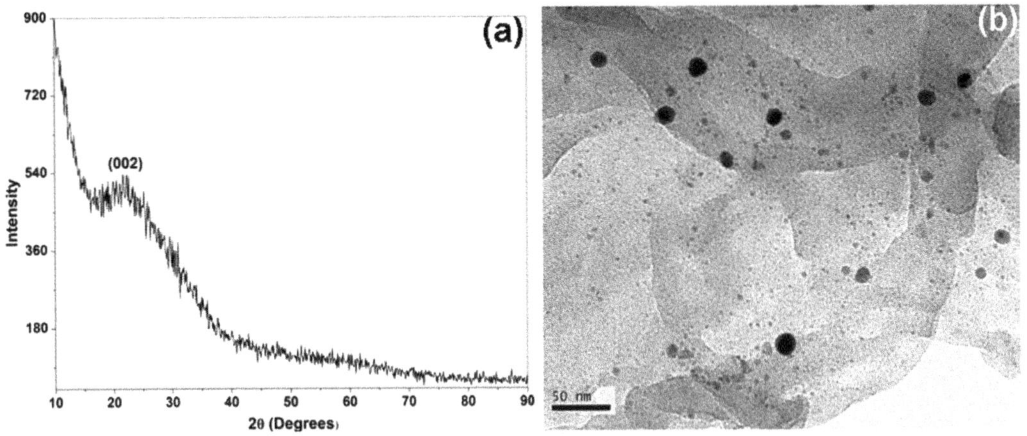

FIGURE 12.3 (a) XRD pattern of starch reduced graphene oxide and (b) TEM image of Ag/rGO nanocomposite.

12.4 APPLICATIONS OF rGO AS AN ANTIBACTERIAL TEMPLATE

The utilization of rGO as an antibacterial agent was investigated by various researchers and found to depend on physicochemical characteristics, such as concentration and size of the nanosheets. Most of the studies indicated that the antibacterial action of rGO was either low or required a higher dosage or longer duration of incubation (Chen, Li, and Chen 2019, Noorunnisa Khanam and Hasan 2019, Norahan et al. 2019). However, the rGO can function as a template or matrix for loading, coating, decorating and functionalizing various antibacterial and bioactive molecules, such as antibiotics, drugs, metals, metal oxides, NPs and ligands and fabrication of nanocomposites. The functionality of rGO as a template for different agents was known to augment the antibacterial activity of nanocomposites against target bacteria, leading to synergistic action (Naseer et al. 2020, Zhang et al. 2020, Zhao et al. 2020, Narayanan, Park, and Han 2021).

The antibacterial and bactericidal action of rGO and rGO nanocomposites were studied against an array of bacteria, including Gram-negative (*E. coli, Pseudomonas aeruginosa, Klebsiella pneumoniae, Proteus mirabilis, Salmonella typhimurium, Vibrio cholerae*) and Gram-positive (*Bacillus cereus, B. subtilis, Listeria monocytogenes, Rhodococcus rhodochrous, Staphylococcus aureus, S. epidermis, S. saprophyticus, Streptococcus pyogenes, Enterococcus faecalis*) bacteria. Various assays, such as disc diffusion, well diffusion, micro and macro broth dilution; absorbance, agar dilution, colony count, growth curve, viability, live/dead staining, biofilm inhibition, reduced nicotinamide adenine dinucletide content, glutathione oxidation, DNA fragmentation, lipid peroxidation, reactive oxygen species (ROS), membrane permeability assays; scanning electron microscopy etc. were used for evaluating the biocidal action of rGO nanocomposites (Table 12.2). The rGO sheets of 140-nm thickness biosynthesized by onion bulb extract hindered the growth of *E. coli, P. aeruginosa, S. aureus* and *S. faecalis* by 90.5, 93.1, 95 and 94%, respectively, at 10 μg/mL and 5 days of incubation (Noorunnisa Khanam and Hasan 2019). The RSF/rGO fibrous mats at 1% prevented the colony growth of *E. coli* and *S. aureus* by 84.9 and 98.6%, respectively. Whereas the respective values for RSF/rGO were 70.3 and 92.8%, after 12 h of treatment (Zhang, Wang, Fan et al. 2021).

The glycopeptide antibiotic, vancomycin-decorated rGO inhibited the growth curve of *S. aureus* and *S. epidermidis* at 2-μg/mL concentration. The vancomycin-decorated rGO film suppressed the bacterial growth in well diffusion assay and reduced the bacterial cell adhesion on the film surface. In a rat infection model, the wound infection induced by *S. aureus* was healed after 14 days of treatment with the composite film confirming its utility as a wound dressing (Xu et al. 2018). The

TABLE 12.2
A Comparison of Antibacterial Action of rGO Nanocomposites towards Various Bacterial Strains, with Reference to Nanoparticle Particle Size and Concentration

Nanocomposite	Size (nm)	Concentration	Bacterial Strains	Antibacterial Action	Reference
rGO	Curve and wave like sheets	20 μg/well	*Escherichia coli* *Bacillus subtilis*	Inhibition zones of 1.9 and 2 mm	Navya Rani et al. (2019)
Vancomycin-decorated rGO	1.45	2 μg/mL	*Staphylococcus aureus* *Staphylococcus epidermidis*	Growth curve inhibition	Xu et al. (2018)
Ciprofloxacin-loaded rGO	–		*Pseudomonas aeruginosa*	Cell rupture, biofilm reduction	Bhattacharya et al. (2021)
Streptomycin-loaded polyethylene glycol-MoS_2/rGO	–	150 μg/mL	*E. coli* *S. aureus*	Structural damage, protein synthesis inhibition, thermal injury, oxidative stress	Zhao et al. (2020)
Ag NPs-decorated rGO	16.3	5.4 μg/disc	*E. coli* *Vibrio cholerae* *Proteus mirabilis* *Salmonella typhi* *S. aureus* *S. epidermidis* *Rhodococcus rhodochrous*	Inhibition zones of 10, 12, 11, 10, 9, 10 and 10 mm	Shanmuganathan et al. (2020)
Ag/rGO	12.7–73.3	60 μg/well	*E. coli* *P. aeruginosa* *Salmonella typhimurium* *P. mirabilis* *S. aureus* *Staphylococcus saprophyticus* *Streptococcus pyogenes* *B. subtilis*	Inhibition zones of 14.6, 23.4, 9.6, 0, 16.4, 23.6, 20.8 and 21.2 mm	Veisi et al. (2019)
Ag/rGO	12.7–73.3	2000, 1000, 4000, 8000, 2000, 1000, 1000 and 1000 μg/mL	*E. coli* *P. aeruginosa* *S. typhimurium* *P. mirabilis* *S. aureus* *S. saprophyticus* *S. pyogenes* *B. subtilis*	MIC	Veisi et al. (2019)
Ag/rGO	12.7–73.3	4000, 1000, 4000, 8000, 4000, 1000, 2000 and 1000 μg/mL	*E. coli* *P. aeruginosa* *S. typhimurium* *P. mirabilis* *S. aureus* *S. saprophyticus* *S. pyogenes* *B. subtilis*	MBC	Veisi et al. (2019)
Ag/rGO	16	100 μg/mL	*E. coli* *P. aeruginosa* *S. aureus*	Growth inhibition of 95.6, 99.9 and 89.3%	Dat et al. (2020)

(Continued)

TABLE 12.2 ***(Continued)***
A Comparison of Antibacterial Action of rGO Nanocomposites towards Various Bacterial Strains, with Reference to Nanoparticle Particle Size and Concentration

Nanocomposite	Size (nm)	Concentration	Bacterial Strains	Antibacterial Action	Reference
Ag/rGO	15	100 µg/mL	*E. coli*	99.9% killing	Chen, Li, and Chen (2019)
Ag/rGO	58.9	15 µg/mL 30 µg/mL	*E. coli* Multidrug resistant *Klebsiella pneumoniae*	100% killing	Tan et al. (2020)
Ag/rGO	6	2 µg/mL	*E. coli*	99.9% killing	Song and Shi (2019)
Ag/PMo/rGO	70	256 µg/mL 512 µg/mL	*E. coli*	MIC MBC	Moghayedi et al. (2020)
Ag/Ag_2S/rGO	37–63	100 µg/mL	*E. coli*	97.7% colony inhibition	Huo et al. (2018)
Chitosan wrapped Ag NPs/rGO sandwich film	15	4.45 At%	*E. coli* *S. aureus*	Inhibition zones of 3.3 and 4 mm	Gu et al. (2021)
PLC Ag/rGO membrane	20.7	2%	*E. coli* *S. aureus*	Colony growth inhibition of 99.4 and 99.5%	Pan et al. (2020)
Cellulose acetate/ polyurethane Ag/rGO scaffolds	8	2.5%	*Pseudomonas* *S. aureus*	Colony growth inhibition of 100 and 96%	Esmaeili et al. (2020)
Cu_2O/rGO	40–60	20 µg/well	*E. coli* *B. subtilis*	Inhibition zones of 3 and 3.5 mm	Navya Rani et al. (2019)
Au/rGO	8–15	50 µg/well	*P. aeruginosa* *K. pneumoniae* *S. aureus* *Bacillus cereus*	Inhibition zones of 18.5, 19.5, 16.3 and 19.6 mm	Saikia et al. (2016)
Chromium and cerium co-doped Fe_3O_4/rGO	22	50 µg/mL 200 µg/mL	*S. aureus*	MIC MBC	Sadeghi Rad et al. (2021)
Ce-TiO_2/rGO	8–14	50 µg/well	*E. coli* *S. aureus*	Inhibition zones of 9.5 and 3 mm	Behera, Barik, and Mohapatra (2021)
Ce-TiO_2/rGO	8–14	30 µg/mL	*E. coli* *S. aureus*	Photoinduced bacterial inactivation of 90 and 100%	Behera, Barik, and Mohapatra (2021)
TiO_2/rGO	17	100 µg/mL	*E. coli*	100% photocatalytic inactivation	Wanag et al. (2018)
CdO/rGO	4–8	3.9, 6.1, 9.6, 4.0, 4.4 and 4.9 µg/mL	*E. coli* *S. typhimurium* *K. pneumoniae* *B. subtilis* *Listeria monocytogenes* *S. aureus*	MIC	Sadhukhan et al. (2019)
rGO/BGNR	60–120	1000 and 500 µg/mL	*P. aeruginosa* *S. aureus*	100% viability loss	Ashok raja et al. (2018)
P/rGO^+	10–70	5%	*E. coli* *Salmonella* *S. aureus* *B. cereus*	Colony count reduction of 74, 79, 95 and 88%	Mirmohseni, Azizi, and Dorraji (2020)

fluoroquinolone antibiotic, ciprofloxacin-loaded rGO caused cell rupture and reduced the biofilm of *P. aeruginosa* (Bhattacharya et al. 2021). The synergistic antibacterial activity of aminoglycoside antibiotic, streptomycin (50 μg/mL)-loaded polyethylene glycol-MoS_2/rGO (150 μg/mL) was studied against *E. coli* and *S. aureus*. Under near-infrared (NIR) irradiation (808 nm, 2 W/cm^2) for 30 min, the antibacterial action was enhanced in terms of bacterial structural damage, protein synthesis inhibition, thermal injury and oxidative stress. The nanocomposites showed low cytotoxicity towards human colon cancer cells and can serve as a photothermal antibacterial agent (Zhao et al. 2020).

The hybrid nanocomposites made up of Ag NPs (16.3 nm) decorated and naringenin reduced GO exhibited inhibition zones towards *E. coli*, *V. cholerae*, *P. mirabilis*, *S. typhi*, *S. aureus*, *S. epidermidis* and *R. rhodochrous* in disc diffusion assay at 5.4 μg/disc. The corresponding inhibition zone values were 10, 12, 11, 10, 9, 10 and 10 mm (Shanmuganathan et al. 2020). The *Pistaciatlantica* leaf extract biosynthesized 12.7–73.3-nm-sized Ag/rGO exhibited inhibition zone values of 14.6, 23.4, 9.6, 0, 16.4, 23.6, 20.8 and 21.2 mm at 60 μg/well against *E. coli*, *P. aeruginosa*, *S. typhimurium*, *P. mirabilis*, *S. aureus*, *S. saprophyticus*, *S. pyogenes* and *B. subtilis* in well diffusion assay, respectively. The corresponding minimum inhibitory concentration (MIC) values were 2000, 1000, 4000, 8000, 2000, 1000, 1000 and 1000 μg/mL towards these bacteria. The respective minimum bactericidal concentration (MBC) values were 4000, 1000, 4000, 8000, 4000, 1000, 2000 and 1000 μg/mL (Veisi et al. 2019). The Ag/rGO nanocomposite integrated with 16-nm-sized Ag NPs suppressed the growth of *E. coli*, *P. aeruginosa* and *S. aureus* by 95.6, 99.9 and 89.3% at 100 μg/mL concentration, respectively (Dat et al. 2020). In a different report, Ag/rGO nanocomposite decorated with Ag NPs of 15 nm size caused 99.9% killing of *E. coli* cells at 100-μg/mL concentration in 3-h duration (Chen, Li, and Chen 2019). The Ag/rGO nanocomposite demonstrated synergetic antibacterial activity against *E. coli* and multidrug resistant *K. pneumoniae* compared with GO, rGO and Ag NPs. Under NIR irradiation (0.30 W/cm^2, 10 min), *E. coli* and *K. pneumoniae* were 100% killed at 15and 30μg/mL of Ag/rGO concentration (Tan et al. 2020). At 2-μg/mL concentration of biosynthesized *S. oneidensis* Ag/rGO nanocomposite decorated with 6nm of Ag NPs, the viability of *E. coli* decreased to 99.9% after 15-min treatment (Song and Shi 2019).

The Ag NPs/phosphomolybdate/rGO (Ag/PMo/rGO) nanocomposite exhibited superior bactericidal action against *E. coli* compared to PMo and rGO. The MIC and MBC values of Ag/PMo/rGO were 256 and 512μg/mL, respectively. Whereas PMo and PMo/rGO showed high MIC values of 3920and 512μg/mL, respectively (Moghayedi et al. 2020). The Ag/Ag_2S/rGO nanocomposite prepared by hydrothermal and UV reduction method decorated with Ag NPs of 37–63-nm size demonstrated 97.7% inhibition of *E. coli* colonies in 24 h, at 100-μg/mL concentration (Huo et al. 2018). The chitosan wrapped Ag NPs/rGO sandwich film exhibited controlled release of silver ions for 14 days. The nanocomposite film impregnated with15-nm-sized Ag NPs showed inhibition zone values of 3.3 and 4 mm in well diffusion towards *E. coli* and *S. aureus*, respectively, at 4.45 At% of Ag. The durable film can be utilized as safe packaging material for shelf life extension of various foods (Gu et al. 2021). The hybrid membranes were fabricated by electrospinning the poly (ε-caprolactone) (PCL) with 20.7-nm-sized Ag NPs anchored on the rGO surface (2%). After 120 min of contact, the fibrous membrane demonstrated enhanced bactericidal activity towards *E. coli* and *S. aureus* by reducing the viability by 99.4 and 99.5%, respectively. While the pristine PCL membrane amended with rGO inactivated the cells of respective bacteria by 72.4 and 65.5% only (Pan et al. 2020). The cellulose acetate/polyurethane nanofibrous scaffolds containing 2.5% nanocomposites of rGO decorated with 8-nm-sized Ag NPs (2.5%) were fabricated by electrospinning. The polymer scaffolds inhibited the colony growth of *Pseudomonas* and *S. aureus* by 100 and 96%, respectively. The nanofibrous mats were biocompatible and promoted *in vivo* healing of artificial wounds (Esmaeili et al. 2020).

The Cu/rGO nanocomposite decorated with Cu NP of 4000 nm was prepared at a copper deposition time of 12 h and the nanocomposite completely (100%) prevented the bacterial growth of *E. coli* and *S. aureus* (Zhang et al. 2020). The hydrothermally synthesized CuO/rGO nanocomposite film exhibited contact mediated complete inactivation of *P. aeruginosa* cells (Alayande, Obaid, and Kim 2020). The cuprous oxide NP (40–60 nm) anchored rGO (Cu_2O/rGO) ceramic nanocomposite biosynthesized by *T. erecta* flower extract showed improved antibacterial activity compared to rGO. The Cu_2O/rGO nanocomposite indicated inhibition zone values of 3 and 3.5 mm against *E. coli* and *B. subtilis* in well diffusion assay at 20 µg/well, respectively. The corresponding values for rGO were 1.9 and 2 mm (Navya Rani et al. 2019). The Au/rGO nanocomposite decorated with Au NP of 8–15 nm, biosynthesized by *Piper pedicellatum* fruit extract, was evaluated for its antibacterial action towards *P. aeruginosa, K. pneumoniae, S. aureus* and *B. cereus*. In the well diffusion assay, inhibition zones of 18.5, 19.5, 16.3 and 19.6 mm were noted towards the nanocomposite for the corresponding bacterial strains at 50µg/well (Saikia et al. 2016).

The nanocomposite of chromium and cerium co-doped Fe_3O_4/rGO showed MIC and MBC values of 50 and 200 µg/mL towards *S. aureus*. The nanocomposite caused bacterial cell membrane damage, deformation and cell structure distortion and leads to cell death (Sadeghi Rad et al. 2021). The Fe_2O_3nano-rod-decorated rGO (Fe_2O_3/rGO) nanocomposite demonstrated a MIC value of 5 µg/mL against ciprofloxacin resistant *S. aureus* (Naseer et al. 2020). The titanium dioxide NPs (17 nm) modified with 1.5% of rGO (TiO_2/rGO) caused 100% photocatalytic inactivation of *E. coli* cells at 100 µg/mL concentration, after 75 min of artificial solar light irradiation (Wanag et al. 2018). The hydrothermally synthesized cerium doped (0.2%) titanium dioxide NP (8–14 nm) deposited rGO (10%) (Ce-TiO_2/rGO) demonstrated inhibition zone values of 9.5 and 3 mm against *E. coli* and *S. aureus* in well diffusion assay under visible light irradiation, at 50 µg/well. At 30 µg/mL concentration of the photoinduced bacterial inactivation was 90 and 100% for *E. coli* and *S. aureus*, respectively (Behera, Barik, and Mohapatra 2021). The rGO@TiO_2 hybrid nano filler reinforced electrospun RSF scaffolds inhibited the colony growth of *S. aureus* by 84.9%. The biocompatible, multifunctional nanocomposite is a promising biomaterial for tissue engineering (Zhang, Wang, Liu et al. 2021). The cadmium oxide NP (4–8 nm)-decorated rGO (CdO/rGO) nanocomposite biosynthesized by glucose exhibited MIC values of 3.9, 6.1, 9.6, 4.0, 4.4 and 4.9 µg/mL towards *E. coli, S. typhimurium, K. pneumoniae, B. subtilis, L. monocytogenes* and *S. aureus*, respectively (Sadhukhan et al. 2019).

The rGO-decorated with 60–120-nm-sized bioactive glass nano rods (rGO/BGNR) caused 100% viability loss for *P. aeruginosa* and *S. aureus* at 1000 and 500 µg/mL, respectively. The nanohybrid biomaterials find applications in the field of bone and dental implants (Ashok raja et al. 2018). The nanohybrids (10–70 nm) made up of cationic rGO intercalated with polyaniline nanofibers (P/rGO^+) caused 74, 79, 95 and 88% reduction in the colony count of *E. coli, Salmonella, S. aureus* and *B. cereus*, respectively at 5% (Mirmohseni, Azizi, and Dorraji 2020). The polyiodide doped starch rGO (SrGO-PI) nanocomposite caused inhibition zones of 22 and 20 mm against *E. coli* and *S. aureus* in well diffusion assay at 2 mg/well (Figure 12.4). The SrGO-PI exhibited MIC and MBC values of 2.5 and 5 mg/mL for both *E. coli* and *S. aureus* strains. At an MBC value of 5 mg/mL, the bacterial lawns of *E. coli* and *S. aureus* grown in petriplates were inhibited entirely, evident from

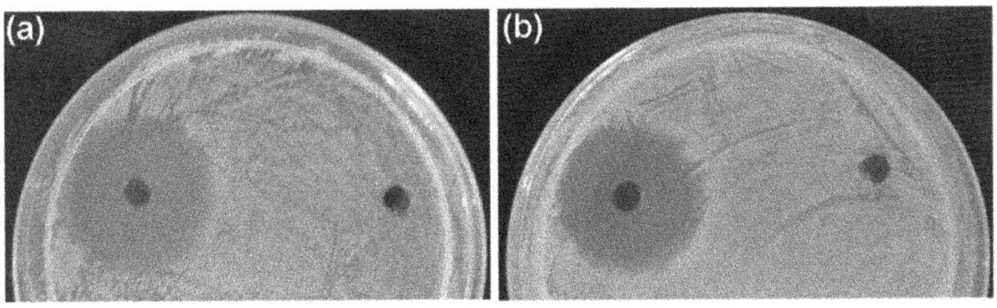

FIGURE 12.4 The inhibition zones of (a) *Escherichia coli* and (b) *Staphylococcus aureus* towards polyiodide doped starch reduced graphene oxide (SrGO-PI) nanocomposite in well diffusion assay at 2 mg/well. Left well: SrGO-PI and right well: SrGO.

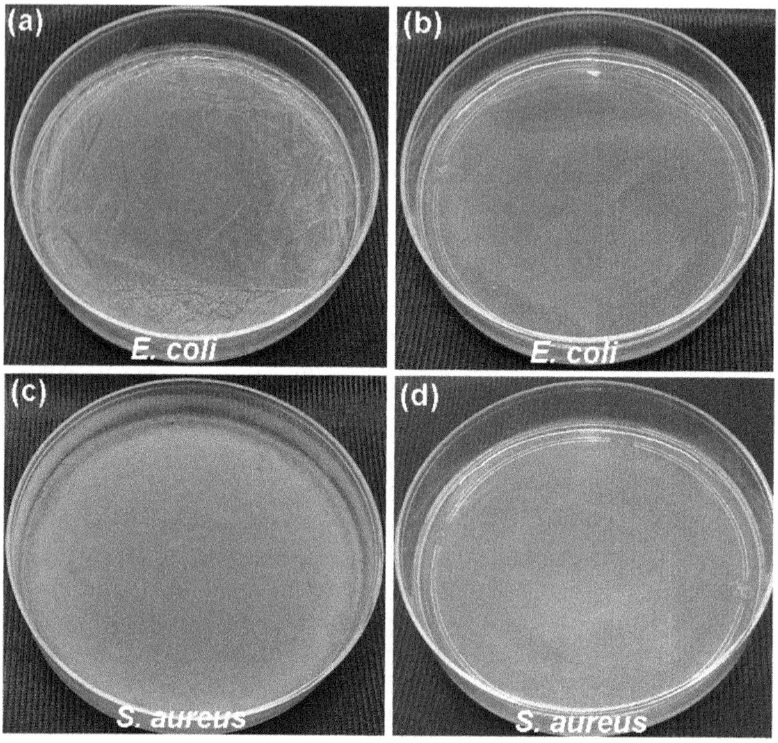

FIGURE 12.5 The inhibition of colony formation by polyiodide doped starch reduced graphene oxide (SrGO-PI) nanocomposite in petriplates, at (a, c) 0 mg/mL and (b, d) 5 mg/mL against *Escherichia coli* and *Staphylococcus aureus.*

no colony formation (Figure 12.5). The IC_{50} value for the corresponding strains was 0.4 mg/mL. Thus, the sustained release of iodine from the nanocomposite can be utilized for food packaging (Narayanan, Park, and Han 2021). The Schiff base assembled on rGO surface (rGO-L2) caused a 67.4 and 84.6% decrease in cell viability of *E. coli* and *S. aureus* at 40 μg/mL, respectively (Omidi, Kakanejadifard, and Azarbani 2017).

12.5 MECHANISM OF ANTIBACTERIAL ACTION

The mode of antibacterial action of structurally intercalated rGO nanocomposite is complex and involves diverse mechanisms. The synergistic antibacterial action of metal NP/rGO nanocomposites is attributed to direct contact with the sharp edges of rGO nanosheets and metal ions release (Ag, Cu, Fe), leading to bacterial cell membrane disruption and ROS generation. The intracellular cytoplasmic contents, including proteins and nucleic acids are leaked outside owing to the loss of membrane integrity. Also, the generated ROS oxidize the various macromolecules of the cell, such as polysaccharides, lipids, proteins and nucleic acids, and hamper their function. The nanocomposite induced oxidative stress and cell membrane damage lead to cell structure distortion and ultimate cell death (Wanag et al. 2018,Song and Shi 2019, Vargas et al. 2019,Naseer et al. 2020, Tan et al. 2020,Zhang et al. 2020, Behera, Barik, and Mohapatra 2021, Sadeghi Rad et al. 2021). In the case of CuO/rGO nanocomposite, contact mediated ROS independent pathway was established in the bacterial inactivation. It is attributed to the direct electron transfer mechanism and increased surface roughness and surface area of the nanocomposite (Alayande, Obaid, and Kim 2020).

12.6 CONCLUSIONS

The biosynthesis of rGO from GO using different biological reductants has gained importance compared to the chemical reduction due to the abundance of low cost, nontoxic, renewable, biodegradable, dual functional biological reductant and stabilizing agents. The rGO served as a template for functionalizing various antibacterial agents and fabricated rGO nanocomposites demonstrated superior antibacterial activity against target bacteria via ROS dependent and ROS independent pathways. The antibacterial rGO nanocomposites are finding applications in water purification, wastewater treatment, anticorrosion coatings, catalyst, photocatalysts, dental implants, drug delivery, photo thermal therapy, cancer treatment, tissue scaffolds, cardiac patch, wound dressings, food packaging, clothing etc.

ACKNOWLEDGMENTS

The author would like to thank Dr M. V. Balarama Krishna, Head, Environmental Science and Nanomaterials Section and Dr Sanjiv Kumar, Head, NCCCM/BARC, for their constant support and encouragement throughout the work.

REFERENCES

Alayande, Abayomi Babatunde, M. Obaid, and In S. Kim. 2020. "Antimicrobial mechanism of reduced graphene oxide-copper oxide (rGO-CuO) nanocomposite films: The case of *Pseudomonas aeruginosa* PAO1."*Materials Science and Engineering* 109:110596. doi: https://doi.org/10.1016/j.msec.2019.110596

Ashok raja, C., S. Balakumar, P. Bargavi, P. Rajashree, B. Anandkumar, R. P. George, and U. Kamachi Mudali. 2018. "Decoration of 1-D nano bioactive glass on reduced graphene oxide sheets: Strategies and *in vitro* bioactivity studies." *Materials Science and Engineering: C* 90:85–94. doi: https://doi.org/10.1016/j.msec.2018.04.040

Behera, Lingaraj, Balaram Barik, and Sasmita Mohapatra. 2021. "Improved photodegradation and antimicrobial activity of hydrothermally synthesized 0.2Ce-TiO_2/RGO under visible light." *Colloids and Surfaces A: Physicochemical and Engineering Aspects* 620:126553. doi: https://doi.org/10.1016/j.colsurfa.2021.126553

Bhattacharya, Proma, Iman Sengupta, Aishee Dey, Sudipto Chakraborty, and Sudarsan Neogi. 2021. "Antibacterial effect of ciprofloxacin loaded reduced graphene oxide nanosheets against *Pseudomonas aeruginosa* strain." *Colloid and Interface Science Communications* 40:100344. doi: https://doi.org/10.1016/j.colcom.2020.100344

Chen, Long, Zhi Li, and Mingguang Chen.2019. "Facile production of silver-reduced graphene oxide nanocomposite with highly effective antibacterial performance." *Journal of Environmental Chemical Engineering* 7 (3):103160. https://doi.org/10.1016/j.jece.2019.103160

Dat, Nguyen Minh, Phung Ngoc Bao Long, Duong Chau Uyen Nhi, Nguyen Nhat Minh, Le Minh Duy, Le Ngoc Quan, Hoang Minh Nam, Mai Thanh Phong, and Nguyen Huu Hieu. 2020. "Synthesis of silver/reduced graphene oxide for antibacterial activity and catalytic reduction of organic dyes." *Synthetic Metals* 260:116260. https://doi.org/10.1016/j.synthmet.2019.116260

Esmaeili, Elaheh, Tarlan Eslami-Arshaghi, Simzar Hosseinzadeh, Elnaz Elahirad, Zahra Jamalpoor, Shadie Hatamie, and Masoud Soleimani. 2020. "The biomedical potential of cellulose acetate/polyurethane nanofibrous mats containing reduced graphene oxide/silver nanocomposites and curcumin: Antimicrobial performance and cutaneous wound healing." *I nternational Journal of Biological Macromolecules* 152:418–427. https://doi.org/10.1016/j.ijbiomac.2020.02.295

Gu, Bin, Qimeng Jiang, Bichong Luo, Chuanfu Liu, Junli Ren, Xiaohui Wang, and Xiaoying Wang. 2021. "A sandwich-like chitosan-based antibacterial nanocomposite film with reduced graphene oxide immobilized silver nanoparticles." *Carbohydrate Polymers* 260:117835. https://doi.org/10.1016/j.carbpol.2021.117835

Huo, Pengwei, Chongyang Liu, Dongyao Wu, Jingru Guan, Jinze Li, Huiqin Wang, Qi Tang, Xiuying Li, Yongsheng Yan, and Shouqi Yuan. 2018. "Fabricated Ag/Ag_2S/reduced graphene oxide composite photocatalysts for enhancing visible light photocatalytic and antibacterial activity." *Journal of Industrial and Engineering Chemistry* 57:125–133. https://doi.org/10.1016/j.jiec.2017.08.015

Ikram, Rabia, Badrul Mohamed Jan, and Waqas Ahmad. 2020. "Advances in synthesis of graphene derivatives using industrial wastes precursors; prospects and challenges." *Journal of Materials Research and Technology* 9 (6):15924–15951. https://doi.org/10.1016/j.jmrt.2020.11.043

Luo, Lan, Lina Xu, and Haibo Zhao. 2017. "Biosynthesis of reduced graphene oxide and its *in-vitro* cytotoxicity against cervical cancer (HeLa) cell lines." *Materials Science and Engineering: C*78:198–202. https://doi.org/10.1016/j.msec.2017.04.031

Maddinedi, Sireesh Babu, Badal Kumar Mandal, Raviraj Vankayala, Poliraju Kalluru, and Sreedhara Reddy Pamanji. 2015. "Bioinspired reduced graphene oxide nanosheets using *Terminalia chebula* seeds extract." *Spectrochimica Acta Part A: Molecular and Biomolecular Spectroscopy* 145:117–124. https://doi.org/10.1016/j.saa.2015.02.037

Maddinedi, Sireesh Babu, Badal Kumar Mandal, Raviraj Vankayala, Poliraju Kalluru, Sai Kumar Tammina, and H. A. Kiran Kumar. 2014. "Casein mediated green synthesis and decoration of reduced graphene oxide." *Spectrochimica Acta Part A: Molecular and Biomolecular Spectroscopy* 126:227–231. https://doi.org/10.1016/j.saa.2014.01.114

Mirmohseni, Abdolreza, Maryam Azizi, and Mir Saeed Seyed Dorraji. 2020. "Cationic graphene oxide nanosheets intercalated with polyaniline nanofibers: A promising candidate for simultaneous anticorrosion, antistatic, and antibacterial applications." *Progress in Organic Coatings* 139:105419. https://doi.org/10.1016/j.porgcoat.2019.105419

Moghayedi, Marjan, Elaheh K. Goharshadi, Kiarash Ghazvini, Hossein Ahmadzadeh, and Majid Namayandeh Jorabchi. 2020. "Antibacterial activity of Ag nanoparticles/phosphomolybdate/reduced graphene oxide nanocomposite: Kinetics and mechanism insights." *Materials Science and Engineering: B* 262:114709. https://doi.org/10.1016/j.mseb.2020.114709

Narayanan, Kannan Badri, Gyu Tae Park, and Sung Soo Han. 2021. "Antibacterial properties of starch-reduced graphene oxide–polyiodide nanocomposite." *Food Chemistry*3 42:128385. https://doi.org/10.1016/j.foodchem.2020.128385

Naseer, Farhan, Erum Zahir, Ekram Y. Danish, Munazza Gull, Syed Noman, and M. Tahir Soomro. 2020. "Superior antibacterial activity of reduced graphene oxide upon decoration with iron oxide nanorods." *Journal of Environmental Chemical Engineering* 8 (5):104424. https://doi.org/10.1016/j.jece.2020.104424

Navya Rani, M., M. Murthy, N. Shyla Shree, S. Ananda, S. Yogesh, and Rangappa Dinesh. 2019. "Cuprous oxide anchored reduced graphene oxide ceramic nanocomposite using *Tagetes erecta* flower extract and evaluation of its antibacterial activity and cytotoxicity." *Ceramics International* 45 (18, Part B): 25020–25026. https://doi.org/10.1016/j.ceramint.2019.04.195

Nguyen, Hang Ngoc, Clemencia Chaves-Lopez, Rodrigo Cardoso Oliveira, Antonello Paparella, and Debora F. Rodrigues. 2019. "Cellular and metabolic approaches to investigate the effects of graphene and graphene oxide in the fungi *Aspergillus flavus* and *Aspergillus niger*." *Carbon* 143:419–429. https://doi.org/10.1016/j.carbon.2018.10.099

Noorunnisa Khanam, P., and Anwarul Hasan. 2019. "Biosynthesis and characterization of graphene by using non-toxic reducing agent from *Allium cepa* extract: Anti-bacterial properties." *International Journal of Biological Macromolecules* 126:151–158. https://doi.org/10.1016/j.ijbiomac.2018.12.213

Norahan, Mohammad Hadi, Mohadeseh Pourmokhtari, Mohammad Reza Saeb, Bita Bakhshi, Mina Soufi Zomorrod, and Nafiseh Baheiraei. 2019. "Electroactive cardiac patch containing reduced graphene oxide with potential antibacterial properties." *Materials Science and Engineering: C* 104:109921. https://doi.org/10.1016/j.msec.2019.109921

Omidi, Sakineh, Ali Kakanejadifard, and Farideh Azarbani. 2017. "Noncovalent functionalization of graphene oxide and reduced graphene oxide with Schiff bases as antibacterial agents." *Journal of Molecular Liquids* 242:812–821. https://doi.org/10.1016/j.molliq.2017.07.074

Pan, Nengyu, Yimin Wei, Mengdi Zuo, Rong Li, Xuehong Ren, and Tung-Shi Huang. 2020. "Antibacterial poly (ε-caprolactone) fibrous membranes filled with reduced graphene oxide-silver." *Colloids and Surfaces A: Physicochemical and Engineering Aspects* 603:125186. https://doi.org/10.1016/j.colsurfa.2020.125186

SadeghiRad, Tannaz, Alireza Khataee, Fatemeh Vafaei, and Shima Rahim Pouran. 2021. "Chromium and cerium co-doped magnetite/reduced graphene oxide nanocomposite as a potent antibacterial agent against *S. aureus*." *Chemosphere* 274:129988. https://doi.org/10.1016/j.chemosphere.2021.129988

Sadhukhan, Sourav, Tapas Kumar Ghosh, Indranil Roy, Dipak Rana, Amartya Bhattacharyya, Rajib Saha, Sanatan Chattopadhyay, Somanjana Khatua, Krishnendu Acharya, and Dipankar Chattopadhyay. 2019. "Green synthesis of cadmium oxide decorated reduced graphene oxide nanocomposites and its electrical and antibacterial properties." *Materials Science and Engineering: C* 99:696–709. https://doi.org/10.1016/j.msec.2019.01.128

Saikia, Indranirekha, Shashanka Sonowal, Mintu Pal, Purna K. Boruah, Manash R. Das, and Chandan Tamuly. 2016. "Biosynthesis of gold decorated reduced graphene oxide and its biological activities." *Materials Letters* 178:239–242. https://doi.org/10.1016/j.matlet.2016.05.011

Shanmuganathan, Rajasree, Gnanasekar Sathishkumar, Kathirvel Brindhadevi, and Arivalagan Pugazhendhi. 2020. "Fabrication of naringenin functionalized-Ag/RGO nanocomposites for potential bactericidal effects." *Journal of Materials Research and Technology* 9 (4):7013–7019. https://doi.org/10.1016/j.jmrt.2020.03.118

Song, Xiaojie, and Xianyang Shi. 2019. "Biosynthesis of Ag/reduced graphene oxide nanocomposites using *Shewanella oneidensis* MR-1 and their antibacterial and catalytic applications." *Applied Surface Science* 491:682–689. https://doi.org/10.1016/j.apsusc.2019.06.154

Tan, Shirui, Xu Wu, Yuqian Xing, Sam Lilak, Min Wu, and Julia Xiaojun Zhao. 2020. "Enhanced synergetic antibacterial activity by a reduce graphene oxide/Ag nanocomposite through the photothermal effect." *Colloids and Surfaces B: Biointerfaces* 185:110616. https://doi.org/10.1016/j.colsurfb.2019.110616

Vargas, Carolina, Raquel Simarro, José Alberto Reina, Luis Fernando Bautista, María Carmen Molina, and Natalia González-Benítez. 2019. "New approach for biological synthesis of reduced graphene oxide." *Biochemical Engineering Journal* 151:107331. https://doi.org/10.1016/j.bej.2019.107331

Veisi, Hojat, Marziyeh Kavian, Malak Hekmati, and Saba Hemmati. 2019. "Biosynthesis of the silver nanoparticles on the graphene oxide's surface using *Pistacia atlantica* leaves extract and its antibacterial activity against some human pathogens." *Polyhedron* 161:338–345. https://doi.org/10.1016/j.poly.2019.01.034

Wanag, Agnieszka, Paulina Rokicka, Ewelina Kusiak-Nejman, Joanna Kapica-Kozar, Rafał J. Wrobel, Agata Markowska-Szczupak, and Antoni W. Morawski. 2018. "Antibacterial properties of TiO_2 modified with reduced graphene oxide." *Ecotoxicology and Environmental Safety* 147:788–793. https://doi.org/10.1016/j.ecoenv.2017.09.039

Wang, Kaixi, Yong Liu, Xiaoying Jin, and Zuliang Chen. 2019. "Characterization of iron nanoparticles/reduced graphene oxide composites synthesized by one step *eucalyptus* leaf extract." *Environmental Pollution* 250:8–13. https://doi.org/10.1016/j.envpol.2019.04.002

Xing, Fu-Yan, Lin-Lin Guan, Yan-Long Li, and Chun-Juan Jia. 2016. "Biosynthesis of reduced graphene oxide nanosheets and their *in vitro* cytotoxicity against cardiac cell lines of *Catla catla*." *Environmental Toxicology and Pharmacology* 48:110–115. https://doi.org/10.1016/j.etap.2016.09.022

Xu, Li Qun, Yan Biao Liao, Ning Li, Yi Jian Li, Jie Yu Zhang, Yun Bing Wang, Xue Feng Hu, and Chang Ming Li. 2018. "Vancomycin-assisted green synthesis of reduced graphene oxide for antimicrobial applications." *Journal of Colloid and Interface Science* 514:733–739. https://doi.org/10.1016/j.jcis.2018.01.014

Xu, Xiaoyue, Xiaoyu Tang, Xiaoxu Wu, and Xiufang Feng. 2019. "Biosynthesis of sorafenib coated graphene nanosheets for the treatment of gastric cancer in patients in nursing care." *Journal of Photochemistry and Photobiology B: Biology* 191:1–5. https://doi.org/10.1016/j.jphotobiol.2018.11.013

Zhang, Chao, Xinru Wang, Suna Fan, Ping Lan, Chengbo Cao, and Yaopeng Zhang. 2021. "Silk fibroin/reduced graphene oxide composite mats with enhanced mechanical properties and conductivity for tissue engineering." *Colloids and Surfaces B: Biointerfaces* 197:111444. https://doi.org/10.1016/j.colsurfb.2020.111444

Zhang, Chao, Xinru Wang, Aihui Liu, Changjiang Pan, Hongyan Ding, and Wei Ye. 2021. "Reduced graphene oxide/titanium dioxide hybrid nanofiller-reinforced electrospun silk fibroin scaffolds for tissue engineering." *Materials Letters* 291:129563. https://doi.org/10.1016/j.matlet.2021.129563

Zhang, Jingxiang, Shengqian Zhu, Kunkun Song, Zhuoyue Wang, Zongpu Han, Keren Zhao, Zengjie Fan, Xi Zhou, and Qiangqiang Zhang. 2020. "3D reduced graphene oxide hybrid nano-copper scaffolds with a high antibacterial performance." *Materials Letters* 267:127527. https://doi.org/10.1016/j.matlet.2020.127527

Zhao, Xingyu, Minmin Chen, Hualin Wang, Li Xia, Min Guo, Suwei Jiang, Qian Wang, Xingjiang Li, and Xuefei Yang. 2020. "Synergistic antibacterial activity of streptomycin sulfate loaded PEG-MoS_2/rGO nanoflakes assisted with near-infrared." *Materials Science and Engineering: C* 116:111221. https://doi.org/10.1016/j.msec.2020.111221

Zheng, Yuhong, Aiwu Wang, Wen Cai, Zhong Wang, Feng Peng, Zhong Liu, and Li Fu. 2016. "Hydrothermal preparation of reduced graphene oxide–silver nanocomposite using *Plectranthus amboinicus* leaf extract and its electrochemical performance." *Enzyme and Microbial Technology* 95:112–117. https://doi.org/10.1016/j.enzmictec.2016.05.010

13 Graphene and Its Composite for Supercapacitor Applications

Priyadharshini Madheswaran and Pazhanivel Thangavelu

CONTENTS

13.1 INTRODUCTION

The ever-growing need of energy coupled with power inadequacy has initiated advanced research on renewable energy sources and its efficient storage. In addition to it, the quick development of business sectors, like portable electronic gadgets, has also made a steady increasing interest for an eco-friendly, superior energy storage device. Supercapacitors are such systems that comprise high power density, high cycling stability (>100 000 cycles) and quick mobility of charged particles[1, 2]. Nowadays, they are used in electronic gadgets, memory backup, etc. A more ongoing use of hybrid vehicles has exhibited its protected and authentic function. Its energy per unit area has considerably been enhanced compared to traditional electrostatic capacitors, the energy of which is still lesser than that of batteries[3, 4]. A large portion of the commercialized supercapacitors has particular energy density under 10 Wh/kg that was three to multiple times lesser than that of batteries. Thus, there exists a great need to explore more enhanced and efficient energy storage devices than that of existing commercialized products[5, 6].

An ultracapacitor is a device which can store charges via two types of mechanisms such as (i) electric double layer capacitance (EDLC) and (ii) pseudocapacitance. Based on the material, either one or both the types of mechanisms were used to store charges. The commonly used electrode materials for the EDLC type are carbon-based materials like activated carbon, graphene, carbon nanotubes [7]. On the other hand, pseudocapacitance electrodes make use of metal oxides, sulphides and conducting polymers. However, the electrode materials still cannot meet the requirement of day-to-day applications. Hence, hybrid electrode materials comprising both the EDLC and redox behaviour with rational optimization and design are expected to meet the requirements of modern society[7, 8].

On the consideration of stated discussion, graphite was regarded as a potential candidate because of its superior electrical and mechanical properties and unique morphology. It is a two-dimensional (2D) single-layered carbon bond with sp^2 hybridization, which raises each and

DOI: 10.1201/9781003169741-13

every graphitic carbons like graphene[9, 10]. Due to its vast applications in different fields, it was also exploited in supercapacitors where its composition with other compounds yielded a better electrochemical performance which was studied here in detail.

13.2 SUPERCAPACITOR PRINCIPLES AND CHARACTERISTICS

As discussed, a supercapacitor comprises two types of mechanisms. In the EDLC type, charges are stored via electrostatic attraction, while in pseudocapacitance, charges are stored via reversible redox mechanism. As same as battery, a supercapacitor consists of two electrodes namely anode and cathode in contact with an electrolytic solution with a separator in between. All the components of supercapacitors contribute to the total specific capacitance of a device. Hence, much attention has been paid to designing an efficient supercapacitor. The ability of the device is majorly characterized on basis of the standards stated in the following: (i) power density considerably larger than commercial energy storage devices along considerably prominent energy per unit area (>10 Wh/kg), (ii) stable cyclability (>10^2 of batteries), (iii) rapid charge-discharge kinetics (< a minute), (iv) lesser self-discharging, (v) reliability and (vi) cost-effective. Also, the period demanded for the current or voltage in a circuit to increase or decrease exponentially by nearly 63% of its amplitude shows that capacitance was an integral multiple of resistance which was the other significant parameter in assessing operation from an ultracapacitor[6, 11, 12].

The anode and cathode were fabricated from a mixture of prepared materials, binder PVdF, activated carbon in 75:5:15 ratio, respectively. Then the mixture was grained using N-methyl pyrrolidone (NMP) solvent until homogeneous paste was formed. The paste was coated on cleaned nickel foam (1 × 1 cm^2) and dried in an oven at 80°C for 24 h. The as-prepared electrode was analysed for electrochemical performance in a 1-M KOH electrolyte solution with a three-electrode system. The synthesized material loaded current collector, Hg/HgO and platinum sheet are applied as working, reference and counter electrodes. The specific capacitance was estimated in charge-discharge measurements using the following relation[7, 9]:

$$\mathbf{C} = \frac{\mathbf{I}\ \Delta \mathbf{t}}{\mathbf{m}\Delta \mathbf{v}}\frac{\mathbf{F}}{\mathbf{g}} \tag{13.1}$$

where 'I' corresponds to current intensity, Δt represents time difference, m was the mass of synthesized sample, and Δv stands for voltage. Hence, the mass ratio of working electrodes was set by a charge-balance equation ($q_+ = q_-$). Also, the following equation represents mass balancing:

$$\frac{\mathbf{m}_+}{\mathbf{m}_-} = \frac{\mathbf{C}_{s-} \times \Delta \mathbf{v}_-}{\mathbf{C}_{s+} \times \Delta \mathbf{v}_+} \tag{13.2}$$

where m, C (F/g) and Δv are the active mass (g), capacitance and potential window, respectively. By using the following equation, specific capacitance (F/g), power density (W/kg) and energy density (Wh/kg) were calculated:

$$\mathbf{E} = \frac{\mathbf{C} \times \Delta \mathbf{v}}{7.2}\frac{\mathbf{Wh}}{\mathbf{kg}} \tag{13.3}$$

$$\mathbf{P} = \frac{3600 \times \mathbf{E}}{\Delta \mathbf{t}}\frac{\mathbf{W}}{\mathbf{kg}} \tag{13.4}$$

where I, Δt, Δv and m were discharge current (mA), charging or discharging time (s), potential and the mass of two electrodes, respectively.

13.3 OTHER GRAPHITIC CARBON

Graphite was a characteristic form of carbon element for almost 50 decades and was broadly used for daily chores, like oil paints lubricants to charcoal pastels. It can excellently transfer thermal energy accompanied by less constrain to electrical energy mobility, which made it an efficient anode material. Furthermore, between the graphite layers, different components may be inserted that may create graphite-based intercalated materials. Moreover, the replacement of carbon in lattice forms of different components, for example, boron in carbon lattice, creates p-type materials. Enlarging the interlayer spacing of graphite was depicted through ionic radicals. Through polarizing anodes or cathodes, ions can be inserted in between the layers of graphite by which the accessibility of electrolyte was enhanced. Asymmetric capacitors comprising different types of electrode materials could combine the properties that facilitate higher efficiency to the system where the synergistic effect of composite materials boosted its overall performance.

Fullerenes with its derived materials are the other graphite elements, regarded as a potential candidate for electrodes in a supercapacitor because of its exceptional dimension that combines faradic capacity attributes along a large surface area. Theoretic forecasts demonstrated that least vacant sub-atomic orbital of fullerene fits tolerance in any case of $6e^-$ decrease. The electrochemical properties of pristine fullerenes are firmly subject to solute–dissolvable interactions[13].

Carbon onions, comprising several graphene circular layers, possess a 350- to 520-m^2/g surface accompanying some nanopores. The available large surface to ionic solution shows these materials as a great possibility of having rich power density to energy storage devices. The recent galvanostatic study on it, at high current density, disclosed the reduction of explicit capacity because of molecular graphiting process and deformity development upon molecule outer boundary. Capacitance of this kind of stuff ranges from 20 to 40 F/g in a 1-M of H_2SO_4 electrolytic arrangement and from 70 to 100 F/g in a 6-M KOH electrolytic arrangement[14–16].

Regardless of the enhanced attributes and ionic transport capacities of listed allotropes of carbon, it was scarcely utilized as an anode and a cathode, for the most part because of lesser capacitance. In carbon allotropes like graphene, it was unclear how ionic particles of solution could enter layered morphology of it, and unique contributions from basal planes and edges to all-out capacity was unknown. Hence, some hypothetical and major examinations during the process of energy storage in this stuff ought to be done in the prospect of progressing energy storage device betterment.

13.4 HETEROATOM-DOPED GRAPHENE

The heteroatom atom-doped graphene involved both EDLCs and pseudocapacitive behaviour and it exhibits electrochemical properties. The efficient faradic attributes manipulation was done by the addition of impurity atoms through the introduction of hetero atoms (nitrogen, sulphur, boron and phosphorous) into layered sp^2 hybridized graphene framework. The activation area was generated via a dopant atom. Electro activity of the surface through offering electronegative site favourable to oxygen fixation plus encouraging the redox process suitable for electro catalytic applications determines the spin density and charge distribution of neighbouring atoms. The generations of additional psuedocapacitance in junction boost its applications as electrodes in energy devices. Activation intensifies the native EDLC of graphene[17]. The increasing conductivity and the pseudocapacitance as well as electro catalytic activity play an important role in the heteroatom-doped graphene for supercapacitor applications. Nitrogen-doped graphene (N-Gr) exhibited that three configurations, namely pyridinic-N, pyrrolic-N and quaternary-N, showed excellent attributes for faradic process. Comparing boron and sulphur-doped graphene the latter could be a better positively charged accelerator to oxygen reduction reaction (ORR) for alkaline medium. The most potent dynamic area on a graphene surface to more beneficial electro catalytic with electro capacitive activity is nitrogen/sulphur. The heteroatom-doped or co-doped graphene required adhering candidates like Nafion to fabricate anode or cathode, which results in rendering powdered final products[18]. Heteroatom-doped graphene has been synthesized by two different methods such as direct synthesis and post-treatment, where the direct synthesis consists of a solvothermal method, chemical vapour deposition (CVD) and arc discharge method and segregation growth. From these, the CVD method is mostly used in direct synthesis. The post-treatment method includes thermal treatment, mechanical exfoliation, hydrothermal method, anodic polarization and plasma process. Compared to direct synthesis, the post-treatment methods are mostly used for the preparation of heteroatom-doped graphene[19]. For instance, Liming Xu[20] successfully prepared the indole-functionalized N-doped-reduced graphene (INFGN) for energy storage applications by vacuum freezing, drying process and hydrothermal method. The pore structure, element composition, thickness and microstructure and electrochemical performance were analysed by Brunauer-Emmett-Teller (BET), atomic force microscopy (AFM), transmission electron microscopy (TEM), scanning electron microscopy (SEM) and electrochemical technologies. The surface morphology of INFGN shows the wrinkle-like structure, and no obvious stacking was found using BET analysis, the value of specific surface area was identified as 135 m^2/g for INFGN. The electrochemical properties of INFGN were analysed by three electrode system. The cyclic voltammetry (CV) can be recorded between the scan rate of −0.2 and 1.04. The quasi-rectangular shape with redox peaks is obtained in the CV curves. The galvanostatic charge discharge (GCD) curves show triangle and symmetrical shape and confirms the electrode are psuedocapacitance behaviour. The electrode of INFGN exhibits the good specific capacitance as 622.3 F/g at 2 A/g and high energy density (21 Wh/kg in 800 W/kg) as the electrolyte of 1-M H_2SO_4. It exhibits excellent cyclic stability 100.5% capacitance retention over 5000 cycles. The INFGN electrodes have good electrochemical properties and used for the supercapacitor applications. The electrochemical behaviour of INFGN electrodes is shown in Figure 13.1.

Taslima Akhter et al.[14] described the N/S incorporated with flexible graphene for supercapacitor application through thermal treatment. The works described the nitrogen and sulphur incorporated with graphene exhibit good specific capacitance of 305 F/g at 100 mV/s and energy density as 28.44 Wh/kg, which also delivers excellent cyclic stability 95.4% capacitance retention over 10 000 cycles. The fabricated electrode has been used for the energy storage application. Followed by S. Suresh Balaji et al.[21] have been successfully synthesized the S-doped graphene (SGO) for energy storage application via supercritical fluid (SCF) processing. The electrochemical analysis of sulphur-doped graphene exhibits specific capacitance of 261 F/g at 1 A/g. The cyclic stability of SGO 90% over 10 000 constant charge discharge in addition possesses better energy density (39 Wh/kg) as the electrolyte of 20% of KOH solution. The prepared SGO electrode is promising material for supercapacitor application.

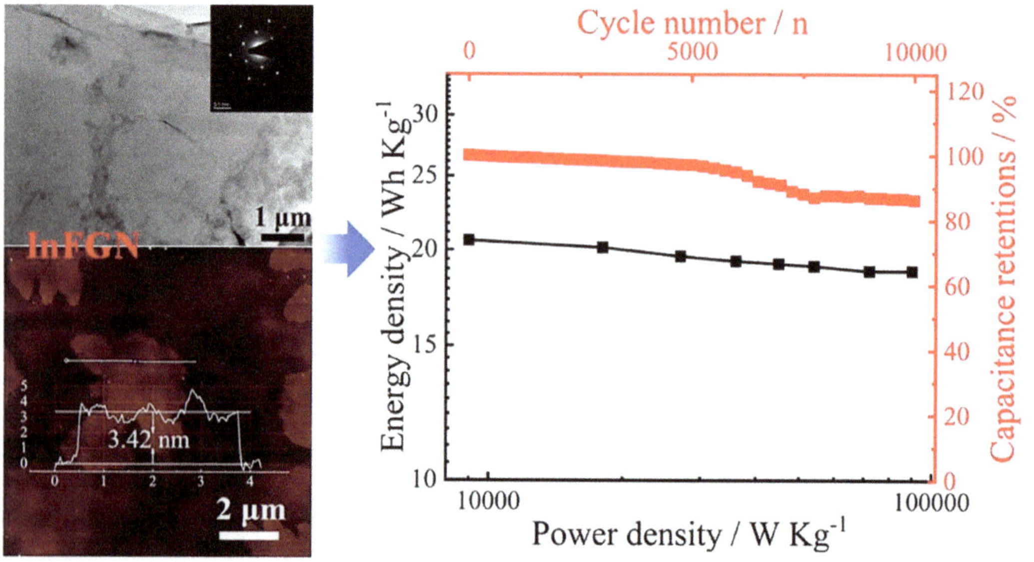

FIGURE 13.1 The overall performance of SSC based on InFGN electrodes with a loading of 10.12 mg/cm^2.

13.5 GRAPHENE – CONDUCTIVE POLYMER HYBRID

Nowadays, the world would face the problem of efficient energy storage device. Here, in various types of materials performed as electrode for supercapacitor, such as conducting polymers, carbon-based materials, metal oxides. The conducting polymers show the pseudocapacitors behaviour and this kind of material have high specific capacitance, large surface area, high energy density and low cyclic stability[19, 22–24]. Graphene is also included in the materials for supercapacitor electrodes due to excellent electrochemical performances, large surface area, superior electrical conductivity and possess mechanical properties. The polymers are widely used in electrode material for supercapacitor and provide specific capacitance, but it exhibits poor cycle stability due to shrinkage and swelling polymers during charge/discharge process. This problem can be solved by polymer composite with graphene. An optimized ratio of graphene composite with conducting polymers will increases the electrochemical performances of the device. The types of conducting polymers used in energy storage devices were polypyrrole (PPy), polyaniline, PEDOT:PSS. Recently, hybrid nanocomposite of conducting polymer/graphene has been most attracted application of sensors, memory devices, catalysis and mostly used for energy storage devices[25, 26]. In polymer/graphene nanocomposite has both psuedocapacitance and EDLC behaviour. The graphene enhanced the electrical conductivity and mechanical stability of polymers. The electrons will transfer in conductivity polymers to graphene during redox reaction with high electrical conductivity. Graphene/polymer nanocomposite has been synthesized through different techniques like mechanical cleavage, ultrasound-induced liquid phase exfoliation (ULIPE), reduction of graphene oxide, CVD and electrochemical and exfoliation. The nanocomposite of graphene/polymer was synthesized by either covalent or noncovalent approaches, by interaction between two components. The prepared nanoparticles are carried out by either a dry sample or liquid media. The graphene synthesized by noncovalent approaches is prepared by physical adsorption via electrostatic or hydrophobic interaction. Similarly, the graphene synthesized by covalent approaches is prepared by chemical reaction[27, 28]. Compared to noncovalent functionalization, covalent approaches are more stable. The hybrid nanocomposite of graphene/polymers is the excellent parameter of supercapacitors application because of high surface and good physiochemical properties. Among different polymers, polypyrrole (PPy) and polyaniline (PANI) were mostly used as composite with graphene as electrodes for supercapacitors. Figure 13.2 depicts the fabrication of symmetric and asymmetric supercapacitors (ASC) using prepared electrodes.

For instance, P. Muhammed Shafi et al.[29] successfully described the $LaMnO_3$/reduced graphene oxide (RGO)/ PANI ternary nanocomposites for high-energy supercapacitor applications. The

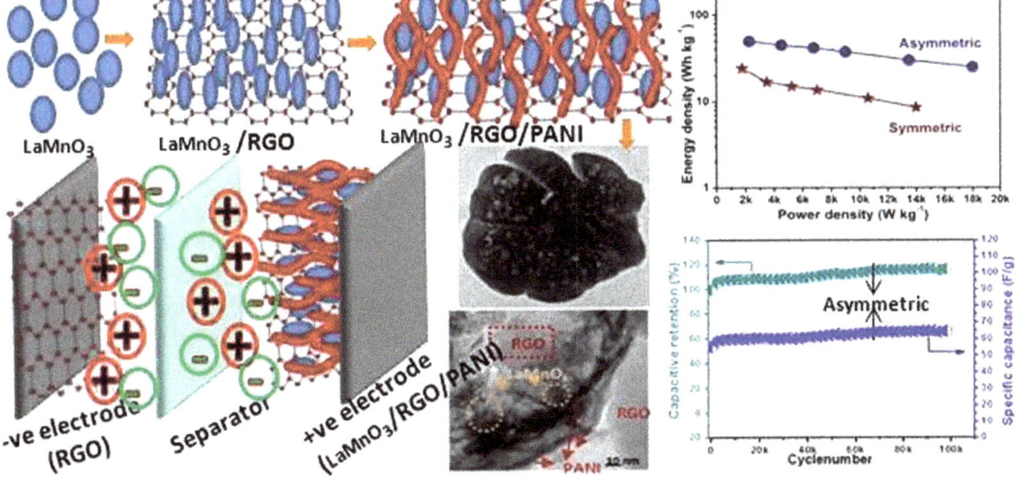

FIGURE 13.2 Fabrication of symmetric and asymmetric supercapacitors of prepared electrodes.

ternary nanocomposite electrode was fabricated by the polymerization route with high-energy density. The structure and internal morphology have been analysed by TEM and further closer inspection was carried out by high resolution transmission electron microscope (HRTEM). The TEM images represent the particle size of prepared composite materials ranging between 20 and 30 nm. The composite material of RGO and PANI is confirmed by HRTEM analysis, which shows an interconnected structure. In this work, they fabricated both symmetric and ASCs as $LaMnO_3$/RGO/polyaniline/$LaMnO_3$/RGO/polyaniline and $LaMnO_3$/RGO/polyaniline/RGO, respectively. However, the composite electrodes have both positive and negative spans and render a large loop region, which represents EDLC behaviour. During a large loop area, the composite electrodes have large specific capacitance. Finally, they concluded that an asymmetric capacitor has the excellent specific capacitance of 111 F/g at 2.5 A/g compared to a symmetric capacitor and also it delivers a maximum energy density of 50 Wh/kg at 2.25K W/kg of power density. Further, ASCs can excel the cyclic stability retention of 117% over 100-K cycles. Following that, Zubair Ahmad et al.[30] described the nanocomposite of graphene oxide/polypyrrole for an energy storage device. Initially, the electrochemical analysis has been carried out by a three-electrode system, using H_2SO_4 electrolytes. Applying BET analysis, the high surface area (61 m^2/g) was calculated, and from the analysis the electrodes exhibit an excellent specific capacitance of 566 F/g at a scan rate of 1 A/g. The outstanding cyclic stability exhibited the electrode as capacity retention at 97% even after 2000 cycles. Although it performs a good electrochemical behaviour, there are still drawbacks that forbid its commercialization. The main challenge is to determine a possible way for cheap bulk yield of graphene/polymer nanocomposites, neither compromising the structures of graphene due to the restacking or aggregation nor its performance. Rational design of a porous structure should also be enhanced instead of bulk collapse and aggregation of composites. In addition to it, despite having great potential, its integration with other electronic devices (e.g., batteries, supercapacitors) stays an objection for practical application.

13.6 GRAPHENE METAL OXIDE HYBRID

The pristine form of graphene is not suitable for charge storage process because of poor electrochemical performance. But the metal oxides (MnO_2, Fe_3O_4, RuO_2, NiO, etc.) have excellent electrochemical properties and are tremendously used in energy storage devices. In transition metal oxide, the electrocatalytic properties are low due to their high electrical resistivity. From these results, it spontaneously affects the rate of capability in charge storage mechanism. To overcome this result, two tactics have been used to improve their electrochemical performances, such as functionalization of electrode (or) fabrication of graphene/metal oxide nanocomposites[24, 31–33]. In comparison of two tactics, the graphene/metal oxide nanocomposites have many advantages like rich energy density, excellent specific capacitance, and a combination of ionic and electronic conductivities. Various methods have been used to prepare the nanocomposite of graphene/metal oxide, such as (i) hydrothermal and solvothermal methods, (ii) electrochemical methods and (iii) CVD method. For instance, Yongmin He et al.[34] prepared the graphene/MnO_2 nanocomposite for energy storage application. He prepared the 3D graphene from nickel foam, and it has flexibility and mechanical strength. Further, MnO_2 has been loaded on the prepared graphene. It exhibits the excellent specific capacitance (130 F/g), large specific area (392 m^2/g) and good cycling performance. Moreover, this type of nanocomposite materials makes the promising electrode for flexible supercapacitors and so on. Further, Xiaonig Tian et al.[35] also described the synthesis of Co_3O_4/N-RGO at 550°C by annealing process. It has a good specific capacitance at 355 F/g at 1 A/g and excellent cycling stability over >90% capacity after 3000 cycles. The interaction between the prepared graphene nanocomposites forms the smallest crystalline size, and due to this, it forms the large pore size and specific surface and makes stable energy storage. The Co_3O_4/N-RGO nanocomposites have better electrochemical performance than the other Co_3O_4 composites. Further, Zhaoling Ma et al.[36] successfully prepared the N-Gr with Fe_2O_3, by one pot facile hydrothermal method. He described the 6.7% of nitrogen incorporated in graphene and it helps to increase the specific capacitance of graphene (267 F/g). The Fe_2O_3 in N-RGO played an important key to increase the electrochemical capacitance and electrical conductivity. The prepared Fe_2O_3/N-RGO is used as promising materials for energy storage applications. Xion-Chen Dong et al.[37] described the cobalt oxide (Co_3O_4) incorporated in 3D graphene nanocomposites, which were developed by two steps (Co_3O_4 nanowires by hydrothermal synthesis and 3D graphene grown by CVD). In Co_3O_4, nanoparticles have light crystallinity, and a uniform diameter forms the nanoparticle 3D graphene skeleton. 3D graphene/Co_3O_4 nanocomposites provide good electrocatalytic and electrochemical properties. The prepared electrode was used for supercapacitor applications, and it exhibits excellent specific capacitance (1100 F/g at 10 A/g) and cycling stability. Madury Chandel et al.[38] successfully prepared Ag and Ni nanoparticles embedded on the graphene oxide (RGO). The nanocomposite of $(Ag_xNi_{(1-x)})$RGO has the hierarchical structure and provides the enhanced electrochemical performances and good cycling stability. The prepared electrodes of $(Ag_xNi_{(1-x)})$RGO show good specific capacitance (897 F/g at 1 A/g) and energy density (80 Wh/kg at 400 W/kg). This nanocomposite is considered the promising electrode material for supercapacitor applications. Furthermore, Amrita De Adhikari et al.[15] described the Cds-CoFe-G nanocomposite for energy storage devices synthesized by chemical process. The hierarchical morphology was analysed by field emission scanning electron microscope (FESEM) and TEM, and it exhibits the excellent specific capacitance (1487 F/g at 5 A/g) of a large surface area due to its ions' diffusions and electron transportation. Further, the nanocomposite materials that have been cycled reversibly in the potential window of 0–1.6 V provide good energy density (528.8 Wh/kg) and high power density (4000 W/kg). Following that, Arpankumar Nayak et al.[39] synthesized the graphene/WO_3 nanocomposite by facile and green solvothermal process. The active electrode material of prepared graphene/WO_3 delivers excellent specific capacitance (465 F/g at 1 A/g) in the electrolyte of 0.1-M H_2SO_4 and a superior cycling stability of 98% after 2000 constant charge discharge. The device also exhibits high energy density (26.7 Wh/kg at 6K W/kg power density). The excellent specific capacitance and good energy density are obtained by the nanocomposite of graphene/WO_3 and it enhanced the promising electrode material for highly supercapacitor applications and Figure 13.3 shows the fabrication of a graphene/WO_3 electrode device.

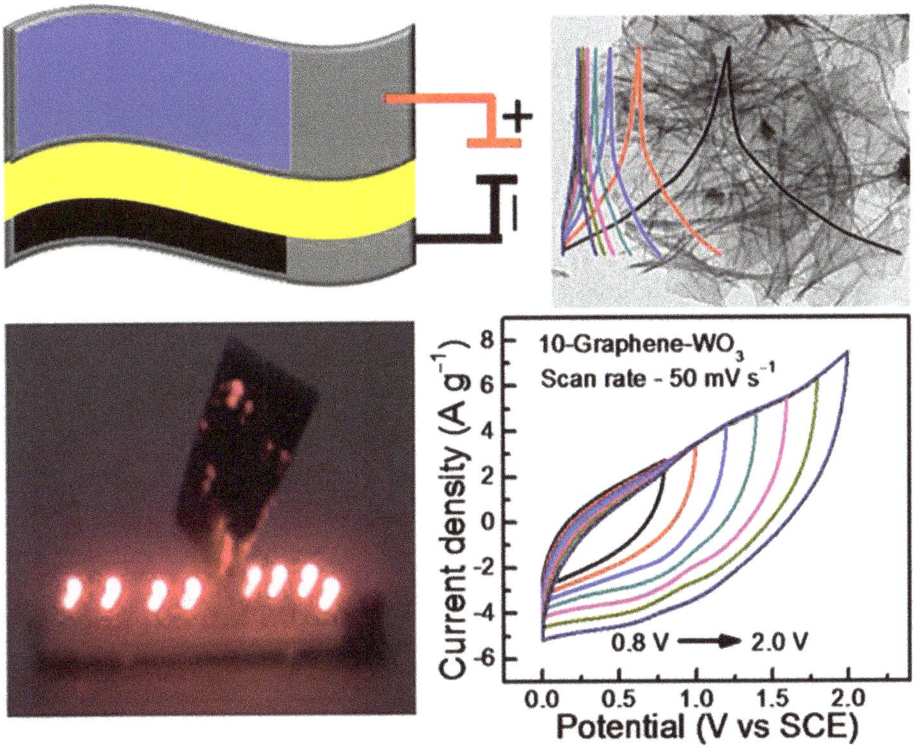

FIGURE 13.3 Graphene/WO_3 electrode device with its CV curve at 50 mV/s.

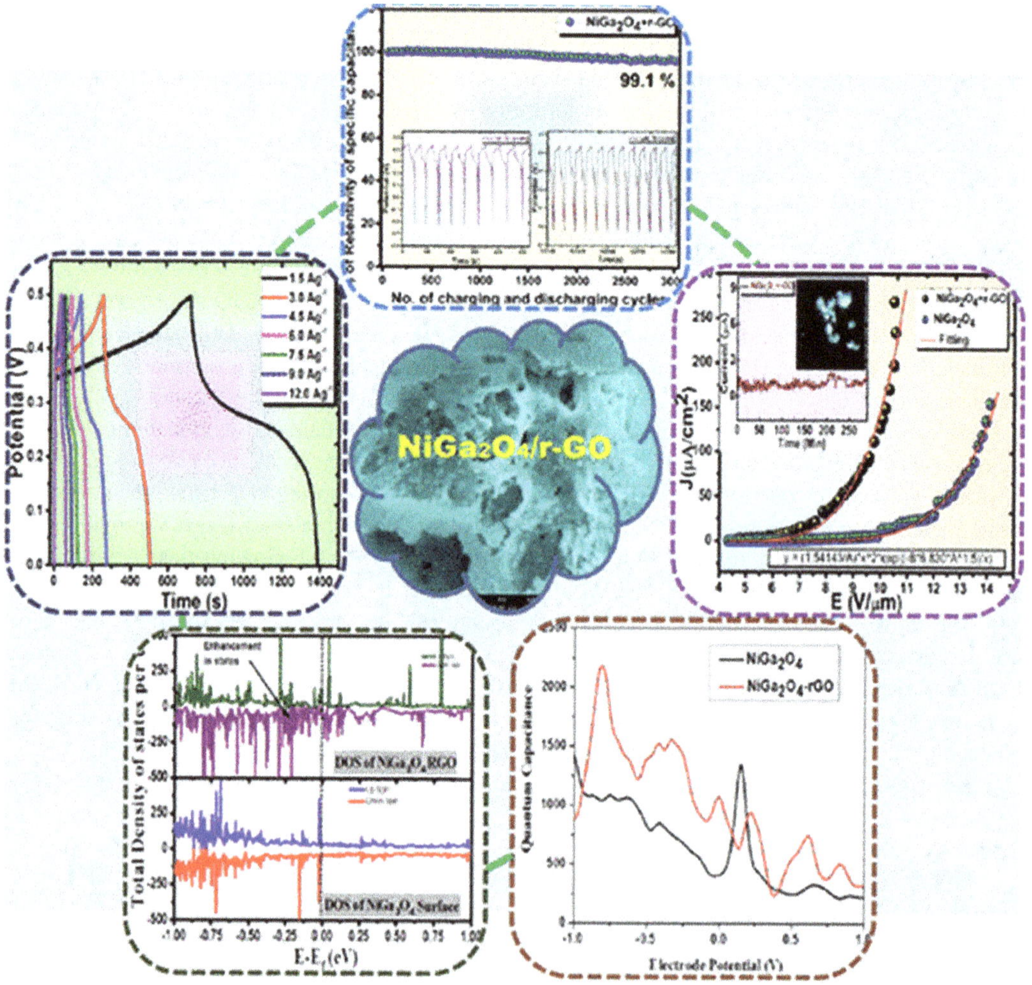

FIGURE 13.4 Electrochemical performance of RGO/$NiGa_2O_4$.

Recently, Subrata Karmakar et al.[40] successfully prepared the RGO reinforced $NiGa_2O_4$ through sol-gel techniques for high supercapacitor performance. From the analysis of XRD, FTIR and Raman deliver the huge cubical parameters for the Ga grown on NiO. The prepared electrodes of $NiGa_2O_4$ exhibit a good specific capacitance of 415 F/g along a cyclic stability of 96% throughout 3000 constant charge discharge. From the other hand, RGO/$NiGa_2O_4$ nanocomposite possesses an excellent specific capacitance of 643 F/g at the same current density along a cyclic stability of 99% throughout 3000 constant charge discharge. These results show that the graphene metal oxide nanocomposite is better than the transition metal oxides, and it suggested that electrodes are used in the highly supercapacitor applications. Figure 13.4 shows the electrochemical performance of RGO/$NiGa_2O_4$.

13.7 GRAPHENE – METAL SULPHIDE HYBRID

In recent time, transition metal oxide is mostly used as an electrode material because of its availability and affordability. Despite its advantages, transition metal oxide has poor electrical conductivity, e.g., MnO_2 (10^{-5}–10^{-6} S/cm). Compared to metal oxide, metal sulphide has better electrical and metallic conductivity. Recently, carbon materials with composites are widely used in the energy storage devices. The carbon-based materials can perform EDLC behaviour. From other carbon materials, RGO had superior physiochemical attributes, including large active region, good energy density, great chemical stability. Therefore, metal sulphide coated on RGO at this combo of nano-composites could meet the requirement of a high performance energy storage device[15, 16, 41]. For instance, Ka-Jing Huang et al.[42] successfully synthesized the RGO embedded on copper sulphide (CuS) by a facile solvothermal method. The morphology structures were analysed by SEM that obtained uniform hallow spheres with a 250-nm diameter. The electrochemical analysis of CV, GCD and electrochemical impedance spectroscopy (EIS) was carried out by three-electrode system using the electrolyte of 6-M KOH. The composite of CuS-RGO confirms that the electrodes are a pseudocapacitor device. CV can be analysed between the scan rate of 2 and 50 mV/s. Further, GCD was used to analyse the specific capacitance of prepared electrodes and recorded between the potential ranges of 0.2 and 0.4 V in the current density. The GCD current shows a pseudocapacitor device, which is well agreement with curves. In GCD, the specific capacitances of CuS and CuS/RGO are determined 917.6 and 2317.8 F/g, respectively, at a current density of 1 A/g. Finally, obtained copper sulphide with grapheme (CuS-RGO) is better than the CuS for high supercapacitor applications. Further, Wei Liu et al.[16] described the transition metal sulphide embedded on graphene sheet for an energy storage device. The transition metal sulphides, such as MoS_2, FeS_2 and CuS_2, are mostly used for the anode materials for electrochemical performances due to good chemical conductivity, low electronegativity and excellent specific capacitance. Here, $CoFeS_2$ was embedded on grapheme nanosheets under hydrothermal condition. The internal morphology and microstructure were analysed by SEM, as shown in Figure 13.5. It clearly shows that a polyhedral structure of FeS_2 is about 1–1.5 μm in size. After the Co species are incorporated on the FeS_2 particles, the polyhedral size decreases (200–300 nm in size) compared to FeS_2 because ions exchange between Co^{2+} ions and FeS_2 nuclei (Figure 13.5(b)). Figure 13.5(c) shows FeS_2/graphene. The composite

FIGURE 13.5 SEM images of (a) pure FeS_2, (b) $CoFeS_2$, (c) graphene/FeS_2 and (d) $CoFeS_2$/graphene nanocomposites.

materials have huge dispersed irregular particles with 80–120 nm in size and grow uniformly on the graphene which forms a sandwitched structure. After Co species are introduced on the composite material (FeS_2/graphene), the $CoFeS_2$/graphene shows a sandwiched structure that becomes smaller (30–60 nm in size) than pure FeS_2 and FeS_2/graphene. Meanwhile, $CoFeS_2$/graphene has a large specific surface due to the decrease of particle size and this nanocomposite is more suitable for electrochemical application. The $CoFeS_2$/graphene nanocomposite exhibits a capacitance of 310 C/g, while the potential ranges between 2 mV/s and a good cyclic stability 62% and 200 mV/s in the aqueous electrolyte of 3-M KOH. In fabricated ASC, delivery provides good energy density (69 Wh/kg at 300 W/kg of power density), which is highly better than the other transition metal sulphides. These electrode materials obtained a long cyclic stability of capacitance retailing 97% over 3000 cycles. From these results, $CoFeS_2$/graphene shows excellent electrochemical performance.

Further, Kaiyang Zhang et al.[43] successfully prepared cobalt nickel sulphides ($CoNiS_2$) on RGO by one-step pyrolysis method. The metal sulphides are promising electrodes for energy storage, but they combine with RGO to improve their capacitance behaviour. Herein ($CoNiS_2$/RGO) nanocomposite has good specific capacitance 1526 F/g at 2.4 A/g. Also, the electrode exhibits more active sites and diffusion in electrode ions, respectively. In ASC, $CoNiS_2$/RGO acts as a positive electrode and AC as negative electrode and delivers a high energy density of 55 Wh/kg at 798 W/kg of power density. Following that, Rajendran Ramachandran et al.[41] derived the zinc sulphide (ZnS) decorated on graphene (ZnS/G) nanocomposites for an energy storage device by a solvothermal method. The morphology of the prepared electrodes was identified by TEM and HRTEM. It has spherical particles (ZnS) with a good agreement with the XRD pattern of ZnS/G nanocomposite. The electrochemical measurements were analysed by electrochemical process by using three-electrode system in an electrolyte solution of 6-M KOH. From CV, the specific capacitance of prepared electrodes (ZnS/G) is derived as 197 F/g at the scan rate of 5 mV/s. It shows 94% efficiency following 1000 constant charge discharge process. The GCD and EIS show high reversible and a conductivity of ZnS/G nanocomposite and suggest that it is more suitable for supercapacitor applications.

13.8 CONCLUSION

Overall literature reports that graphene was established as a potential candidate for electrode materials in supercapacitors. Regardless of producing vast articles on graphene and its oxide, an eco-friendly and low budget strategy for the growth of graphene is still in high demand. Governance across quality with an amount of outcome on blending graphene needs a total comprehension of material sciences with a science of its various synthetic procedures. Various basic issues, like total exfoliation of graphite, stabilizing layers in different solutions and holding the pure attributes of pristine 2D graphene that should have been dealt earlier to synthetic exfoliation technique would practically be utilized in synthesizing it.

The synergistic effect of other substances comprises faradic nature with pure graphene, for example, metal oxide, sulphide and polymer with graphene that are potential candidates in supercapacitors. The prospective examination of endeavours ought to be put on upgrading the interaction among the constituents of composites to enhance the redox process along the junction in order to accomplish effective redox reaction, notwithstanding EDL capacity. Utilizing these composites may form pillars with graphene sheets, which may be decent exploration bearing towards using one of kind properties of graphene for electrochemical applications.

REFERENCES

1. Yu, A.; Davies, A.; Chen, Z. Electrochemical supercapacitors. Electrochem. Technol. Energy Storage Convers. 2012, *1*, 317–382.
2. Simon, P.; Gogotsi, Y.; Dunn, B. Where do batteries end and supercapacitors begin? Science. 2014, *343*(6176), 1210–1211.

3. Matheswaran, P.; Karuppiah, P.; Chen, S.M.; Thangavelu, P. A binder-free $Ni_2P_2O_7/Co_2P_2O_7$ nanograss array as an efficient cathode for supercapacitors. New J. Chem. 2020, *44*(30), 13131–13140.
4. Priyadharshini, M.; Pazhanivel, T.; Bharathi, G. Carbon quantum dot incorporated nickel pyrophosphate as alternate cathode for supercapacitors. ChemistrySelect 2020, *5*(8), 2643–2652.
5. Augustyn, V.; Simon, P.; Dunn, B. Pseudocapacitive oxide materials for high-rate electrochemical energy storage. Energy Environ. Sci. 2014, *7*(5), 1597–1614.
6. Zhang, L.L.; Zhao, X.S. Carbon-based materials as supercapacitor electrodes. Chem. Soc. Rev. 2009, *38*(9), 2520–2531.
7. Matheswaran, P.; Karuppiah, P.; Thangavelu, P. Contribution of different charge storage mechanisms in cobalt pyrophosphate–based supercapattery. Ionics (Kiel). 2021, *27*(80), 1769–1780.
8. Matheswaran, P.; Karuppiah, P.; Chen, S.M.; Thangavelu, P.; Ganapathi, B. Fabrication of g-C_3N_4 nanomesh-anchored amorphous $NiCoP_2O_7$: tuned cycling life and the dynamic behavior of a hybrid capacitor. ACS Omega 2018, *3*(12), 18694–18704.
9. Zhang, X.; Samorì, P. Graphene/polymer nanocomposites for supercapacitors. ChemNanoMat 2017, *3*(6), 362–372.
10. Li, K.; Guo, M.; Yan, Y.; Zhan, K.; Yang, J.; Zhao, B.; Li, J. Ultrasmall $Co_2P_2O_7$ nanocrystals anchored on nitrogen-doped graphene as efficient electrocatalysts for the oxygen reduction reaction. New J. Chem. 2019, *43*(17), 6492–6499.
11. Snook, G.A.; Kao, P.; Best, A.S. Conducting-polymer-based supercapacitor devices and electrodes. J. Power Sources 2011, *196*(1), 1–12.
12. Yan, J.; Wang, Q.; Wei, T.; Fan, Z. Recent advances in design and fabrication of electrochemical supercapacitors with high energy densities. Adv. Energy Mater. 2014, *4*(4), 1300816.
13. Frackowiak, E.; Béguin, F. Carbon materials for the electrochemical storage of energy in capacitors. J. Carbon 2001, *39*, 937–950.
14. Akhter, T.; Islam, M.M.; Faisal, S.N.; Haque, E.; Minett, A.I.; Liu, H.K.; Konstantinov, K.; Dou, S.X. Self-assembled N/S codoped flexible graphene paper for high performance energy storage and oxygen reduction reaction. ACS Appl. Mater. Interfaces 2016, *8*(3), 2078–2087.
15. De Adhikari, A.; Oraon, R.; Tiwari, S.K.; Saren, P.; Lee, J.H.; Kim, N.H.; Nayak, G.C. CdS-$CoFe_2O_4$@ reduced graphene oxide nanohybrid: an excellent electrode material for supercapacitor applications. Ind. Eng. Chem. Res. 2018, *57*(5), 1350–1360.
16. Liu, W.; Niu, H.; Yang, J.; Cheng, K.; Ye, K.; Zhu, K.; Wang, G.; Cao, D.; Yan, J. Ternary transition metal sulfides embedded in graphene nanosheets as both the anode and cathode for high-performance asymmetric supercapacitors. Chem. Mater. 2018, *30*(3), 1055–1068.
17. Wang, H.; Maiyalagan, T.; Wang, X. Review on recent progress in nitrogen-doped graphene: synthesis, characterization, and its potential applications. ACS Catal. 2012, *2*(5), 781–794.
18. Wang, X.; Sun, G.; Routh, P.; Kim, D.H.; Huang, W.; Chen, P. Heteroatom-doped graphene materials: syntheses, properties and applications. Chem. Soc. Rev. 2014, *43*(20), 7067–7098.
19. Wen, Y.; Huang, C.; Wang, L.; Hulicova-Jurcakova, D. Heteroatom-doped graphene for electrochemical energy storage. Chin. Sci. Bull. 2014, *59*(18), 2102–2121.
20. Xu, L.; Zhang, Y.; Zhou, W.; Jiang, F.; Zhang, H.; Jiang, Q.; Jia, Y.; Wang, R.; Liang, A.; Xu, J.; Duan, X. Fused heterocyclic molecule-functionalized N-doped reduced graphene oxide by non-covalent bonds for high-performance supercapacitors. ACS Appl. Mater. Interfaces 2020, *12*(40), 45202–45213.
21. Balaji, S.S.; Anandha Raj, J.; Karnan, M.; Sathish, M. Supercritical fluid assisted synthesis of S-doped graphene and its symmetric supercapacitor performance evaluation using different electrolytes. Synth. Met. 2019, *255*(July), 116111.
22. Wu, P.; Hu, H.Y.; Xie, N.; Wang, C.; Wu, F.; Pan, M.; Li, H.F.; Di Wang, X.; Zeng, Z.; Deng, S.; Dai, G.P. A N-doped graphene-cobalt nickel sulfide aerogel as a sulfur host for lithium-sulfur batteries. RSC Adv. 2019, *9*(55), 32247–32257.
23. He, J.; Chen, Y.; Lv, W.; Wen, K.; Xu, C.; Zhang, W.; Qin, W.; He, W. Three-dimensional CNT/graphene-Li_2S aerogel as freestanding cathode for high-performance Li-S batteries. ACS Energy Lett. 2016, *1*(4), 820–826.
24. Wang, J.; Wu, Z.; Hu, K.; Chen, X.; Yin, H. High conductivity graphene-like MoS_2/polyaniline nanocomposites and its application in supercapacitor. J. Alloys Compd. 2015, *619*, 38–43.
25. Ma, G.; Peng, H.; Mu, J.; Huang, H.; Zhou, X.; Lei, Z. In situ intercalative polymerization of pyrrole in graphene analogue of MoS_2 as advanced electrode material in supercapacitor. J. Power Sources 2013, *229*, 72–78.
26. Huang, K.-J.; Wang, L.; Liu, Y.-J.; Wang, H.-B.; Liu, Y.-M.; Wang, L.-L. Synthesis of polyaniline/2-dimensional graphene analog MoS_2 composites for high-performance supercapacitor. Electrochim. Acta 2013, *109*, 587–594.

27. Zhang, J.; Shi, L.; Liu, H.; Deng, Z.; Huang, L.; Mai, W.; Tan, S.; Cai, X. Utilizing polyaniline to dominate the crystal phase of Ni(OH)<inf>2</inf> and its effect on the electrochemical property of polyaniline/Ni(OH)<inf>2</inf> composite. J. Alloys Compd. 2015, *651*, 126–134.
28. Lee, C.C.; Omar, F.S.; Numan, A.; Duraisamy, N.; Ramesh, K.; Ramesh, S. An enhanced performance of hybrid supercapacitor based on polyaniline-manganese phosphate binary composite. J. Solid State Electrochem. 2017, *21*(11), 3205–3213.
29. Shafi, P.M.; Ganesh, V.; Bose, A.C. $LaMnO_3$/RGO/PANI ternary nanocomposites for supercapacitor electrode application and their outstanding performance in all-solid-state asymmetrical device design. ACS Appl. Energy Mater. 2018, *1*(6), 2802–2812.
30. Ahmad, Z.; Kim, W.; Kumar, S.; Yoon, T.H.; Lee, J.S. Nanocomposite supercapacitor electrode from sulfonated graphene oxide and poly(pyrrole-(biphenyldisulfonic acid)-pyrrole). ACS Appl. Energy Mater. 2020, *3*(7), 6743–6751.
31. Pazhamalai, P.; Krishnamoorthy, K.; Sudhakaran, M.S.P.; Kim, S.J. Fabrication of high-performance aqueous Li-ion hybrid capacitor with $LiMn_2O_4$ and graphene. ChemElectroChem 2017, *4*(2), 396–403.
32. Yan, X.; Xu, R.; Guo, J.; Cai, X.; Chen, D.; Huang, L.; Xiong, Y.; Tan, S. Enhanced photocatalytic activity of Cu_2O/g-C_3N_4 heterojunction coupled with reduced graphene oxide three-dimensional aerogel photocatalysis. Mater. Res. Bull. 2017, *96*, 18–27.
33. Wang, R.; Xu, C.; Sun, J.; Gao, L. Three-dimensional Fe_2O_3 nanocubes/nitrogen-doped graphene aerogels: nucleation mechanism and lithium storage properties. Sci. Rep. 2014, *4(1) 1-7.*
34. He, Y.; Chen, W.; Li, X.; Zhang, Z.; Fu, J.; Zhao, C.; Xie, E. Freestanding three-dimensional graphene/ MnO_2 composite networks as ultralight and flexible supercapacitor electrodes. ACS Nano 2013, *7*(1), 174–182.
35. Tian, X.; Sun, X.; Jiang, Z.; Jiang, Z.J.; Hao, X.; Shao, D.; Maiyalagan, T. Exploration of the active center structure of nitrogen-doped graphene for control over the growth of Co_3O_4 for a high-performance supercapacitor. ACS Appl. Energy Mater. 2018, *1*(1), 143–153.
36. Ma, Z.; Huang, X.; Dou, S.; Wu, J.; Wang, S. One-pot synthesis of Fe_2O_3 nanoparticles on nitrogen-doped graphene as advanced supercapacitor electrode materials. J. Phys. Chem. C 2014, *118*(31), 17231–17239.
37. Dong, X.C.; Xu, H.; Wang, X.W.; Huang, Y.X.; Chan-Park, M.B.; Zhang, H.; Wang, L.H.; Huang, W.; Chen, P. 3D graphene-cobalt oxide electrode for high-performance supercapacitor and enzymeless glucose detection. ACS Nano 2012, *6*(4), 3206–3213.
38. Chandel, M.; Makkar, P.; Ghosh, N.N. Ag–Ni nanoparticle anchored reduced graphene oxide nanocomposite as advanced electrode material for supercapacitor application. ACS Appl. Electron. Mater. 2019, *1*(7), 1215–1224.
39. Nayak, A.K.; Das, A.K.; Pradhan, D. High performance solid-state asymmetric supercapacitor using green synthesized graphene-WO_3 nanowires nanocomposite. ACS Sustain. Chem. Eng. 2017, *5*(11), 10128–10138.
40. Karmakar, S.; Mistari, C.D.; Shajahan, A.S.; More, M.A.; Chakraborty, B.; Behera, D. Enhancement of pseudocapacitive behavior, cyclic performance, and field emission characteristics of reduced graphene oxide reinforced $NiGa_2O_4$ nanostructured electrode: a first principles calculation to correlate with experimental observation. J. Phys. Chem. C 2021, *125*(14), 7898–7912.
41. Ramachandran, R.; Saranya, M.; Kollu, P.; Raghupathy, B.P.C.; Jeong, S.K.; Grace, A.N. Solvothermal synthesis of zinc sulfide decorated graphene (ZnS/G) nanocomposites for novel supercapacitor electrodes. Electrochim. Acta 2015, *178*, 647–657.
42. Huang, K.J.; Zhang, J.Z.; Liu, Y.; Liu, Y.M. Synthesis of reduced graphene oxide wrapped-copper sulfide hollow spheres as electrode material for supercapacitor. Int. J. Hydrogen Energy 2015, *40*(32), 10158–10167.
43. Zhang, K.; Wei, Y.; Huang, J.; Xiao, Y.; Yang, W.; Hu, T.; Yuan, K.; Chen, Y. A generalized one-step in situ formation of metal sulfide/reduced graphene oxide nanosheets toward high-performance supercapacitors. Sci. China Mater. 2020, *63*(10), 1898–1909.

Index

Note: Locators in *italics* represent figures and **bold** indicate tables in the text.

D

E

F

L

M

N

O

P

Q

R

U

V

W

X

Z

Milton Keynes UK
Ingram Content Group UK Ltd.
UKHW051026210124
436410UK00001B/1

9 780367 770914